D1564863

27
Topics in Medicinal Chemistry

Aims and Scope

Drug research requires interdisciplinary team-work at the interface between chemistry, biology and medicine. Therefore, the new topic-related series Topics in Medicinal Chemistry will cover all relevant aspects of drug research, e.g. pathobiochemistry of diseases, identification and validation of (emerging) drug targets, structural biology, drugability of targets, drug design approaches, chemogenomics, synthetic chemistry including combinatorial methods, bioorganic chemistry, natural compounds, high-throughput screening, pharmacological in vitro and in vivo investigations, drug-receptor interactions on the molecular level, structure-activity relationships, drug absorption, distribution, metabolism, elimination, toxicology and pharmacogenomics.

In general, special volumes are edited by well known guest editors.

In references Topics in Medicinal Chemistry is abbreviated Top Med Chem and is cited as a journal.

More information about this series at http://www.springer.com/series/7355

Amanda L. Garner
Editor

RNA Therapeutics

With contributions by

P. Agarwala · R. Brenk · J.L. Childs-Disney · M.G. Costales · M.D. Disney · A.L. Garner · T. Hermann · G.M. Karp · D.A. Lorenz · S. Maiti · B.L. Miller · K.E. Murphy-Benenato · N.A. Naryshkin · S. Pandey · M.G. Stanton · T. Wehler · M.G. Woll · C.J. Woo

Editor
Amanda L. Garner
University of Michigan
Department of Medicinal Chemistry
Ann Arbor, Michigan
USA

ISSN 1862-2461 ISSN 1862-247X (electronic)
Topics in Medicinal Chemistry
ISBN 978-3-319-68090-3 ISBN 978-3-319-68091-0 (eBook)
DOI 10.1007/978-3-319-68091-0

Library of Congress Control Number: 2017953775

Printed on acid-free paper

This Springer imprint is published by Springer Nature
The registered company is Springer International Publishing AG
The registered company address is: Gewerbestrasse 11, 6330 Cham, Switzerland

Preface

The central dogma is the hallmark of genetic transmission and molecular biology. The fundamental processing of DNA to RNA, via transcription, and RNA to protein, via splicing and translation, drives all cellular activities and its imbalance leads to disease. While current efforts in drug discovery tend to take a more protein-centric view of the central dogma, whereby the great majority of approved drugs target proteins, we may be limiting ourselves by doing so. Following completion of the human genome project, it was revealed that only ~2% of our genome encodes for proteins, and the overwhelming majority is comprised of often highly conserved RNAs. Since that time, RNA has been shown to play a crucial role in nearly all of human biology from transcriptional regulation, splicing, translation, RNA function, and catalysis. These efforts are carried out by a pantheon of RNA molecules ranging in size and structure, from small ~22 nucleotide microRNAs (miRNAs) to >200 nucleotide long noncoding RNAs (lncRNAs) to highly structured RNA catalysts, ribozymes. Accordingly, we have witnessed an explosion in discoveries connecting these RNAs with human diseases, making the search for RNA-targeted therapeutics ever more pressing. This connection has peaked the interest of not only academic medicinal chemists but also the biotech and pharmaceutical industries. To embark upon this new frontier of RNA-targeted drug discovery, research efforts span the development of new methodologies for tackling the challenging molecular recognition problem of targeting RNA with small molecules to traditional antisense oligonucleotide-based approaches to the new and fascinating concept of using mRNAs as drugs. In this volume of *Topics in Medicinal Chemistry*, I have selected a series of reviews representing the current state of the art in the field in the areas of RNA-targeted small molecule discovery, RNA drug targets, and the emerging field of mRNA therapeutics.

The first four chapters describe enabling tools and methodologies for the discovery of small molecule RNA ligands. This goal represents a difficult medicinal chemistry problem due to the challenges associated with the discovery of such molecules, namely RNA's highly electronegative nature and dynamic structure. Nonetheless, inroads are being made due to the creative work of many leading groups in the field. In the first chapter by Costales, Childs-Disney, and Disney, a computational platform for the rational design of sequence-selective small molecules targeting RNA will be described. This will be followed up by a chapter from Miller describing an alternative approach for this goal by using dynamic combinatorial chemistry. Both of these tactics, however, require the use of known RNA-binding motifs and scaffolds for rational design. In order to discover new chemical space for targeting RNA, high-throughput screening methods are needed. In the third chapter, Wehler and Brenk discuss applications of structure-based drug discovery for targeting RNA. The fourth chapter by Lorenz and Garner reviews assay technologies used for the discovery of small molecules targeting a specific class of RNA, microRNAs, that play a key role in the fine-tuning of gene expression.

The following four chapters describe promising RNA drug targets and current medicinal chemistry efforts for these targets. Some of the first attempts at RNA-targeted drug discovery focused on the RNA viruses, human immunodeficiency virus (HIV), and hepatitis C virus (HCV). Significant work continues in these areas and will be described by Hermann in the fifth chapter. Many genetic disorders, such as spinal muscular atrophy and Duchenne muscular dystrophy, have aberrant pre-mRNA splicing implicated in their pathology. Woll, Naryshkin, and Karp will discuss the use of splice-switching oligonucleotides and small molecules as therapeutic modalities in the sixth chapter. The seventh chapter from Pandey, Agarwala, and Maiti will describe therapeutic applications of targeting a specific RNA structural class, namely G-quadruplexes, which are found in many disease-relevant RNAs. For the eighth chapter, Woo provides a review on a new class of RNA, long noncoding RNAs, which have been shown to function in the regulation of transcription and epigenetics and serve as an up-and-coming area for RNA-targeted drug discovery.

To date, drug development has focused almost entirely on small molecules or biologics, such as oligonucleotides and antibodies, but these efforts are not without challenges and failures. As the final chapter, Stanton and Murphy-Benenato describe how advances in RNA chemistry and drug delivery have made the promise of therapeutic mRNAs, which can generate any protein-of-interest in vivo, an impending reality.

As outlined here and throughout this volume, an interdisciplinary group of scientists will be required to effectively tackle RNA-targeted drug discovery. It is my hope that this volume will not only serve as a compilation of current knowledge for those well versed in the field of RNA but also inspire our next generation of researchers to find new ways to contribute to this exciting field.

Ann Arbor, MI, USA
May 2017

Amanda L. Garner

Contents

Computational Tools for Design of Selective Small Molecules Targeting RNA: From Small Molecule Microarrays to Chemical Similarity Searching 1
Matthew G. Costales, Jessica L. Childs-Disney, and Matthew D. Disney

A Modular Approach to the Discovery and Affinity Maturation of Sequence-Selective RNA-Binding Compounds 17
Benjamin L. Miller

Structure-Based Discovery of Small Molecules Binding to RNA 47
Thomas Wehler and Ruth Brenk

Approaches for the Discovery of Small Molecule Ligands Targeting microRNAs 79
Daniel A. Lorenz and Amanda L. Garner

Viral RNA Targets and Their Small Molecule Ligands 111
Thomas Hermann

Drugging Pre-mRNA Splicing 135
Matthew G. Woll, Nikolai A. Naryshkin, and Gary M. Karp

Targeting RNA G-Quadruplexes for Potential Therapeutic Applications 177
Satyaprakash Pandey, Prachi Agarwala, and Souvik Maiti

The Therapeutic Targeting of Long Noncoding RNA 207
Caroline J. Woo

Messenger RNA as a Novel Therapeutic Approach 237
Matthew G. Stanton and Kerry E. Murphy-Benenato

Index 255

Top Med Chem (2018) 27: 1–16
DOI: 10.1007/7355_2016_21

Published online: 10 May 2017

Computational Tools for Design of Selective Small Molecules Targeting RNA: From Small Molecule Microarrays to Chemical Similarity Searching

Matthew G. Costales, Jessica L. Childs-Disney, and Matthew D. Disney

Abstract RNA is an important drug target, yet few lead compounds elicit their effects by acting on RNA outside of the bacterial ribosome. Herein, we describe various synergistic strategies to identify small molecules that target RNA and how computational approaches can be utilized for lead optimization. In particular, we describe the development of small molecule microarray approaches applied towards RNA and its application to identify small molecule binders and for the facile study of antibiotic resistance mechanisms for known or novel lead antibacterials. Additionally, a microarray-based library-versus-library screen, which probes millions of combinations, is described that identifies RNA motif binding partners preferred by small molecules. Lead compounds can be designed by searching for these privileged interactions in a disease-causing RNA. Computational chemistry can be used to optimize these compounds. For example, lead compounds that target the r(CCUG) repeats expansions that cause myotonic dystrophy type 2 (DM2) were lead optimized by using structure-based design. Specifically, the compounds were developed to allow an in situ click chemistry approach in which a disease-affected cell synthesizes its own drug on-site by using the disease-causing biomolecule as a cellular catalyst. In another lead optimization strategy, chemical similarity searching was employed to lead optimize small molecules that target the r(CUG) repeat expansion that causes myotonic dystrophy type 1 (DM1). These studies allowed for the identification of an in vivo active small molecule that targets r(CUG) and improves disease-associated defects. As more studies are completed to understand the role of RNA in disease biology, the number of potential RNA targets will increase. In order to leverage these important investigations to develop compounds that target RNA, approaches that allow one to identify and optimize small molecules for selectivity and potency must be carefully considered.

M.G. Costales, J.L. Childs-Disney, and M.D. Disney (✉)
Departments of Chemistry and Neuroscience, The Scripps Research Institute, 130 Scripps Way, Jupiter, FL 33458, USA
e-mail: Disney@scripps.edu

Keywords Chemical biology, Drug design, High throughput screening, Nucleic acids, RNA

Contents

1 Introduction 2
2 RNA Targeting and the Transcriptome: Structured RNAs Abound! 3
3 RNA Targets: Binding to Isolated Fragments Mimics Binding to the Biologically Active Structure 4
4 Small Molecule Microarrays for Studying Ligand Binding 5
5 Two-Dimensional Combinatorial Screening: Facile Profiling of Libraries of RNA Motifs for Binding to Libraries of Small Molecules 6
6 Using Results of Two-Dimensional Combinatorial Screening to Computationally Define RNA Targets in the Transcriptome 7
7 Structure-Aided Drug Design: Development of an On-Site Drug Synthesis Approach 9
8 Chemical Similarity Searching: A Ligand-Based Computational Approach to Rapidly Lead Optimize Small Molecules Targeting RNA 9
9 Summary and Outlook 11
References 12

1 Introduction

RNA has been long viewed as an important drug target in both bacterial and human systems. The earliest examples of small molecules that target RNA include antibiotics binding to various sites in the bacterial ribosome [1] and more recently riboswitches [2, 3]. Additionally, antisense and other oligonucleotide-based approaches can be designed to target RNA through Watson-Crick base pairing [4, 5]. Complementary oligonucleotides bind a target RNA and modulate its function through cleavage of the targeted transcript or by steric blockage [6, 7]. These approaches have been invaluable to treat bacterial infections, to study translational machinery, and to target human RNAs to provide pre-clinical compounds that can potentially treat disease [8].

The use of small molecules to target human RNAs has been limited, but there have been a few important examples that include affecting pre-mRNA splicing by binding to an mRNA [9–11], reducing various modes of toxicity associated with RNA repeat expansions [12–18], and precisely inhibiting microRNA biogenesis by targeting microRNA hairpin precursors [19–21]. Small molecules may be particularly advantageous to target structured regions that provide binding pockets, in contrast to antisense oligonucleotides, which prefer to target unstructured regions [22, 23]. Further, small molecules can, in principle, be more easily optimized by using medicinal chemistry approaches than antisense oligonucleotides that have strict chemical requirements not only for base pairing but also for targeted mRNA cleavage [24].

In this review, we describe various approaches to identify lead small molecules that target RNA. An emphasis is placed on various bead- and microarray-based screening approaches to identify and study lead RNA-targeting small molecules [25–28]. Computational approaches are described that can be used to optimize lead small molecules for bioactivity once a lead is identified. These methods include structure-based design, where small molecules targeting RNA repeat expansions were optimized by using a novel on-site drug synthesis approach [29], and chemical similarity searching, which was used to lead optimize compounds targeting the RNA repeat expansions that cause myotonic dystrophy Type 1 (DM1) and Huntington's disease [16, 30]. In the coming years, these approaches are likely to advance a paradigm in which cell permeable small molecules can be quickly designed to target a disease-causing RNA.

2 RNA Targeting and the Transcriptome: Structured RNAs Abound!

Since the early days in RNA biology, it was well known that RNAs, including ribosomal (rRNA) and transfer (tRNAs) RNAs, adopt a folded structure [31]. Indeed, crystal structures of tRNAs provided the first glimpses into these defined structures [32–35]. It became apparent from these and other studies that RNA folding is much more complex than that observed in Watson-Crick pairing of DNA, owing to the fact that RNA is single stranded. Thus, it folds onto itself to minimize its free energy, forming various noncanonically paired regions such as internal loops, hairpin loops, bulges, and multibranch loops (Fig. 1) [36–39].

RNA folding has been studied ever since, including investigations on phylogenetic comparison of rRNA to identify the kingdoms of life and even to define the new ones [40, 41]. Many biophysical chemists also sought to develop methods to predict RNA folding. Work of Tinoco [31], Turner [42], and others developed free energy folding rules for RNA, enabled by understanding the folding free energy of the various motifs that RNA adopts. These rules were then adapted to algorithms, such as those by Zuker [43], to compute the lowest free energy RNA structures as well as suboptimally folded structures. During the course of these studies, it was found that biologically active folds of RNA are not necessarily the lowest free energy structure but can be

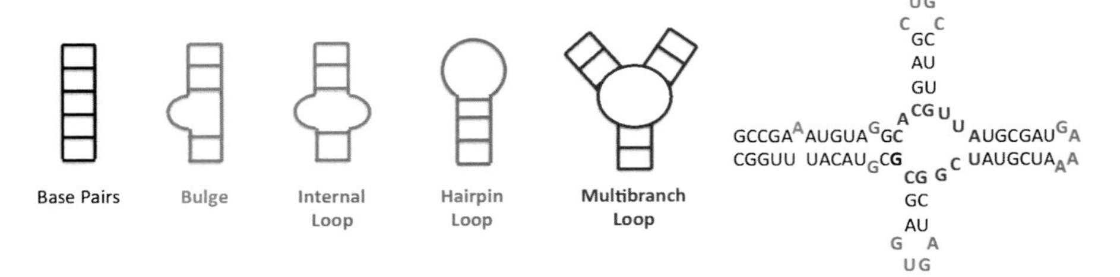

Fig. 1 Secondary structural elements formed when single stranded RNA folds on itself to minimize its free energy. RNA can form canonical base pairs (GC, AU, and GU pairs) or noncanonically paired regions including bulges, internal loops (both symmetric and asymmetric), hairpin loops, and multibranch loops. The different elements are *color coded* in the secondary structure on the *right*

a "suboptimally" folded one [43]. Thus, experimental and computational approaches were developed to identify the biologically active structure from predicted ensembles, including chemical and enzymatic mapping to probe RNA structure [44] and the implementation of algorithms to utilize these data to refine structure prediction [45]. Chemical probes of RNA structure include the Gilbert–Maxam sequencing reagents (dimethyl sulfate (DMS), 1-cyclohexyl-3-(2-(4-morpholinyl)ethyl) carbodiimide tosylate (CMCT), and kethoxal) [46–48], but more recently the powerful selective 2′-hydroxyl acylation analyzed by primer extension (SHAPE) approach [49].

A renaissance in RNA structural biology was put forward by X-ray crystallography [50] and NMR spectroscopy [51, 52], which elucidated structural details of RNA folds of small RNA fragments or, in rare cases, the structure of full length biological RNAs. The determination of the crystal structure by the Doudna Laboratory of domains of *Tetrahymena thermophila* group 1 intron advanced the field of RNA structural biology considerably [53]. At a similar time, major efforts were put into studying the structure of the ribosomal particle, culminating in atomic or near atomic resolution structures of segments and the whole ribosome by the work or Yonath, Ramakrishnan, Noller, Steitz, Moore, Cate, and many others [54–60]. One of the most interesting aspects of these studies was that the phylogenic derived predicted secondary structure nearly perfectly matched the experimentally determined structure! Additionally, the work of Weeks by using RNA structure prediction and SHAPE showed that computational methods constrained by experiments could also recapitulate the secondary structure of rRNA [61].

One major advance emanating from studies on the ribosome included defining the binding pockets of various antibiotics [55, 56, 58]. These studies defined or confirmed the mechanism of action of these important compounds and highlighted the importance of chemical probes of RNA function. That is, antibiotics locked the ribosomal particle into a conformation that allowed it to be studied. Further, structure-based drug design was enabled providing several clinically useful antibacterials [62].

3 RNA Targets: Binding to Isolated Fragments Mimics Binding to the Biologically Active Structure

In many of the above examples, small molecule screens were completed to identify and study small molecule binding to isolated regions of a biologically active RNA target. The works of Wong [63], Puglisi [64, 65], Pilch [66, 67], Westhof [68], Hermann [69], Tor [69, 70], Chow [71, 72], Mobashery [71, 72], and many others have used the binding of aminoglycoside antibacterials to an isolated A-site to infer binding to the whole bacterial ribosome. For example, a structure determined by NMR spectroscopy from the Puglisi lab indicated that paromomycin binding to the A-site altered the dynamic structure of an A-rich internal loop in the A-site upon compound binding

[64]. Chemical mapping of ribosomal particles bound to paromomycin showed that dynamic switching affected decoding of tRNAs and explained the antibacterial activity of these small molecules.

Studies by Pilch, Hermann, and Tor advanced these studies in isolated systems and showed that the ability of aminoglycosides to affect dynamics of this site was a better predictor of antibacterial efficacy than simple binding [67, 69, 70]. Ramakrishnan completed structural studies on the 30S ribosome particle and showed that the binding of aminoglycosides in the whole particle displayed the dynamic switching features initially described in the isolated systems [55]. Furthermore, the studies on HIV TAR and other fragments of biological RNAs have shown that studying the ligand binding capacity of isolated pieces could be used to score compounds for binding and infer biological activity [73].

4 Small Molecule Microarrays for Studying Ligand Binding

The Schreiber group developed the small molecule microarray approach by arraying small molecules onto surfaces in a spatially defined manner [74]. Compounds on the array bound a labeled protein target, indicated by signal at specific positions. Thus, this approach provided a facile method to score compounds for binding to targets.

Inspired by these studies, Disney and coworkers developed a small molecule microarray approach to profile the binding of small molecules to RNA targets. These studies were used to profile aminoglycosides for binding to isolated A-sites from both the human and bacterial ribosome and identify small molecules that were not substrates for resistance-causing enzymes [25, 26]. These studies were further developed to study the effect of aminoglycoside modification by resistance-causing enzymes on the ability of compounds to bind to the bacterial A-site [75]. Indeed by using various modification enzymes and labeled substrates, the ability of a compound to be modified by a resistance-causing enzyme could be profiled quickly. Known work by the Chow [76] and Mobashery [77] groups showed that enzymatic modification significantly diminished ligand binding affinity to A-sites. Indeed, decreased binding of array-modified aminoglycosides to A-sites was also observed in our microarray approach in high throughput fashion [75]. The use of the microarray method to study binding of RNA targets has been extended by us and others and is gaining widespread use.

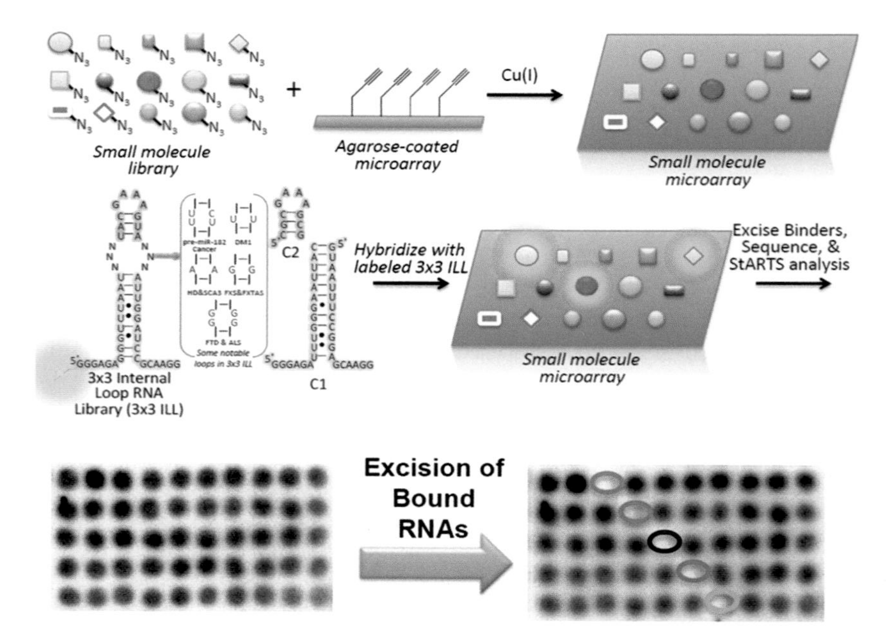

Fig. 2 Schematic of Two-*D*imensional *C*ombinatorial *S*creening (2DCS), a library-vs.-library screening approach, to identify the RNA motifs preferred by small molecules [28]. Briefly, small molecules are site-specifically immobilized onto a functionalized agarose-coated microarray, for example, by using click chemistry. The arrays are hybridized with a labeled RNA library that displays discrete secondary structural elements such as internal loops that are likely to be found in biological RNAs. Hybridization is completed under conditions of high stringency using oligonucleotides that mimic regions that are common to all members of the RNA library (C1 and C2). Bound RNAs, as visualized by deposition of signal on the array surface, are excised from the array, amplified, and sequenced. The RNA-small molecule pairs are then deposited into a database that can be used as a facile lead compound identification strategy (Inforna)

5 Two-Dimensional Combinatorial Screening: Facile Profiling of Libraries of RNA Motifs for Binding to Libraries of Small Molecules

Given the developments of profiling a single or few RNA targets for binding to array-immobilized ligands, we sought to develop a novel method to profile large-scale molecular recognition events via microarray (Fig. 2). To accomplish this goal, we developed an RNA library that displayed randomized motifs and a microarray-based platform that allowed for screening of chemical and RNA motif space simultaneously [28]. Thus, in Two-*D*imensional *C*ombinatorial *S*creening (2DCS), a library of small molecules is spatially arrayed onto an agarose microarray surface followed by hybridization of the surface with RNA motif libraries in the presence of competitors, which constrain selected interactions to the randomized region in the RNA library

(Fig. 2) [27]. This approach and a bead-based equivalent [78] have defined the RNA motif binding preferences for a wide number of small molecule compound types, defining rules for molecular recognition. For example, bead-based studies defined that RNA internal loops that contain pyrimidine-rich regions are preferred binding sites for 6′-acylated kanamycin A derivatives [78].

6 Using Results of Two-Dimensional Combinatorial Screening to Computationally Define RNA Targets in the Transcriptome

The results of 2DCS were used to identify new druggable targets in the transcriptome. For example, overlap was identified between the RNA motif binding preferences of a 6′-acylated kanamycin A derivative harboring an alkyne and the RNA internal loops that are present in the r(CCUG) repeat expansion (r(CCUG)^{exp}) that causes myotonic dystrophy type 2 (DM2). In DM2, r(CCUG)^{exp} is present in intron 1 of the zinc finger 9 (*ZNF9*) pre-mRNA [79] and folds into a hairpin structure forming repeating 5′C$\underline{CU}$G/3′G$\underline{UC}$C motifs in the stem. r(CCUG)^{exp} sequesters and inactivates various proteins, in particular muscleblind-like 1 protein (MBNL1), causing deregulation of alternative pre-mRNA splicing [80].

Because r(CCUG)^{exp} folds into an array of 2×2 nucleotide pyrimidine-rich internal loops, a modular approach was employed (Fig. 3a). Polyvalent scaffolds displaying the 6′-acylated kanamycin with varied spacing were synthesized and studied to identify the compound that best mimicked the distance between loops in r(CCUG)^{exp} [81]. In this case, a peptoid scaffold was used as peptoids are easy to synthesize and allow for precise control of the spacing between RNA-binding modules (by the number of propylamine spacers inserted between ligand conjugation sites) and valency/density (by controlling the number of azide-containing modules installed on the backbone) [81]. Additionally, peptoids are cell permeable [83]. By using a screen to identify compounds that can inhibit a complex between MBNL1 and r(CCUG)$_{12}$, a mimic of r(CCUG)^{exp}, it was identified that compounds with four propylamine spacers, or **nK-4** (where **nK** indicates the number of kanamycin-binding modules and **4** indicates the number of propylamine spacers), were optimal (Fig. 3) [81]. Furthermore, binding analysis showed that the designed, cell permeable small molecules bound with an order of magnitude greater affinity and selectivity than MBNL1 [81]. Thus, cell permeable small molecules can have more favorable properties for binding RNA over evolved proteins.

The effect of the compounds on improving DM2-associated defects was also assessed in cell culture. Three designer compounds, monomer **K**, dimer **2K-4**, and trimer **3K-4**, improve DM2-associated pre-mRNA splicing defects [82]. The potencies of the compounds increased as a function of valency, providing IC$_{50}$s in the low micromolar regime [82]. These studies identified one of the first designer compounds that improve disease-affected defects in cells.

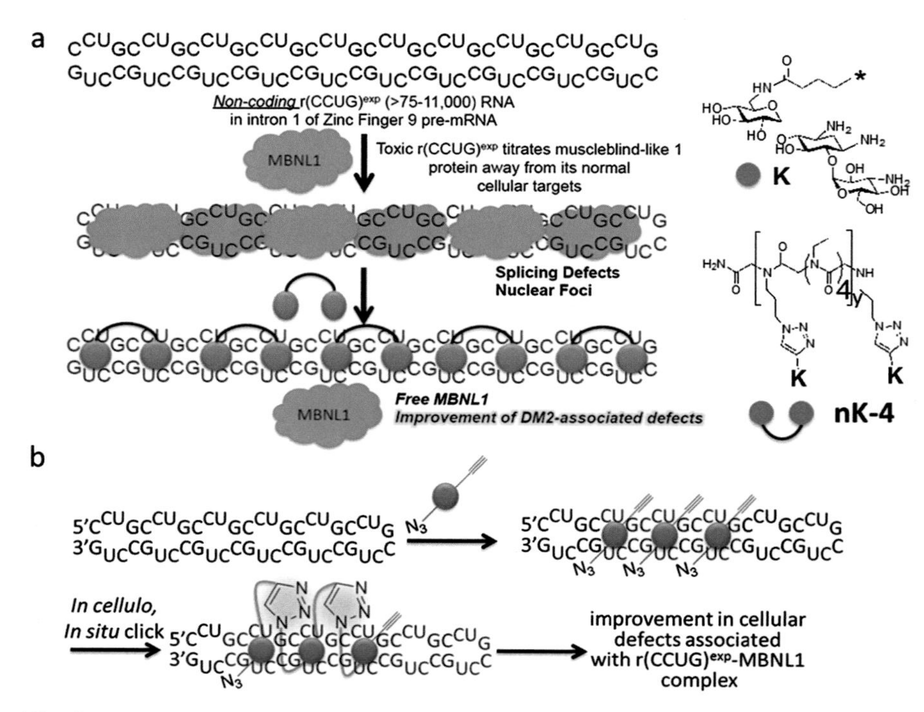

Fig. 3 Approaches to target the expanded r(CCUG) repeats (r(CCUG)^{exp}) that cause myotonic dystrophy type 2 (DM2). (**a**) The DM2 repeat folds into a hairpin structure with periodically repeating 5′CCUG3′/3′GUCC5′ internal loops. We identified a 6′-acylated kanamycin A derivative (**K**) that binds the loops with high affinity [27]. By identifying the optimal distance between **K**s to target adjacent loops simultaneously (**nK-4**), affinity, selectivity, and activity in cellular models were greatly improved [81, 82]. (**b**) Structure-aided design using a structure of **2K-4** bound to r(CCUG) repeats enabled design of a "clickable" **K** derivative with precisely positioned alkyne and azide groups [82]. When the clickable compound binds adjacent loops in r(CCUG)^{exp}, a multivalent compound is synthesized on-site. Notably, oligomeric **K** species are not observed in cells that do not express r(CCUG)^{exp}. That is, the disease-affected cell serves as a round bottomed flask and r(CCUG)^{exp} serves as the catalyst to synthesize its own inhibitor [29]

The Zimmerman group has also developed small molecules that target r(CCUG)^{exp} [84, 85]. These designer compounds inhibit formation of DM2-associated nuclear foci at micromolar concentrations.

7 Structure-Aided Drug Design: Development of an On-Site Drug Synthesis Approach

We sought to use structure-aided drug design to improve the activity of our **nK-4**, DM2-targeting compounds. To enable these studies, we determined a structure of a mimic of r(CCUG)^{exp} bound to **2K-4**, a dimeric compound that targets two 5′CCUG/3′GUCC motifs in r(CCUG)^{exp} simultaneously [82]. This structure enabled an on-site drug synthesis approach (Fig. 3b) [29] based on the click chemistry methodology of Sharpless and colleagues that was initially described by using acetyl choline esterase as a reaction vessel to synthesize its own inhibitor in vitro [86]. That is, two compounds equipped with an azide and/or alkyne bind adjacent sites on the target, which bring the otherwise unreactive groups into close proximity. Once in close proximity, the two groups click together to form a stable triazole linkage.

Studies of **2K-4** and derivatives thereof in their bound state of r(CCUG)^{exp} showed that an azide moiety at the 6″ position on kanamycin A would be in close proximity to an alkyne moiety displayed from 6′-N-5-hexynoate kanamycin A when bound to adjacent internal loops (Fig. 3b) [82]. Indeed, it was shown that incubation of these two compounds with a mimic of r(CCUG)^{exp}, but not other nucleic acids, allowed for dimer formation in vitro [29].

We next studied if the synthesis of larger and thus more potent oligomers could be catalyzed by r(CCUG)^{exp}. We synthesized a singular compound 6′-*N*-5-hexynoate-6″-azido-kanamycin A that contains appropriately spaced click reactive groups. The compound does not react intramolecularly in vitro but reacts upon exposure to r(CCUG)^{exp} to form oligomers [29]. Importantly, 6′-*N*-5-hexynoate-6″-azido-kanamycin A forms oligomeric species in cells that express r(CCUG)^{exp} (but not cells that do not) and potently improves DM2-associated defects including deregulation of pre-mRNA splicing and formation of foci [29]. Thus, on-site probe synthesis was accomplished by using a disease-affected cell as a round bottomed flask and a disease-causing RNA as a cellular catalyst. Many extensions of using this approach are currently being explored.

8 Chemical Similarity Searching: A Ligand-Based Computational Approach to Rapidly Lead Optimize Small Molecules Targeting RNA

In addition to targeting the r(CCUG)^{exp} repeat that causes DM2, we have also actively developed small molecules that target the r(CUG)^{exp} repeat expansion that causes DM1. In DM1, an expanded r(CUG) repeat (r(CUG)^{exp}) is present in the 3′ untranslated region (UTR) of the dystrophia myotonica protein kinase (*DMPK*) mRNA [87]. Like r(CCUG)^{exp}, the DM1 repeat expansion also sequesters MBNL1 and deregulates pre-mRNA splicing [88]. Although we have used the above strategies for DM1 that were described for DM2, in this module we describe the use of chemical

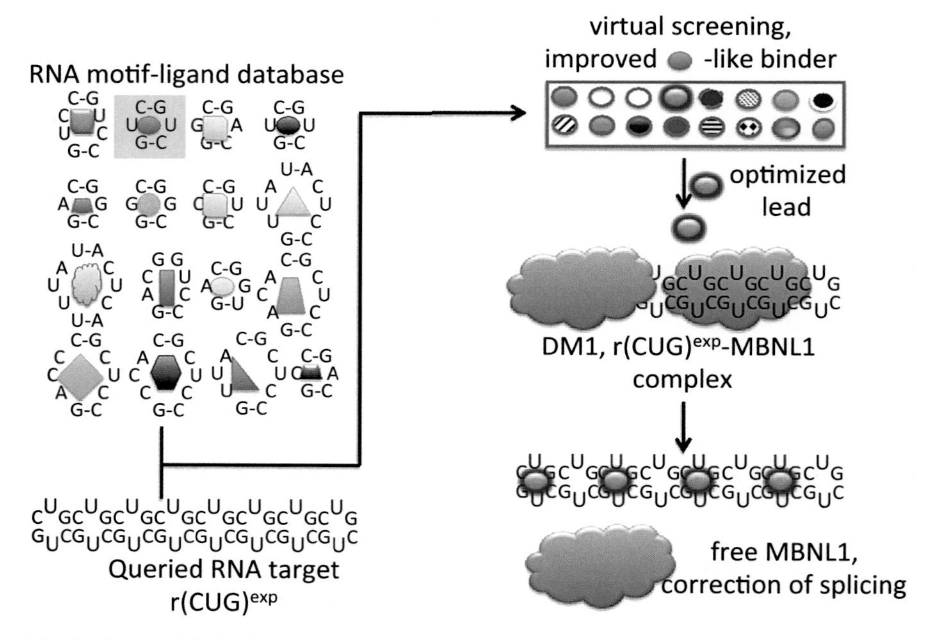

Fig. 4 Chemical similarity searching has been used to optimize lead compounds targeting expanded RNA repeats that were inactive in cells to small molecules that are active in cellular and animal models [16, 30]

similarity searching to quickly lead optimize compounds that target r(CUG)^{exp}, affording compounds that improve various disease-causing defects in cellular and animal models of disease [16].

A lead *bis*-benzimidazole compound was identified that targets r(CUG)^{exp}; however, application of this compound to cellular models of disease had no effect [16]. Thus, chemical similarity searching was used to identify compounds that are similar to the lead compound but with differences that may improve bioactivity (Fig. 4). Indeed, by searching libraries of small molecules for compounds that are similar to the *bis*-benzimidazole as determined by shape Tanimoto scores, a library of 500 compounds was constructed [16]. An in vitro assay for inhibition of the r(CUG)^{exp}–MBNL1 complex found that a subset of compounds from the library possessed as much as 6-fold improved activity over the lead starting compound [16]. The most potent compound identified, **H1**, improved various DM1-associated defects in cellular models (Fig. 5) [16].

A mouse model is also available for DM1 in which r(CUG)^{exp} is knocked in and expressed via the human skeletal actin promoter [89]. DM1 mice display disease-associated phenotypes including pre-mRNA splicing defects and myotonia. Addition of 100 mg/kg **H1** delivered by *i.p.* injection improved pre-mRNA splicing defects [16]. These studies validated that small molecules targeting this RNA can be active in vivo.

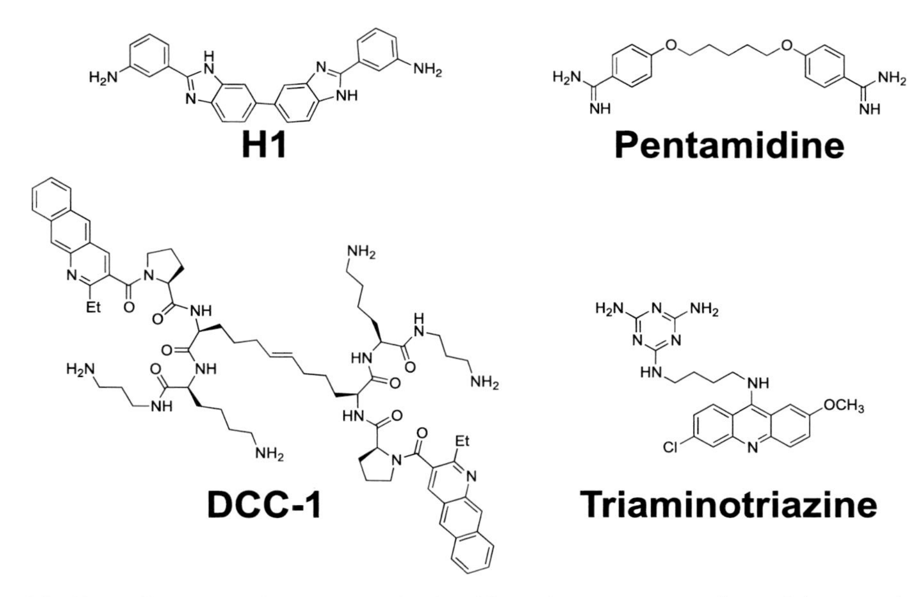

Fig. 5 Small molecules that have been developed by various groups to ameliorate defects associated with myotonic dystrophy type 1 (DM1) [12, 14, 16, 18]

Other excellent work has also been conducted on targeting r(CUG)^{exp} repeats. This includes work by the Miller, Zimmerman, and Berglund groups (Fig. 5). In the case of Miller, dynamic combinatorial chemistry and lead optimization was used to identify compounds that target the r(CUG)^{exp} repeats and improve disease-associated defects, including **DCC-1** (Fig. 5) [90]. Work primarily by the Zimmerman group used the structure of 1 $\times$ 1 nucleotide UU internal loops in r(CUG)^{exp} to develop triaminotriazine molecules that can hydrogen bond with the U bases and stack on the adjacent CG base pairs [12, 91]. These compounds have activity in cells and also in a *Drosophila* model of disease [92]. Work of the Berglund group has focused on the identification of compounds, ranging from pentamidine [18] to actinomycin D [93] (Fig. 5), that affect the transcription of r(CUG)^{exp} repeats; both compounds improve defects in a DM1 mouse model. Thus, there is much activity in this area and various strategies are being brought to bear on targeting this repeating transcript. The observation that multiple modalities work and have activity in vivo is exciting for the field.

9 Summary and Outlook

As many diseases are associated with its dysfunction and aberrant expression, RNA has become an increasingly more relevant drug target. Breakthroughs in sequencing and structural determination have led to the identification of disease-causing RNAs and their structures. A library-vs.-library microarray screening platform has provided a wealth of data defining privileged RNA motif–small molecule interactions [28]. By searching for these motifs in biologically active RNA folds, a facile lead identification strategy has been developed [20]. Both chemical similarity searching [16, 30] and structure-aided design [29] have been used effectively to optimize lead small molecules that target expanded repeating RNAs. The latter enabled an on-site synthesis approach to target r(CCUG)^{exp} that has vast potential to progress the field of precision medicine [29]. Challenges remain to improve current lead compounds; however, work by various groups will continue to accelerate developments in the field of small molecule modulation of RNA and methods to design selective RNA-targeting ligands.

References

1. von Ahsen U, Noller HF (1993) Footprinting the sites of interaction of antibiotics with catalytic group I intron RNA. Science 260:1500–1503
2. Clyde K, Harris E (2006) RNA secondary structure in the coding region of dengue virus type 2 directs translation start codon selection and is required for viral replication. J Virol 80:2170–2182
3. Elghonemy S, Davis WG, Brinton MA (2005) The majority of the nucleotides in the top loop of the genomic 3′ terminal stem loop structure are cis-acting in a West Nile virus infectious clone. Virology 331:238–246
4. Chan JH, Lim S, Wong WS (2006) Antisense oligonucleotides: from design to therapeutic application. Clin Exp Pharmacol Physiol 33:533–540
5. Rayburn ER, Zhang R (2008) Antisense, RNAi, and gene silencing strategies for therapy: mission possible or impossible? Drug Discov Today 13:513–521
6. Sharp PA, Zamore PD (2000) RNA interference. Science 287:2431–2433
7. Swayze EE, Bhat B (2008) The medicinal chemistry of oligonucleotides. In: Crooke ST (ed) Antisense drug technology: principles, strategies, and applications, 2nd edn. CRC Press, Boca Raton, FL
8. Kole R, Krainer AR, Altman S (2012) RNA therapeutics: beyond RNA interference and antisense oligonucleotides. Nat Rev Drug Discov 11:125–140
9. Rishton GM, LaBonte K, Williams AJ, Kassam K, Kolovanov E (2006) Computational approaches to the prediction of blood-brain barrier permeability: a comparative analysis of central nervous system drugs versus secretase inhibitors for Alzheimer's disease. Curr Opin Drug Discov Devel 9:303–313
10. Atlas R, Behar L, Elliott E, Ginzburg I (2004) The insulin-like growth factor mRNA binding-protein IMP-1 and the Ras-regulatory protein G3BP associate with tau mRNA and HuD protein in differentiated P19 neuronal cells. J Neurochem 89:613–626
11. Luo Y, Disney MD (2014) Bottom-up design of small molecules that stimulate exon 10 skipping in mutant *MAPT* pre-mRNA. Chembiochem 15:2041–2044
12. Jahromi AH, Nguyen L, Fu Y, Miller KA, Baranger AM, Zimmerman SC (2013) A novel CUG^{exp}•MBNL1 inhibitor with therapeutic potential for myotonic dystrophy type 1. ACS Chem Biol 8:1037–1043
13. Childs-Disney JL, Hoskins J, Rzuczek SG, Thornton CA, Disney MD (2012) Rationally designed small molecules targeting the RNA that causes myotonic dystrophy type 1 are potently bioactive. ACS Chem Biol 7:856–862

14. Gareiss PC, Sobczak K, McNaughton BR, Palde PB, Thornton CA, Miller BL (2008) Dynamic combinatorial selection of molecules capable of inhibiting the (CUG) repeat RNA-MBNL1 interaction in vitro: discovery of lead compounds targeting myotonic dystrophy (DM1). J Am Chem Soc 130:16254–16261
15. Lee MM, Childs-Disney JL, Pushechnikov A, French JM, Sobczak K, Thornton CA, Disney MD (2009) Controlling the specificity of modularly assembled small molecules for RNA via ligand module spacing: targeting the RNAs that cause myotonic muscular dystrophy. J Am Chem Soc 131:17464–17472
16. Parkesh R, Childs-Disney JL, Nakamori M, Kumar A, Wang E, Wang T, Hoskins J, Tran T, Housman DE, Thornton CA, Disney MD (2012) Design of a bioactive small molecule that targets the myotonic dystrophy type 1 RNA via an RNA motif-ligand database & chemical similarity searching. J Am Chem Soc 134:4731–4742
17. Pushechnikov A, Lee MM, Childs-Disney JL, Sobczak K, French JM, Thornton CA, Disney MD (2009) Rational design of ligands targeting triplet repeating transcripts that cause RNA dominant disease: application to myotonic muscular dystrophy type 1 and spinocerebellar ataxia type 3. J Am Chem Soc 131:9767–9779
18. Warf MB, Nakamori M, Matthys CM, Thornton CA, Berglund JA (2009) Pentamidine reverses the splicing defects associated with myotonic dystrophy. Proc Natl Acad Sci U S A 106:18551–18556
19. Velagapudi SP, Disney MD (2014) Two-dimensional combinatorial screening enables the bottom-up design of a microRNA-10b inhibitor. Chem Commun 50:3027–3029
20. Velagapudi SP, Gallo SM, Disney MD (2014) Sequence-based design of bioactive small molecules that target precursor microRNAs. Nat Chem Biol 10:291–297
21. Bose D, Jayaraj G, Suryawanshi H, Agarwala P, Pore SK, Banerjee R, Maiti S (2012) The tuberculosis drug streptomycin as a potential cancer therapeutic: inhibition of miR-21 function by directly targeting its precursor. Angew Chem Int Ed Engl 51:1019–1023
22. Chen RX, Xia YH, Xue TC, Ye SL (2012) Suppression of microRNA-96 expression inhibits the invasion of hepatocellular carcinoma cells. Mol Med Rep 5:800–804
23. Mathews DH, Burkard ME, Freier SM, Wyatt JR, Turner DH (1999) Predicting oligonucleotide affinity to nucleic acid targets. RNA 5:1458–1469
24. Moreno PMD, Pêgo AP (2014) Therapeutic antisense oligonucleotides against cancer: hurdling to the clinic. Front Chem 2:87
25. Disney MD, Seeberger PH (2004) Aminoglycoside microarrays to explore interactions of antibiotics with RNAs and proteins. Chemistry 10:3308–3314
26. Disney MD, Magnet S, Blanchard JS, Seeberger PH (2004) Aminoglycoside microarrays to study antibiotic resistance. Angew Chem Int Ed Engl 43:1591–1594
27. Disney MD, Labuda LP, Paul DJ, Poplawski SG, Pushechnikov A, Tran T, Velagapudi SP, Wu M, Childs-Disney JL (2008) Two-dimensional combinatorial screening identifies specific aminoglycoside-RNA internal loop partners. J Am Chem Soc 130:11185–11194
28. Childs-Disney JL, Wu M, Pushechnikov A, Aminova O, Disney MD (2007) A small molecule microarray platform to select RNA internal loop-ligand interactions. ACS Chem Biol 2:745–754
29. Rzuczek SG, Park H, Disney MD (2014) A toxic RNA catalyzes the in cellulo synthesis of its own inhibitor. Angew Chem Int Ed Engl 53:10956–10959
30. Kumar A, Parkesh R, Sznajder LJ, Childs-Disney JL, Sobczak K, Disney MD (2012) Chemical correction of pre-mRNA splicing defects associated with sequestration of muscleblind-like 1 protein by expanded r(CAG)-containing transcripts. ACS Chem Biol 7:496–505
31. Tinoco I Jr, Bustamante C (1999) How RNA folds. J Mol Biol 293:271–281
32. Shi H, Moore PB (2000) The crystal structure of yeast phenylalanine tRNA at 1.93 A resolution: a classic structure revisited. RNA 6:1091–1105
33. Jovine L, Djordjevic S, Rhodes D (2000) The crystal structure of yeast phenylalanine tRNA at 2.0 Å resolution: cleavage by Mg^{2+} in 15-year old crystals. J Mol Biol 301:401–414
34. Robertus JD, Ladner JE, Finch JT, Rhodes D, Brown RS, Clark BFC, Klug A (1974) Structure of yeast phenylalanine tRNA at 3 Å resolution. Nature 250:546–551

35. Quigley GJ, Teeter MM, Rich A (1978) Structural analysis of spermine and magnesium ion binding to yeast phenylalanine transfer RNA. Proc Natl Acad Sci U S A 75:64–68
36. Batey RT, Rambo RP, Doudna JA (1999) Tertiary motifs in RNA structure and folding. Angew Chem Int Ed Engl 38:2326–2343
37. Mathews DH, Sabina J, Zuker M, Turner DH (1999) Expanded sequence dependence of thermodynamic parameters improves prediction of RNA secondary structure. J Mol Biol 288:911–940
38. Mathews DH, Disney MD, Childs JL, Schroeder SJ, Zuker M, Turner DH (2004) Incorporating chemical modification constraints into a dynamic programming algorithm for prediction of RNA secondary structure. Proc Natl Acad Sci U S A 101:7287–7292
39. Leontis NB, Stombaugh J, Westhof E (2002) The non-Watson-Crick base pairs and their associated isostericity matrices. Nucleic Acids Res 30:3497–3531
40. Woese CR, Fox GE (1977) Phylogenetic structure of the prokaryotic domain: the primary kingdoms. Proc Natl Acad Sci U S A 74:5088–5090
41. Eloe-Fadrosh EA, Paez-Espino D, Jarett J, Dunfield PF, Hedlund BP, Dekas AE, Grasby SE, Brady AL, Dong H, Briggs BR, Li W-J, Goudeau D, Malmstrom R, Pati A, Pett-Ridge J, Rubin EM, Woyke T, Kyrpides NC, Ivanova NN (2016) Global metagenomic survey reveals a new bacterial candidate phylum in geothermal springs. Nat Commun 7:10476
42. Mathews DH, Moss WN, Turner DH (2010) Folding and finding RNA secondary structure. Cold Spring Harb Perspect Biol 2:a003665
43. Zuker M, Jaeger JA, Turner DH (1991) A comparison of optimal and suboptimal RNA secondary structures predicted by free energy minimization with structures determined by phylogenetic comparison. Nucleic Acids Res 19:2707–2714
44. Ziehler WA, Engelke DR (2001) Probing RNA structure with chemical reagents and enzymes. Curr Protoc Nucleic Acid Chem Chapter 6:Unit 6.1
45. Stormo GD (2006) An overview of RNA structure prediction and applications to RNA gene prediction and RNAi design. Curr Protoc Bioinformatics Chapter 12:Unit 12.11
46. Maxam AM, Gilbert W (1977) A new method for sequencing DNA. Proc Natl Acad Sci U S A 74:560–564
47. Tijerina P, Mohr S, Russell R (2007) DMS footprinting of structured RNAs and RNA-protein complexes. Nat Protoc 2:2608–2623
48. Gopinath SCB (2009) Mapping of RNA–protein interactions. Anal Chim Acta 636:117–128
49. Mortimer SA, Weeks KM (2007) A fast-acting reagent for accurate analysis of RNA secondary and tertiary ttructure by SHAPE chemistry. J Am Chem Soc 129:4144–4145
50. Doudna JA, Cate JH (1997) RNA structure: crystal clear? Curr Opin Struct Biol 7:310–316
51. Fürtig B, Richter C, Wöhnert J, Schwalbe H (2003) NMR spectroscopy of RNA. Chembiochem 4:936–962
52. Ulyanov NB, Mujeeb A, Du Z, Tonelli M, Parslow TG, James TL (2006) NMR structure of the full-length linear dimer of stem-loop-1 RNA in the HIV-1 dimer initiation site. J Biol Chem 281:16168–16177
53. Doudna JA, Cech TR (1995) Self-assembly of a group I intron active site from its component tertiary structural domains. RNA 1:36–45
54. Ban N, Nissen P, Hansen J, Moore PB, Steitz TA (2000) The complete atomic structure of the large ribosomal subunit at 2.4 Å resolution. Science 289:905–920
55. Carter AP, Clemons WM, Brodersen DE, Morgan-Warren RJ, Wimberly BT, Ramakrishnan V (2000) Functional insights from the structure of the 30S ribosomal subunit and its interactions with antibiotics. Nature 407:340–348
56. Hansen JL, Ippolito JA, Ban N, Nissen P, Moore PB, Steitz TA (2002) The structures of four macrolide antibiotics bound to the large ribosomal subunit. Mol Cell 10:117–128
57. Nissen P, Hansen J, Ban N, Moore PB, Steitz TA (2000) The structural basis of ribosome activity in peptide bond synthesis. Science 289:920–930
58. Pioletti M, Schlunzen F, Harms J, Zarivach R, Gluhmann M, Avila H, Bashan A, Bartels H, Auerbach T, Jacobi C, Hartsch T, Yonath A, Franceschi F (2001) Crystal structures of complexes of the small ribosomal subunit with tetracycline, edeine and IF3. EMBO J 20:1829–1839

59. Wimberly BT, Brodersen DE, Clemons WM Jr, Morgan-Warren RJ, Carter AP, Vonrhein C, Hartsch T, Ramakrishnan V (2000) Structure of the 30S ribosomal subunit. Nature 407:327–339
60. Yusupov MM, Yusupova GZ, Baucom A, Lieberman K, Earnest TN, Cate JHD, Noller HF (2001) Crystal structure of the ribosome at 5.5 Å resolution. Science 292:883–896
61. Deigan KE, Li TW, Mathews DH, Weeks KM (2009) Accurate SHAPE-directed RNA structure determination. Proc Natl Acad Sci U S A 106:97–102
62. Wimberly BT (2009) The use of ribosomal crystal structures in antibiotic drug design. Curr Opin Investig Drugs 10:750–765
63. Wong CH, Hendrix M, Priestley ES, Greenberg WA (1998) Specificity of aminoglycoside antibiotics for the A-site of the decoding region of ribosomal RNA. Chem Biol 5:397–406
64. Fourmy D, Yoshizawa S, Puglisi JD (1998) Paromomycin binding induces a local conformational change in the A-site of 16 S rRNA. J Mol Biol 277:333–345
65. Recht MI, Fourmy D, Blanchard SC, Dahlquist KD, Puglisi JD (1996) RNA sequence determinants for aminoglycoside binding to an A-site rRNA model oligonucleotide. J Mol Biol 262:421–436
66. Kaul M, Barbieri CM, Pilch DS (2006) Aminoglycoside-induced reduction in nucleotide mobility at the ribosomal RNA A-site as a potentially key determinant of antibacterial activity. J Am Chem Soc 128:1261–1271
67. Barbieri CM, Kaul M, Pilch DS (2007) Use of 2-aminopurine as a fluorescent tool for characterizing antibiotic recognition of the bacterial rRNA A-site. Tetrahedron 63:3567–6574
68. Kondo J, Pachamuthu K, Francois B, Szychowski J, Hanessian S, Westhof E (2007) Crystal structure of the bacterial ribosomal decoding site complexed with a synthetic doubly functionalized paromomycin derivative: a new specific binding mode to an a-minor motif enhances in vitro antibacterial activity. ChemMedChem 2:1631–1638
69. Blount KF, Zhao F, Hermann T, Tor Y (2005) Conformational constraint as a means for understanding RNA-aminoglycoside specificity. J Am Chem Soc 127:9818–9829
70. Wang H, Tor Y (1998) RNA-aminoglycoside interactions: design, synthesis, and binding of "amino-aminoglycosides" to RNA. Angew Chem Int Ed Engl 37:109–111
71. Llano-Sotelo B, Azucena EF Jr, Kotra LP, Mobashery S, Chow CS (2002) Aminoglycosides modified by resistance enzymes display diminished binding to the bacterial ribosomal aminoacyl-tRNA site. Chem Biol 9:455–463
72. Haddad J, Kotra LP, Llano-Sotelo B, Kim C, Azucena EF Jr, Liu M, Vakulenko SB, Chow CS, Mobashery S (2002) Design of novel antibiotics that bind to the ribosomal acyltransfer site. J Am Chem Soc 124:3229–3237
73. Lind KE, Du Z, Fujinaga K, Peterlin BM, James TL (2002) Structure-based computational database screening, in vitro assay, and NMR assessment of compounds that target TAR RNA. Chem Biol 9:185–193
74. Hergenrother PJ, Depew KM, Schreiber SL (2000) Small-molecule microarrays: covalent attachment and screening of alcohol-containing small molecules on glass slides. J Am Chem Soc 122:7849–7850
75. Disney MD, Barrett OJ (2007) An aminoglycoside microarray platform for directly monitoring and studying antibiotic resistance. Biochemistry 46:11223–11230
76. Chow CS, Mahto SK, Lamichhane TN (2008) Combined approaches to site-specific modification of RNA. ACS Chem Biol 3:30–37
77. Zhang W, Fisher JF, Mobashery S (2009) The bifunctional enzymes of antibiotic resistance. Curr Opin Microbiol 12:505–511
78. Disney MD, Childs-Disney JL (2007) Using selection to identify and chemical microarray to study the RNA internal loops recognized by 6′-*N*-acylated kanamycin A. Chembiochem 8:649–656
79. Liquori CL, Ricker K, Moseley ML, Jacobsen JF, Kress W, Naylor SL, Day JW, Ranum LP (2001) Myotonic dystrophy type 2 caused by a CCTG expansion in intron 1 of *ZNF9*. Science 293:864–867

80. Konieczny P, Stepniak-Konieczna E, Sobczak K (2014) MBNL proteins and their target RNAs, interaction and splicing regulation. Nucleic Acids Res 42:10873–10887
81. Lee MM, Pushechnikov A, Disney MD (2009) Rational and modular design of potent ligands targeting the RNA that causes myotonic dystrophy 2. ACS Chem Biol 4:345–355
82. Childs-Disney JL, Yildirim I, Park H, Lohman JR, Guan L, Tran T, Sarkar P, Schatz GC, Disney MD (2014) Structure of the myotonic dystrophy type 2 RNA and designed small molecules that reduce toxicity. ACS Chem Biol 9:538–550
83. Kwon Y-U, Kodadek T (2007) Quantitative evaluation of the relative cell permeability of peptoids and peptides. J Am Chem Soc 129:1508–1509
84. Wong CH, Fu Y, Ramisetty SR, Baranger AM, Zimmerman SC (2011) Selective inhibition of MBNL1-CCUG interaction by small molecules toward potential therapeutic agents for myotonic dystrophy type 2 (DM2). Nucleic Acids Res 39:8881–8890
85. Nguyen L, Lee J, Wong CH, Zimmerman SC (2014) Small molecules that target the toxic RNA in myotonic dystrophy type 2. ChemMedChem 9:2455–2462
86. Lewis WG, Green LG, Grynszpan F, Radic Z, Carlier PR, Taylor P, Finn MG, Sharpless KB (2002) Click chemistry in situ: acetylcholinesterase as a reaction vessel for the selective assembly of a femtomolar inhibitor from an array of building blocks. Angew Chem Int Ed Engl 41:1053–1057
87. Mahadevan M, Tsilfidis C, Sabourin L, Shutler G, Amemiya C, Jansen G, Neville C, Narang M, Barcelo J, O'Hoy K, et al (1992) Myotonic dystrophy mutation: an unstable CTG repeat in the 3′ untranslated region of the gene. Science 255:1253–1255
88. Miller JW, Urbinati CR, Teng-umnuay P, Stenberg MG, Byrne BJ, Thornton CA, Swanson MS (2000) Recruitment of human muscleblind proteins to (CUG)n expansions associated with myotonic dystrophy. EMBO J 19:4439–4448
89. Mankodi A, Logigian E, Callahan L, McClain C, White R, Henderson D, Krym M, Thornton CA (2000) Myotonic dystrophy in transgenic mice expressing an expanded CUG repeat. Science 289:1769–1773
90. Ofori LO, Hoskins J, Nakamori M, Thornton CA, Miller BL (2012) From dynamic combinatorial 'hit' to lead: in vitro and in vivo activity of compounds targeting the pathogenic RNAs that cause myotonic dystrophy. Nucleic Acids Res 40:6380–6390
91. Arambula JF, Ramisetty SR, Baranger AM, Zimmerman SC (2009) A simple ligand that selectively targets CUG trinucleotide repeats and inhibits MBNL protein binding. Proc Natl Acad Sci U S A 106:16068–16073
92. Wong C-H, Nguyen L, Peh J, Luu LM, Sanchez JS, Richardson SL, Tuccinardi T, Tsoi H, Chan WY, Chan HYE, Baranger AM, Hergenrother PJ, Zimmerman SC (2014) Targeting toxic RNAs that cause myotonic dystrophy Type 1 (DM1) with a bisamidinium inhibitor. J Am Chem Soc 136:6355–6361
93. Siboni RB, Nakamori M, Wagner SD, Struck AJ, Coonrod LA, Harriott SA, Cass DM, Tanner MK, Berglund JA (2015) Actinomycin D specifically reduces expanded CUG repeat RNA in myotonic dystrophy models. Cell Rep 13:2386–2394

Top Med Chem (2018) 27: 17–46
DOI: 10.1007/7355_2016_23

Published online: 10 May 2017

A Modular Approach to the Discovery and Affinity Maturation of Sequence-Selective RNA-Binding Compounds

Benjamin L. Miller

Abstract This chapter describes a strategy developed at the University of Rochester that relies on a two-step process for the generation of high-affinity, sequence-selective RNA-binding compounds with target-relevant biological activity. First, a natural product-inspired dynamic combinatorial library (DCL) is employed to rapidly produce "hit" compounds able to bind the target RNA. Second, a process of analog synthesis is employed to enhance affinity, bioavailability, and sequence selectivity. This strategy has been used to successfully produce compounds able to bind target RNAs with high (nanomolar) affinity, and target-relevant activity in cellular assays and in vivo. In particular, approaches to RNA targets of critical importance in Myotonic Dystrophy (a triplet repeat RNA-mediated disease) and in the life cycle of HIV will be discussed.

Keywords Bioisosteres, Dynamic combinatorial chemistry, Frameshifting, HIV, Myotonic dystrophy, Natural products

Contents

1 Introduction 18
2 Origins of the Method 18
3 Initial Demonstrations with Coordination Complexes 19
4 Towards more Complex Library Analysis: Resin-Bound Dynamic Combinatorial Chemistry 20
5 Library Design 21
6 Case Studies in the RNA World 23
7 Triplet Repeat RNA 23
8 HIV Frameshift-Stimulating RNA 24
9 Library Selections 25

B.L. Miller (✉)
Department of Dermatology, University of Rochester Medical Center, Rochester, NY 14642, USA
e-mail: benjamin_miller@urmc.rochester.edu

9.1 Initial Results: HIV-1 FSS RNA [48] ... 27
9.2 $(CUG)^{exp}$ RNA [49] ... 28
10 Building from RBDCC: Enhancing Affinity, Selectivity, and Stability via Medicinal Chemistry ... 29
10.1 Replace Disulfides with Non-labile Bioisosteres ... 29
10.2 Enhance Affinity Through Pi-System Extension ... 31
10.3 Constrain Conformation via Amide N-Methylation ... 31
10.4 Enhancing Affinity, Selectivity, and Bioavailability for $(CUG)^{exp}$ ... 35
11 Beyond Dimers: Accelerating Compound Development Through Orthogonal and Bidirectional Exchange ... 38
11.1 Orthogonal Exchange Chemistry ... 39
11.2 Dynamic Linker Incorporation ... 39
12 Conclusions ... 41
References ... 42

1 Introduction

Until recently, RNA was largely ignored as a target for molecular discovery. Only in recent years have biologists and chemists studying RNA learned that, rather than a simple intermediary between DNA and proteins, RNA has a rich functional landscape all its own. Recognition of RNA's regulatory functions, structural diversity, and central role in many human diseases gave impetus to the goal of developing broadly applicable strategies for the design of molecules capable of sequence- and structure-selective recognition of RNA. While the field remains a long way from being able to rapidly generate highly selective non-nucleotide based molecules for any arbitrary RNA sequence (as is now largely a solved problem for many DNA sequences [1]), researchers working in many groups around the globe have made significant strides towards this goal [2–4]. This volume contains several chapters written by other researchers active in this field. This chapter will focus on an approach developed at the University of Rochester over the past two decades.

At the outset of our efforts in this area, there were few examples of successful RNA-targeted molecular discovery. As such, directed ligand design was not feasible, and establishing a high-throughput screening facility was beyond available resources. Therefore, we sought an alternative approach. In particular, methods that might allow a biological target to select *and amplify* its own ligand were deemed worthy of pursuit. This led us (and, as it turned out, several other groups simultaneously) to develop the concept of Dynamic Covalent Chemistry, or Dynamic Combinatorial Chemistry.

2 Origins of the Method

Dynamic processes have been at the heart of chemistry since its inception. There have been numerous reviews of the development of Dynamic Combinatorial Chemistry (DCC) [5–7], and as such that history will not be reviewed in detail here. One seminal contribution that should be highlighted, however, is a 1986 paper by Rideout describing the in situ assembly of hydrazone cytotoxins from component hydrazines and carbonyl compounds [8]. This early demonstration of a self-assembled molecule gaining biological activity, in conjunction with templated synthetic chemistry described by Feldman and coworkers [9], spontaneous resolution of dynamic equilibria of enantiomers under thermodynamic control [10], and self-assembled, self-selected coordination networks reported by the Lehn group [11], led us to the following hypothesis (Fig. 1). If one were to mix molecular fragments capable of reversible assembly, this would generate an initial equilibrium mixture, or dynamic combinatorial library (DCL). Adding a target of interest (for example, an RNA sequence) to this dynamic mixture would shift the equilibrium according to Le Chatelier's principle, as transiently formed compounds bound to the target in proportion to their affinity. After a suitable re-equilibration time, one could identify the highest affinity compound based on the change in concentrations for library members. In the end, this analytical challenge was perhaps the most difficult of the DCL process. Early contributions by other groups included important publications from the Sanders group in which transesterification of cholic acid derivatives was used to produce self-selecting mixtures of receptors [12], from Still describing the dynamic selection of macrocyclic receptors for resin-immobilized peptides via disulfide exchange [13], and the Lehn group's identification of small molecules binding carbonic anhydrase via imine exchange followed by sodium cyanoborohydride reduction to halt equilibration [14].

3 Initial Demonstrations with Coordination Complexes

An attractive aspect of DCC with regard to its application to nucleic acid-targeted compound discovery is that the technique is inherently modular, and therefore amenable to the extended binding surface presented by many nucleic acids. Initial efforts in the Rochester group focused on the recognition of DNA using ligand exchange in transition metal coordination complexes. Use of self-assembled libraries of transition metal complexes was inspired both by Nature (in the context of zinc finger proteins, sequence-selective DNA binding proteins which gain structural stability by coordination to $Zn^{(2+)}$) [15], by the aforementioned work of Lehn, and by early studies from the Schepartz group. In the Schepartz effort, salicylaldehyde was condensed with amines

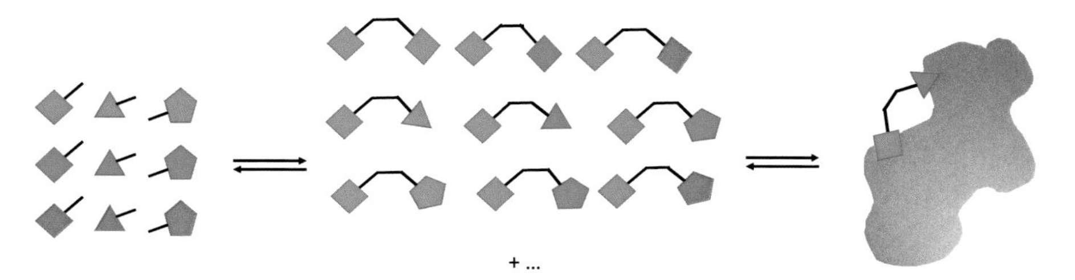

Fig. 1 Dynamic Combinatorial Chemistry (DCC)

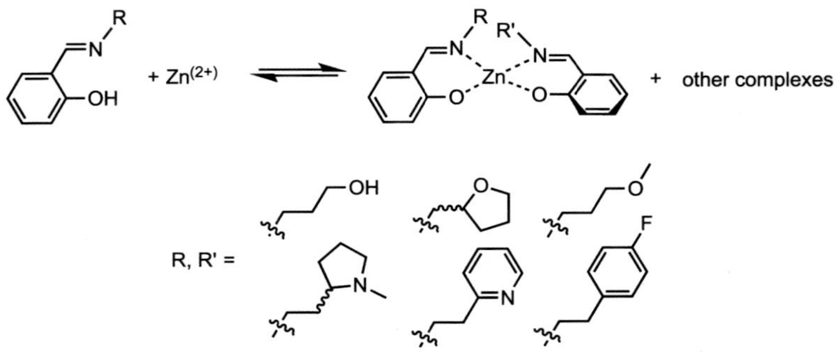

Fig. 2 Metal salicylaldimines employed in a DNA-targeted dynamic combinatorial library (DCL)

to form "self-assembled ionophores," selective for different ions [16]. With these precedents, we began by synthesizing a small library of six salicylaldimines (Fig. 2), and incubating those with resin-bound homopolymeric (d(T)•d(A)) DNA in the presence and absence of $Zn^{(2+)}$ ions. Analyzing material eluted from the column (i.e., library components not bound to the DNA resin) revealed a decrease in concentration of an N-methyl pyrrolidine-containing ligand, consistent with its $Zn^{(2+)}$-dependent selection by the DNA. Analysis of binding of this ligand to DNA in the presence and absence of zinc confirmed both its affinity and requirement for the presence of $Zn^{(2+)}$ [17, 18]. A subsequent RNA-targeted study using $Cu^{(2+)}$ as the coordinating metal similarly yielded selection from a small library of salicylamides; here, equilibrium dialysis was employed in the selection scheme [19].

While these proof-of-principle experiments with metal salicylamides and salicylaldimines demonstrated the ability of DCC to yield both DNA- and RNA-binding compounds, transitioning the method to the identification of actually *useful* structures would require modifications to the approach. First, as discussed above, DCC as originally developed in our lab and others required the examination of library selections pre- and post-incubation with a target by HPLC or HPLC-mass spectrometry. While there have been rare examples of extending this methodology to large (>1,000 compound) libraries [20], the analytical challenge of doing so prevented general use of DCC with libraries of sufficient numerical and structural diversity to prove broadly useful in the context of RNA-targeted compound discovery. Second, while interesting, the coordination complexes that were an initial focus of our work did not have a clear path to use in cells and in vivo due to their solution lability. Note that several other outstanding DNA-targeted DCC experiments have been described by both Balasubramanian [21] and Lehn [22], in each case using DCC to identify G-quadruplex targeted compounds.

4 Towards more Complex Library Analysis: Resin-Bound Dynamic Combinatorial Chemistry

In order to simplify the process of identifying active compounds from a complex DCL, we followed a path that had been successful in the context of oligonucleotide analysis: phase separation. Immobilization of oligonucleotides on spatially segregated spots in microarrays enabled a host of high-throughput biological experiments [23]. Therefore, we reasoned that by immobilizing library components in a microarray format, and allowing a second set of identical library components to react with the array-immobilized materials, one could rapidly synthesize and screen (using a fluorophore-tagged target) large dynamic libraries of dimers. Unfortunately, while attractive conceptually, this idea faced an insurmountable hurdle: given the typical spot size of a microarray, there would be insufficient material in each spot to outcompete formation of compounds in solution, given sufficient solution monomer concentration for rapid exchange. Therefore, we focused on an alternative that provided the advantages of phase segregation in combination with larger amounts of material: library synthesis on resin beads. In this "Resin-Bound Dynamic Combinatorial Chemistry" (RBDCC) strategy (Fig. 3), each bead would carry a unique DCL monomer, synthesized using standard split-pool methods [24]. When allowed to undergo dynamic exchange with a mixture of DCL monomers in solution, a sub-library forms on each bead. Selection under equilibrating conditions allows the distribution of compounds on each bead to change, but, importantly, the original monomer synthesized on bead remains invariant. After halting dynamic exchange, resin beads carrying the highest affinity combinations may be identified by fluorescence microscopy, removed from the remaining pool, and analyzed by mass spectrometry.

5 Library Design

RBDCC provides what is essentially a "theme and variations" approach to compound discovery. The "theme" in this case is the underlying structure of the library and an exchange reaction, varied by monomer diversity. A potentially useful "theme" for nucleic acid-targeted library design was suggested by the structure of bisintercalating peptide antibiotics. Represented by molecules such as Echinomycin (**1**) and Triostin A (**2**), these compounds achieve selective recognition of DNA through a combination of relatively nonspecific binding interactions via intercalation, in combination with

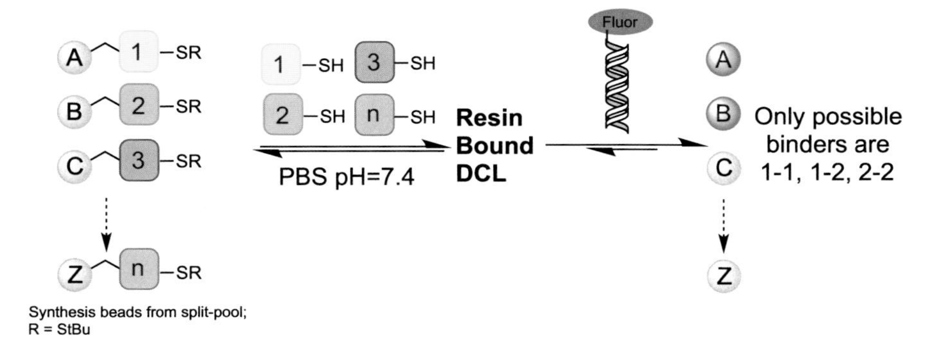

Fig. 3 Resin-bound dynamic combinatorial chemistry (RBDCC)

specific interactions with nucleic acid bases via the peptide backbone and side chains [25]. One could in principle fully mimic these natural products by applying both trans-esterification and disulfide exchange to create bicyclic structures. Thus far, however, we have focused on linear dimers equilibrated via disulfide exchange. Initially examined in depth by the Whitesides group [26] and implemented in the context of DCC independently by Still and Sanders (vide supra), disulfide exchange is ideal for DCC experiments because of its ability to proceed under mild aqueous conditions, compatibility with a broad range of chemical functionality, and, critically, by the ability to halt equilibration easily. At buffer pH values near or above the pKa of an alkyl thiol, exchange occurs via rapid S_N2-like displacement of thiolate. If the pH is dropped below the thiol pKa, this reaction is effectively halted allowing identification of the selected compound. In practice, this means that exchange occurs with reasonable facility at physiological pH (7.4) under standard buffering conditions and is halted below pH 7.

Initial testing of the RBDCC concept was accomplished by synthesis of a simple library of nine dipeptide monomers (Fig. 4) [27]. Cysteine provided the thiol. The library was screened against two duplex DNA sequences, one known to bind triostin A (**DNA_1**: 5′-TCTAGACGTC-3′) [28], and one preferentially bound by the related synthetic compound TANDEM (**DNA_2**: 5′-CCATGATATC-3′) [29]. Library screening was initially conducted in solution phase, by analogy to "traditional" DCC screens. While HPLC traces showed clear perturbation of the equilibrium composition of the library in the presence of **DNA_2**, identification of the highest affinity compound was not possible given the complexity of the equilibrium. **DNA_1** did not produce any obvious changes in the library HPLC chromatogram.

Next, the RBDCC concept was tested. Nine reaction vessels were prepared containing resin beads bearing each of the nine monomer compounds. A solution containing all nine monomers was then added to each tube, along with fluorescently tagged **DNA_2**. After allowing the samples to reach equilibrium (the length of time needed for this was determined using model peptide monomers), the exchange reaction was halted, and beads were thoroughly washed with buffer to remove unbound DNA. Examination of resin beads from each tube by fluorescence microscopy unambiguously revealed that those bearing monomer **7** were highly fluorescent, while remaining beads had only modest or no fluorescence at the same exposure. This result suggested

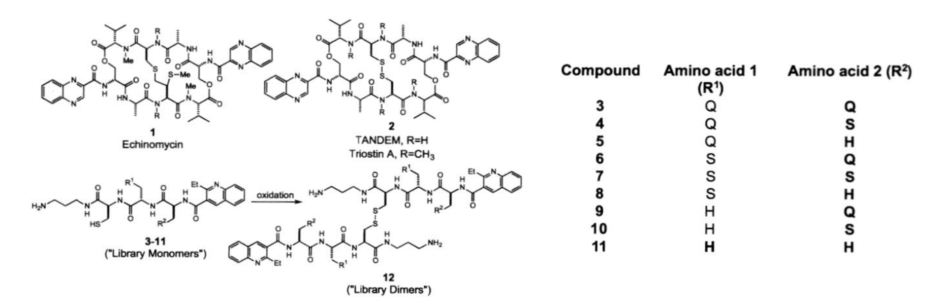

Compound	Amino acid 1 (R^1)	Amino acid 2 (R^2)
3	Q	Q
4	Q	S
5	Q	H
6	S	Q
7	S	S
8	S	H
9	H	Q
10	H	S
11	H	H

Fig. 4 Proof of concept DNA-targeted resin-bound dynamic combinatorial library (RBDCL)

that homodimer **7-7** was the highest affinity binder. Confirmation of this observation was provided by NMR titrations and by equilibrium dialysis: while **7-7** was found to have a dissociation constant (K_D) of 2.8 μM, the non-selected compound **10-10** was found to bind **DNA_2** with a K_D of 10.8 μM. These results confirmed that the RBDCC protocol could differentiate between compounds of differing affinities. It remained to expand RBDCC to a larger library size, conduct selection experiments under true complex mixture conditions rather than in tubes separated by resin-bound monomer, and apply the concept to an RNA target.

6 Case Studies in the RNA World

The demonstration experiments using a simple library undergoing selection against a DNA sequence verified that the RBDCC method was capable of yielding nucleic acid binding compounds. Could one expand this to a larger library, and, more importantly, would that yield compounds that had sufficient selectivity and affinity for nucleic acid targets to be useful (following a modicum of medicinal chemistry) for downstream biological studies? Over the past decade, we have demonstrated that the answer to both questions is "yes." We were particularly eager to apply this methodology to RNA-targeted compound discovery. Published work from our group has focused on two RNA targets, described in detail below. As early work proceeded in parallel, and results obtained from one target influenced experiments focused on the other, we will first discuss the targets themselves, and then discuss the screening and medicinal chemistry efforts in tandem.

7 Triplet Repeat RNA

Type I Myotonic Dystrophy (DM1) is a debilitating disease characterized by several symptoms including muscle wasting, myotonia (an inability to relax voluntary muscle after contraction), cataracts, cardiac defects, and testicular atrophy [30]. The disease results from the expansion of CTG repeats within the 3′ UTR of the *DMPK* gene (Fig. 5); its severity correlates with the number of CTG repeat units present [31, 32]. Central to DM1 pathogenesis is that when these repeat units are transcribed into CUG repeat RNA (referred to as $(CUG)^{exp}$), they yield transcripts capable of binding a splicing factor known as muscleblind, or MBNL1. Bound up as complex RNA-protein foci within cell nuclei, these structures prevent MBNL1 from carrying out its normal splicing function. As targets of muscleblind include proteins essential for function of muscle cells, muscle dysfunction ensues. One attractive strategy for addressing DM1 pharmacologically, therefore, would be to produce compounds able to specifically bind $(CUG)^{exp}$ RNA in competition with MBNL1. Extensive efforts by the Thornton group demonstrated early on that this could be accomplished with antisense oligonucleotides, and a construct descended from those early studies is currently in clinical trials. In the small- to medium-sized molecule regime, several innovative approaches to $(CUG)^{exp}$

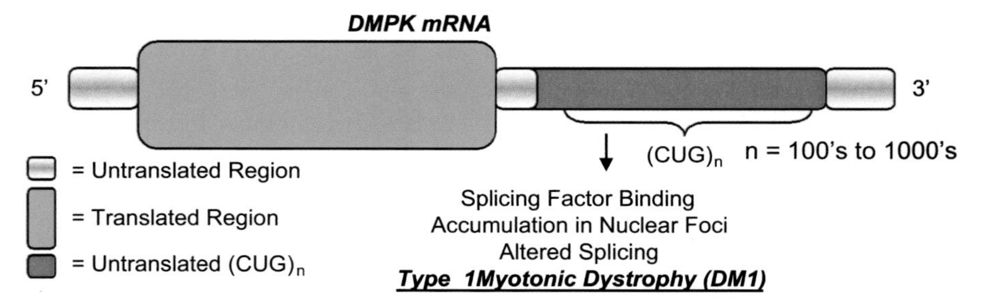

Fig. 5 Expansion of CUG repeats in the 3′ untranslated region of *DMPK* leads to Type 1 Myotonic Dystrophy (DM1)

RNA recognition have been described in addition to the work we summarize here (selected examples include: Pushechnikov et al., Arambula et al., and Warf et al. [33–35]). A closely related disease, type 2 Myotonic Dystrophy or DM2, likewise results from nucleotide repeat expansion, but the sequence in this case at the RNA level is expanded tetranucleotide repeats, or $(CCUG)^{exp}$ [36]. Structurally, $(CUG)^{exp}$ RNA forms hairpins regardless of length, made up of C*G base pairs and U*U bulges [37]. This makes it ideally suited for a modular approach to compound discovery, and multivalent recognition.

8 HIV Frameshift-Stimulating RNA

Nature uses a fascinating range of processes to control the production of proteins from genetic information. One of the most surprising is the process of recoding, or "frameshifting" [38]. This allows for the product of an mRNA to be changed at the level of translation, and typically relies on modulation of ribosomal behavior by specific RNA structures. While several types of frameshifting are possible [39], we will focus here on a process resulting in a 1-nucleotide backwards slippage of the ribosome, also known as a −1 programmed ribosomal frameshifting (−1 PRF). This process is used to regulate the ratio of two polyproteins critical to the life cycle of HIV: Gag, which undergoes further processing to yield HIV structural proteins (capsid, matrix, and nucleocapsid) and Pol, which is processed to yield several viral enzymes (HIV protease, integrase, and reverse transcriptase). Pol is produced only as a fusion with Gag, and only if a −1 PRF event occurs [40]. The process goes like this: 90–95% of the time, a ribosome processing the *gag-pol* mRNA produces Gag, reaches a stop codon, and falls off. However, with a frequency of 5–10%, during this process the ribosome pauses and shifts backwards 1 nucleotide on a so-called "slippery sequence" of repeating uridines, taking it out of the reading frame that includes the Gag stop codon, and into the Pol reading frame. This enables the production of Gag-Pol. −1 PRF in HIV is thought to be regulated by interaction of the ribosome with an exceptionally stable ($T_m = >90°$) [41] stem-loop structure known as the "HIV-1 Frameshift

Stimulatory Sequence," or HIV-1 FSS, just downstream of the slippery sequence (Fig. 6). Dual plasmid experiments verified that the ratio of Gag to Gag-Pol is critical to the infectivity of HIV-1; variation of the ratio in either direction results in the production of virus that is dramatically less infective [43]. The importance of both the slippery sequence and the FSS is reflected in their resistance to mutation [44]. As such, small-molecule mediated interference with HIV-1 −1 PRF has been a recognized goal for some time, since researchers hypothesized that sensitivity of the virus to FSS mutation would reduce its ability to evolve resistance to compounds targeting this sequence. The first report of a compound able to influence frameshifting appeared in 1998 [45]. Here, bis-benzamide **13** was found to inhibit HIV-1 via what was believed to be a frameshift-dependent mechanism. Subsequent work by the Butcher group, however, revealed that **13** actually only binds the FSS weakly ($K_D = 360\ \mu M$) and is toxic at a concentration far below this [46]. Successful inhibition of HIV via modulation of the frameshift process would therefore require substantially higher affinity and specificity for FSS binding.

9 Library Selections

Experiments targeting both the HIV-1 FSS and CUG RNAs proceeded in identical fashion, building on the previously described DNA-targeting proof-of-concept work. A library of 150 tripeptide "monomers" incorporating cysteine residues and a

Fig. 6 The HIV-1 frameshift element consists of a "slippery sequence" and highly stable upper stem loop. Other elements of the secondary structure shown may be different in the context of the full HIV-1 genome [42]

heterocyclic cap was synthesized using split-pool techniques on solid phase (Fig. 7). A photocleavable linker [47] was employed to allow for downstream cleavage of library products via photolysis, for subsequent identification by mass spectrometry. Library diversity was provided by varying two amino acids, the heterocyclic cap, and the position of the cysteine. The 50 combinations of varied amino acids and heterocyclic cap were selected so as to provide both a diversity of functionality (charged/uncharged) and to be uniquely identifiable by mass. The position of the cysteine residue was encoded by the size of the resin bead used, with small (155 μm), medium (225 μm), and large (300 μm) resin employed for encoding cysteine in the first, second, and third position, respectively. Resin loading was normalized across the three bead sizes by selective acetylation. Following synthesis and deprotection, half the resin was cleaved, providing a mixture of 150 monomers for use in solution, and an identical set of 150 monomers on bead. Resin-bound monomers were checked via fluorescence microscopy to ensure that there were no intrinsically fluorescent species. Likewise, library monomers on bead were incubated with fluorescently tagged RNA targets to confirm that monomers alone had negligible affinity. Next, the RBDCC experiments themselves were conducted. As with the DNA-targeted proof-of-concept example, libraries were allowed to equilibrate with RNA under exchange conditions for a period of days before exchange was halted, unbound material eluted from the system, and resin beads washed. Library beads carrying dimer compounds able to bind the RNA were then identified via fluorescence microscopy. Individual beads were photolyzed using a handheld UV lamp to cleave bound materials, and analyzed via mass spectrometry. As RBDCC employs stoichiometry in which the bead-bound monomer is in excess (driving the equilibrium towards binding to the resin), mass spectral identification only reveals the identity of monomers, not complete compounds. While this requires a

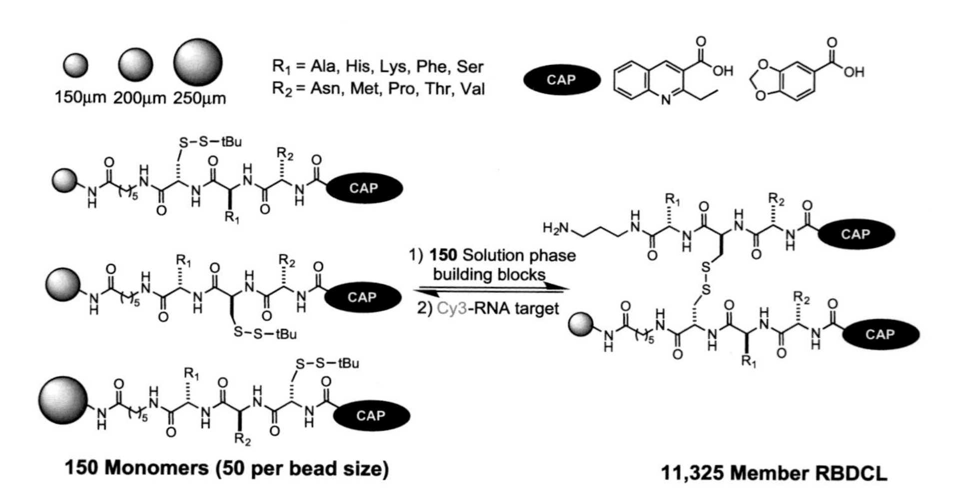

Fig. 7 RBDCL employed in RNA-targeted experiments

secondary screen or testing of individual combinations of monomers post-screen, it nonetheless provided a dramatic simplification relative to full mixture-based screening.

9.1 Initial Results: HIV-1 FSS RNA [48]

Replicate screens conducted with the HIV-1 FSS RNA revealed three candidate monomers involved in binding interactions, representing nine possible combinations (counting symmetry breaking via bead attachment) (Fig. 8). To confirm the results of the initial screens, a secondary screen was conducted using only these three monomers on bead, and the same three monomers in solution, in nine separate reactions. This experiment confirmed that the RBDCC process had indeed identified binding compounds, and also highlighted the competitive nature of RBDCC: while the heterodimer of **14** and **15** captured fluorescently tagged RNA when **14** was on bead and allowed to undergo exchange with **15** in solution, it did not when **15** was on bead and **14** was in solution. This suggested that the formation of homodimer **14-14** in solution outcompeted the heterodimer for binding. Indeed, the reaction in which **14-14** was formed on bead yielded the strongest fluorescence, suggesting that this compound had the highest affinity. Resynthesis of these compounds followed by SPR analysis (here, with the compound immobilized on the SPR chip and RNA flowed in solution, all subsequent SPR experiments used a mode in which the RNA was immobilized) confirmed that **16** bound the HIV-1 FSS with a K_D of 4.1 ± 2.4 μM. While sequence

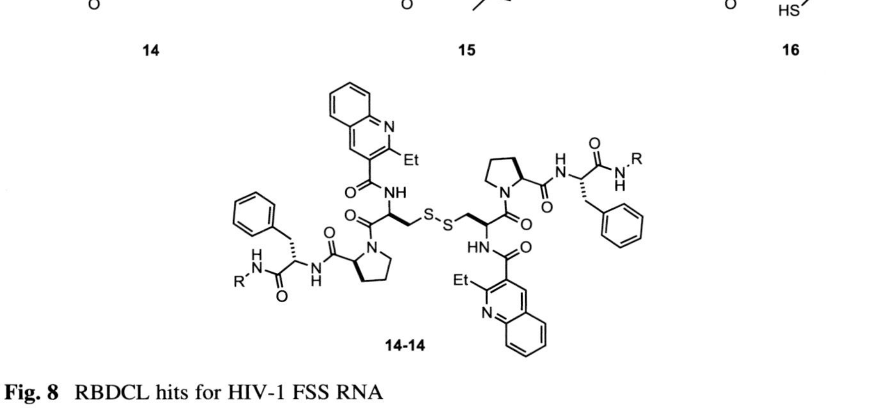

Fig. 8 RBDCL hits for HIV-1 FSS RNA

selectivity was not explored in detail, two closely related hairpins bound with K_Ds more than tenfold weaker than the FSS.

9.2 *(CUG)^exp RNA [49]*

Screening the library against CUG repeat RNA proceeded in an analogous fashion, here using Cy3-tagged $(CUG)_{10}$ as the target. Analysis of replicate screens revealed four monomers selected (Fig. 9); no overlap was observed between HIV-1 FSS RNA and $(CUG)_{10}$ RNA selections. In this case, all ten possible unique combinations of the four monomers were synthesized, purified, and tested. A slot blot assay was employed to provide an initial indication of affinity. This assay suggested that several of the compounds had similar affinities, and all four monomers were able to participate in at least one moderate-affinity binder. Results from this assay confirmed that none of the beads selected in the RBDCC process yielded false positive hits.

Next, the ability of compounds to inhibit MBNL1 binding to $(CUG)^{exp}$ RNA was tested. Here, an enzyme fragment complementation assay was employed. In this system, $(CUG)_{109}$ RNA was first immobilized in wells of a 96-well plate, and allowed to bind MBNL1 fused to an enzyme donor peptide. When this binds an "enzyme acceptor" (a beta-galactosidase construct), it produces a functional enzyme able to process a substrate producing a chemiluminescent signal [50]. Competition for MBNL1 binding by compounds was anticipated to interfere with enzyme complementation. Indeed, while compounds provided a spectrum of activity, all were able to inhibit MBNL1 binding with IC_{50} values similar to the K_D values measured by slot blot.

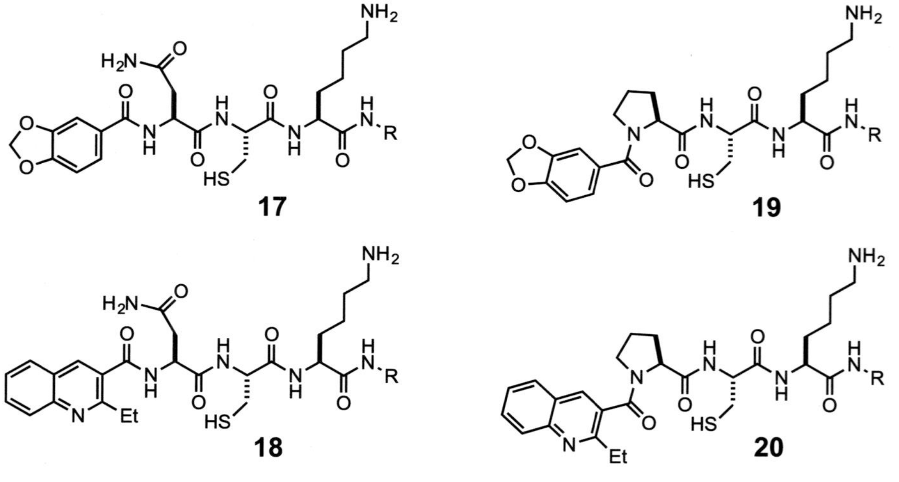

17 19 18 20

R = $(CH_2)_3NH_2$ or $(CH_2)_5C(O)NH_2$

Fig. 9 Monomers selected from RBDCL screening of Cy3-$(CUG)_{10}$

Control compounds (not selected from the library and not able to bind $(CUG)^{exp}$) did not interfere with the luminescence reaction. One unexplained observation is that not all MBNL1 binding could be abrogated.

It has been noted that results from screening experiments are frequently contaminated by the presence of aggregators [51, 52]. To ensure that this was not an issue with our RBDCL screen, we examined compounds **15-15**, **20-20**, **19-20**, and **15-20** by dynamic light scattering (DLS). This technique provides both the apparent size of particles present in a solution as well as particle polydispersity. As a positive control, Congo Red (a dye known to aggregate [50]) was found to form particles with an average hydrodynamic radius of 69.25 nm. In contrast, only compound **15-15** formed particles (hydrodynamic radius approximately 30 nm), but only at concentrations above 100 μM. Addition of a detergent (Triton X-100, 0.01%) to the solution did not measurably alter the aggregation of **15-15**. Experiments analogous to these have been used throughout subsequent research by the Rochester group on all RNA-targeted compounds.

10 Building from RBDCC: Enhancing Affinity, Selectivity, and Stability via Medicinal Chemistry

Experiments employing the 11,325-member RBDCC library to identify disulfides binding CUG repeat RNA and the HIV-1 FSS were successful, yielding compounds with moderate (low micromolar) affinity, and binding selectivity over tRNA and target-related RNA sequences. Particularly impressive in the case of CUG-targeted compounds was the ability to inhibit $(CUG)^{exp}$-MBNL1 binding in vitro, with IC_{50}s comparable to the K_D. Moving beyond this, however, to compounds realistically useful for further study in cells and in vivo would require improvements in affinity, selectivity, and most importantly stability: while the lability of the disulfide bond is critical to the exchange process that makes RBDCC possible, it is a distinct liability for bioassay. To address these issues, efforts thus far have focused on the general concepts outlined in Fig. 10.

10.1 Replace Disulfides with Non-labile Bioisosteres

Research to date has focused on hydrocarbon linkages (alkane and alkene), although other strategies are also possible. While not a perfect mimic of the disulfide linkage (the –C–C=C–C– olefin dihedral angle is either 180° or 0°, in contrast to the 108° lowest-energy angle adopted by the –C–S–S–C– disulfide; Fig. 11), we were encouraged by the literature reports that previous replacement of disulfides with olefins had proven successful [54–57]. A particularly notable example of this was work from the Nicolaou group, in which selectivities of disulfide- and olefin-linked vancomycin

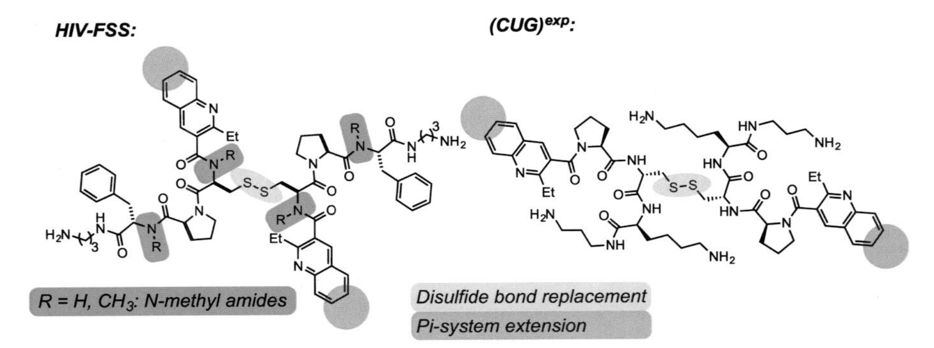

Fig. 10 Strategies used to convert RBDCC hits to compounds suitable for bioassay

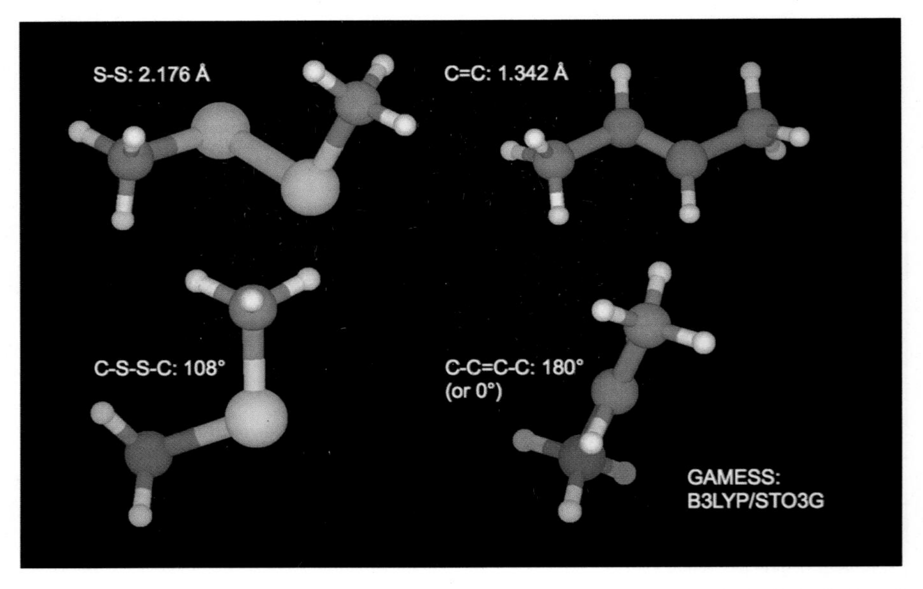

Fig. 11 Geometric differences between equilibrium geometries for a model disulfide (*left*) and olefin (*right*). Optimized structures were calculated using GAMESS [53] using density functional theory (B3LYP/STO3G)

dimers were shown to be similar [58]. In the series of compounds targeting (CUG)^{exp} RNA, replacement of the disulfide with an olefin also allowed us to examine the effect of linker length on compound affinity, with useful results.

10.2 Enhance Affinity Through Pi-System Extension

An early hypothesis was that, much like the bisintercalating peptide antibiotics that served as a model for the core structure used in our RNA-targeted RBDCC library, selectivity was likely to be a function of the peptidic portion of the molecule, while affinity largely resulted from the heterocycle. Therefore, we reasoned that extension of the heterocycle's pi-surface would produce a better intercalator, and hence a higher-affinity molecule. The obvious structural modification therefore was to extend the quinolone moiety to a benzo[G]quinoline. Surprisingly, despite the extensive work done on derivatives of acridine [59], a closely related heterocycle, virtually no prior work had occurred on benzo[G]quinoline. Of course, it was important to be mindful of the general observation that intercalative binding is nonselective, and that there was the potential that this modification would enhance affinity, but reduce selectivity.

10.3 Constrain Conformation via Amide N-Methylation

N-methylation is employed both by Nature and in the design of peptidomimetics as a strategy for conformational constraint, resistance to proteolysis, and enhancement of bioavailability. Cyclosporine, an FDA-approved natural product used as an immunosuppressive agent gains its biostability and ability to cross cell membranes in part through amide N-methylation. Use of amide N-methylation in the context of peptidomimetics has been summarized by Kessler and colleagues, who have been at the forefront of the field [60]. In the case of RNA-targeted compounds, we hypothesized that selective amide N-methylation would enhance affinity through conformational bias.

With the above hypotheses as a framework, efforts focused on the HIV FSS began by selective modification of compound **14-14**. Replacement of the 2-ethylquinoline with either 2-methyl- (**21**) or unsubstituted (**22**) quinoline had only a modest effect on affinity, while both a 2-methylpyridine analog (**23**) and the simple peptide (**24**) had no measurable binding ability. Initial attempts to enhance binding affinity through use of a larger heterocycle including synthesis of **25** and **26** were unsuccessful, however (Fig. 12).

Replacement of the disulfide linkage in **16** with an olefin was accomplished via cross-metathesis between a resin-bound peptide incorporating allylglycine in place of cysteine, and an identical peptide in solution. While in principle this reaction should be feasible entirely in solution, in practice cleanup of the reaction was considerably simpler in the resin-bound case. Cross-metathesis yielded both the *E* and *Z* isomers **27** and **28**, which were readily separable by preparative HPLC. For saturated analog **29**, olefin cross-metathesis was followed by on-bead hydrogenation catalyzed by Wilkinson's catalyst in a 1:10 mixture of methanol:methylene chloride. All three compounds bound the FSS RNA with dissociation constants (as measured by surface plasmon resonance) in the 100 nM to low micromolar regime; a roughly threefold difference in affinity separated the tightest binder (*Z* isomer) from the weakest binder

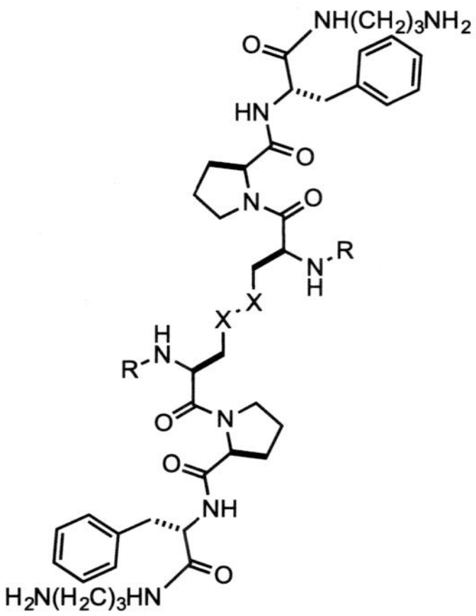

Compound	X-X	R	K_D (□M)
14-14	S-S	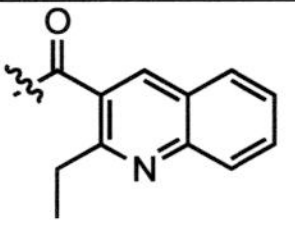	0.35 ± 0.11[a]
21	S-S	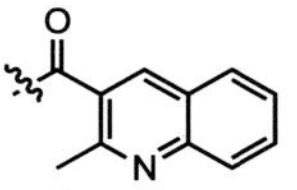	0.43 ± 0.05[a]
22	S-S	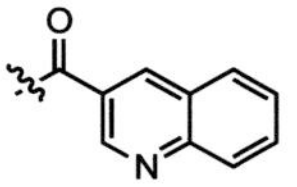	0.65 ± 0.04[a]
23	S-S	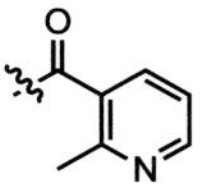	No Binding[a]
24	S-S	H	No Binding[a]
25	S-S	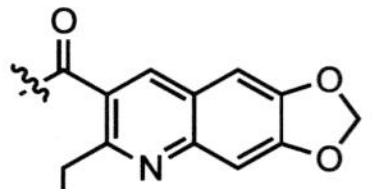	1.42 ± 0.17[a]
26	S-S	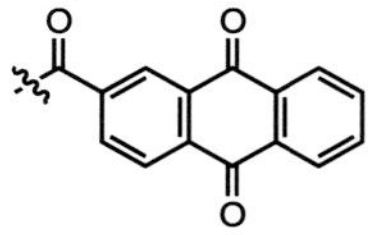	0.23 ± 0.03[a]
27	CH=CH, *E*	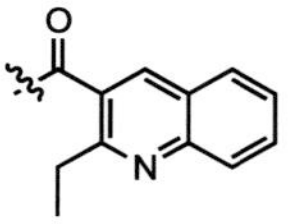	0.33 ± 0.02[b]
28	CH=CH, *Z*	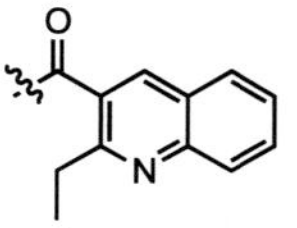	0.18 ± 0.02[b]
29	CH_2-CH_2	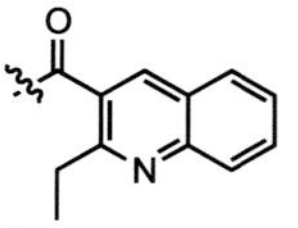	1.27 ± 0.11[b]
30	CH=CH, *E*	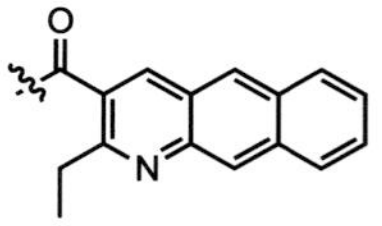	0.096[b]
31	CH=CH, *Z*	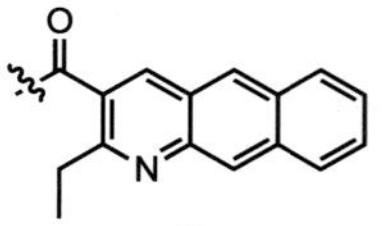	0.071[b]
32	CH=CH, *E*	H	No binding

Fig. 12 First- and second-generation analogs of HIV FSS-binding compounds: (a) K_D measured by fluorescence titration; (b) K_D measured by surface plasmon resonance

(the saturated hydrocarbon). Competition fluorescence titrations revealed no change in affinity for FSS RNA in the presence of a 20-fold excess of total yeast tRNA, indicative of high selectivity [61].

Following this initial set of analogs, a series of compounds was synthesized incorporating a benzo[G]quinoline in place of quinoline [62]. As expected, these compounds (**30** and **31**) displayed enhanced affinity relative to their quinoline congeners. Confirming earlier results with disulfide-containing peptides, a compound lacking both benzo[G]quinoline moieties (**32**) had no observable affinity for FSS RNA, while compounds bearing only a single benzo[G]quinoline had dramatically reduced binding ability. Likewise, although the "half-compound" **33** bound weakly, its affinity was completely ablated in the presence of excess tRNA, indicating a nonselective binding mode. These results confirmed the importance of all major aspects of the initially selected structure. Parallel fluorescence titration experiments suggested that the increase in affinity for benzo[G]quinoline containing compounds did come at a cost in selectivity, as compounds were found to bind a homologous DNA sequence with affinities only two- to fourfold weaker than FSS RNA.

Analogs incorporating both the olefin bioisostere for disulfide and the benzo[G] quinoline moiety proved to have the high affinity needed for moving the program into cellular assays. The presence of the benzo[G]quinoline structure also simplified early tests of bioavailability, as its fluorescence provided a readily observable "tag" for the presence of the compound. While the ability of compounds to cross cell membranes was not quantitated at this stage, all benzo[G]quinoline containing analogs were found to pass into the cytosol of HEK 293T cells as observed by fluorescence microscopy. Toxicity was measured using a WST-1 assay in HEK293T cells as well; this assay employs a reagent producing a colored dye in response to mitochondrial metabolism [63]. Compounds were found to have toxicities (TD_{50}) ranging from 10 to >60 μM.

Next, a dual-luciferase assay was employed to examine the effect of compounds on frameshifting [64]. In this assay, plasmids are used encoding two luciferase enzymes (*Renilla* and firefly) fused together. These enzymes catalyze a reaction producing luminescence of differing wavelengths. Thus, when transiently transfected into HEK 293T cells, luminescence readout on a plate reader of the relevant wavelengths provides a measure of the ratio of active *Renilla* and firefly luciferase present. A control (0-frame) plasmid is employed in which both enzymes are in-frame with each other, producing a baseline level of *Renilla*–firefly fusion protein. In the experimental plasmid, firefly luciferase is in the −1 reading frame from *Renilla*, separated by a frameshift-stimulating sequence. For our experiments, we employed a plasmid incorporating the HIV-1 FSS, and a plasmid incorporating the frameshift-stimulating RNA sequence from HTLV-2 as a separate selectivity control. Thus, compounds able to enhance frameshifting via binding to the HIV-1 FSS would be expected to increase the amount of active *Renilla*–firefly fusion, and thereby the ratio of firefly luminescence to *Renilla*. This is indeed what was observed with compounds **30** and **31**. A concentration-dependent increase in frameshift was found in the HIV FSS-controlled system, while no analogous increase in frameshifting was observed for the system under the control of the HTLV-2 FSS.

Of course, the primary remaining question was whether this activity would translate into an effect on virus. To address that issue, we employed an infectivity assay with pseudotyped (single-cycle) HIV. In this assay, HEK 293T cells are transfected with plasmids encoding a modified HIV genome (in which the *env* and *nef* genes are replaced with green fluorescent protein, GFP), and a plasmid encoding the vesicular stomatitis virus (VSV-G) envelope protein [65]. This yields pseudotyped HIV virus particles that are able to infect target cells, but not replicate. Here, the TZM-bl cell line is used as the target for infection; this is a derivative of the HeLa cell line in which an enzyme has been introduced to provide a luminescence readout for HIV infection [66]. This assay enables quantification of the amount of virus produced (simply by observing GFP fluorescence in producer cells) and its ability to infect human cells (by visualizing luminescence in target cells). Virus particles harvested from producer cells are also quantitated and normalized via a p24 ELISA assay prior to being introduced to target cells.

Using this assay, compounds **30** and **31** were found to produce a dose-dependent decrease in the amount of virus produced, and a dramatic decrease in the infectivity of pseudotyped virus particles. It is not clear why virus production is inhibited by FSS-targeting compounds. Concentrations at which this occurs are below the TD_{50}, and cells appear to grow normally, so although a global effect on translation is possible it is unlikely that this is the primary cause. To gain a further understanding of compound effect on virus infectivity, harvested virus particles were western blotted for reverse transcriptase (RT) and p24. This provides a measure of changes in Gag:Gag-Pol ratio, as p24 is present in both polyproteins, while RT is only produced in Gag-Pol. We observed that, concomitant with the decrease in infectivity of virus particles, the ratio of Gag-Pol to Gag increased as a function of increased compound concentration. In contrast, the control compound **32** produced no change in infectivity, and no significant change in Gag:Gag-Pol ratio. While not definitive for compounds **30** and **31** operating via a frameshift-dependent mechanism, these results are consistent with that hypothesis. The band pattern observed in western blots was inconsistent with the one produced on treatment with the protease inhibitor Indinavir, arguing against an interaction of FSS-targeting compounds with protease (subsequent experiments confirmed this; vide infra).

As a next step in enhancing binding and biocompatibility, we prepared a series of analogs incorporating N-methyl amides (Fig. 13) [67]. As mentioned previously, amide N-methylation is one of the simplest peptide modifications providing conformational restraint, enhanced bioavailability, and enhanced biostability. Eight analogs were synthesized, incorporating two or four methylated amides. Only symmetrical methylation patterns were examined at this stage. Methylation was accomplished using a variation of methodology first described by Miller and Scanlan [68], in which the growing peptide chain is first N-nitrobenzenesulfonyl (Nosyl) protected, then methylated via TMS-diazomethane. Deprotection of the Nosyl group allows for further extension using standard peptide coupling reagents.

Given that a potential outcome of this study was a dramatic reduction in affinity (i.e., if conformational restriction provided by the N-methyl amides uniformly disfavored binding conformers), we were gratified to observe a significant improvement

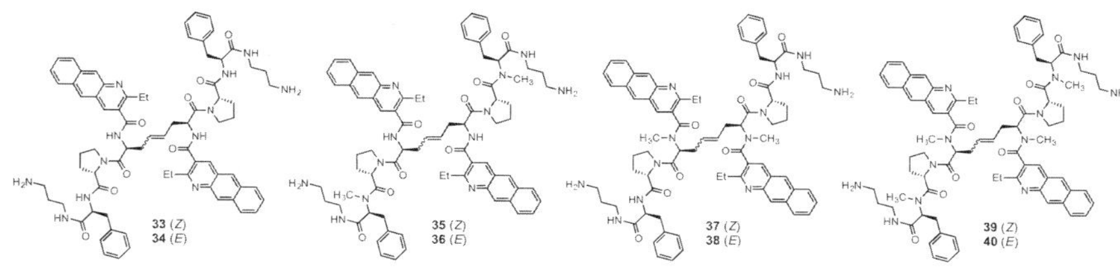

Compound	Sequence, K_D (nM)				
	FSS-RNA	FSS-DNA	Yeast tRNA	FSS RNA : FSS DNA Selectivity	FSS RNA : tRNA Selectivity
34	100 ± 30	230 ± 20	220 ± 30	2.3 : 1	2.2 : 1
33	66 ± 30	120 ± 20	330 ± 30	1.8 : 1	5 : 1
38	2.7 ± 1.0	81 ± 30	120 ± 40	30 : 1	44 : 1
37	2.7 ± 0.7	65 ± 10	180 ± 20	24 : 1	67 : 1
36	3.0 ± 0.1	79 ± 10	90 ± 10	26 : 1	30 : 1
35	1.1 ± 0.5	60 ± 8.0	67 ± 10	55 : 1	61 : 1
40	5.0 ± 2.0	34 ±10	91 ± 20	6.8 : 1	18 : 1

Fig. 13 N-methyl analog series. K_D values were measured by fluorescence titration

in binding affinity for all compounds, as measured by fluorescence titration. Of interest was the fact that this enhancement was driven by an order-of-magnitude increase in the association rate constant (k_{on}). This observation is consistent with the conformational restriction hypothesis, as it suggests that the ground state conformational ensemble is better biased in favor of a conformer or conformers able to bind the FSS RNA. As with previous analogs, toxicity and cell permeability were assessed in HEK 293T cells. Toxicities were similar to those measured for the un-methylated compounds. Flow cytometry was used to quantitate the relative amounts of each compound able to cross cell membranes. We observed that this roughly followed the calculated logP of the series, with more hydrophobic compounds attaining a higher concentration in the interior of the cell. N-methylated compounds were also highly active in the pseudo-typed HIV infectivity assay.

10.4 Enhancing Affinity, Selectivity, and Bioavailability for $(CUG)^{exp}$

In parallel to work on enhancing the affinity and selectivity of HIV-1 FSS-binding compounds, the Rochester group prepared and tested a similar analog series targeting $(CUG)^{exp}$ RNA [69]. The primary focus here was on derivatives of compound **20-20**, one of the best-performing structures selected from the initial Cy3-$(CUG)_{10}$ RBDCC screen. Compounds **41-49** (Fig. 14) were synthesized. In addition to replacing the 2-ethylquinoline with benzo[G]quinoline and replacing disulfide with an olefin linkage (compounds **42** and **43**), E and Z isomers (**48** and **49**) of a compound incorporating an extended olefin linker were prepared, along with both isomers of a "scrambled" sequence (**46** and **47**). Binding constants were measured for these compounds by SPR against a series of (CUG) repeat RNAs, and compared with both DM2 ($(CCUG)^{exp}$) RNA and off-target sequences (Fig. 15).

As for HIV-1 FSS binders, $(CUG)^{exp}$ binding analogs gained substantial affinity through substitution of 2-ethylquinoline with benzo[G]quinoline. Compounds **42** and **43** bound $(CUG)_2$-$(CUG)_{10}$ with similar affinity, and although neither displayed selectivity over (CCUG) repeats or (CAG), the E isomer (**43**) was found to have modest selectivity for $(CUG)^{exp}$ over duplex (CAG/CUG). Removal of benzo [G]quinolines (**44** and **45**) reduced affinity for $(CUG)_{10}$, but did not ablate it entirely, in contrast to previously described results from HIV FSS-binding compounds lacking benzo[G]quinoline. Likewise, scrambled compounds **46** and **47** were found to have reduced affinity for (CUG) repeats and actually were somewhat selective for duplex (CAG/CUG) and an off-target RNA (the HIV-1 FSS). Most intriguing, however, was the performance of the extended linker compounds **48** and **49**. These were strongly selective for $(CUG)^{exp}$ sequences over both $(CCUG)^{exp}$ and off-target RNAs; further analysis of compound **49** showed selectivity for longer repeats $(CUG)_{10}$ over short repeats. On- and Off-rates were derived from SPR binding curves, with the off-rate

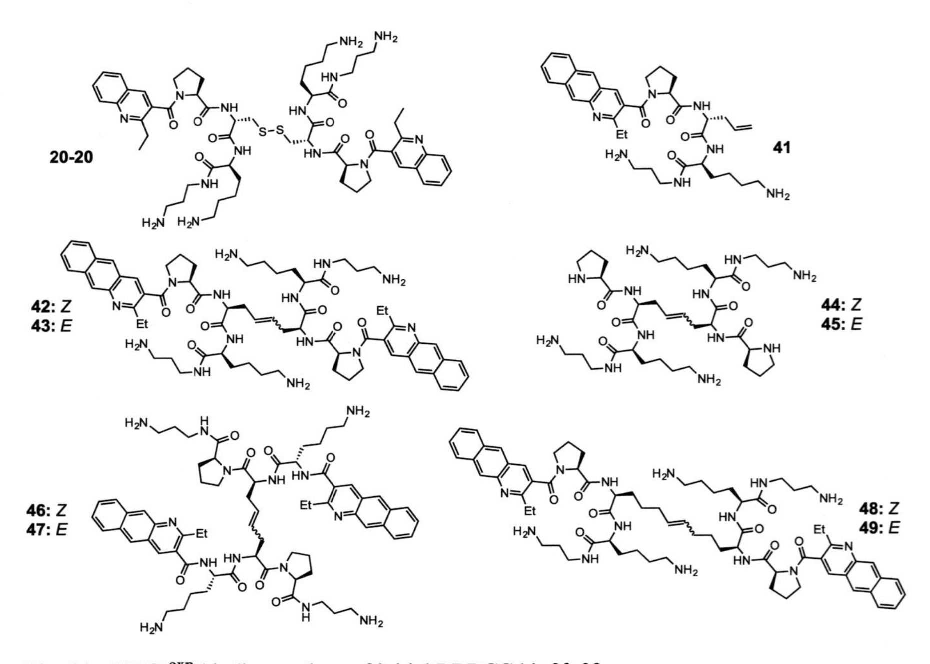

Fig. 14 $(CUG)^{exp}$-binding analogs of initial RBDCC hit **20-20**

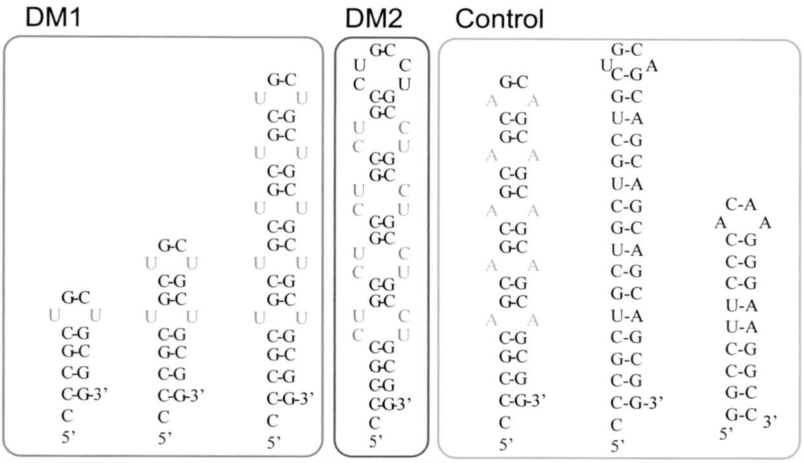

DM1 DM2 Control

Compound	RNA Sequence						
	$(CUG)_2$	$(CUG)_4$	$(CUG)_{10}$	$(CCUG)_{10}$	$(CAG)_{10}$	Duplex	HIV-1 FSS
4	32 (n=2)	22.5 (n=4)	39 (n=10)	38 (n=10)	49 n=10	67	1240
5	21.5	38	68 (n=10)	15	55.5	307.5	759
6	ND	ND	263	21900	160.5	336.5	116
7	ND	ND	422	36100	312.5 (n=3)	ND	302.5 (n=1)
8	ND	ND	133.5	294	474.5	94	ND
9	ND	ND	120.5	138	263	77	67
10	ND	ND	81.5	278.5	ND	ND	ND
11	668	60 (n=1)	22.5 (n=5)	210	862	No binding	2000
Neomycin	ND	ND	409	1955	ND	ND	ND

Fig. 15 Binding constants (apparent K_D, nM) and stoichiometries (n) for $(CUG)^{exp}$ targeted compounds as measured by SPR

providing the strongest indicator of selectivity (i.e., the compound with the longest residence time on the target sequence had the highest selectivity). Thermodynamic binding constants for all compounds to $(CUG)_{10}$ were confirmed by fluorescence titration and yielded values consistent with SPR.

SPR experiments were also employed to determine stoichiometry for selected compounds and sequences. Given that the benzo[G]quinoline was hypothesized to act as an intercalator, we were surprised to find stoichiometries incompatible with this binding mode, according to neighbor exclusion rules [70]. Thus, these results suggest either a groove-binding mode for both benzo[G]quinolines or a mixed mode in which one intercalates and the other binds in the groove. Structural analysis will be required to fully understand this observation.

Toxicities of compounds **42** and **43** were tested in mouse myoblasts, as this cell line would be used in further experiments testing efficacy in a DM1 disease model. Toxic doses for each compound (TD_{50}) were found to be on the order of 200 μM. Fluorescence microscopy also indicated compounds localized in the nucleus, in contrast to the uniform distribution observed for HIV-1 FSS RNA-binding compounds. To test the ability of $(CUG)^{exp}$ compounds to displace MBNL1 in a cellular context, their activity in mouse myoblasts carrying a luciferase gene fused to $(CUG)_{800}$ in the 3′-UTR was examined. In this assay, MBNL1 binding to the long $(CUG)^{exp}$ repeat prevents transport of the luciferase mRNA out of the nucleus and subsequent translation. Compound-dependent disruption of the $(CUG)^{exp}$–MBNL1 interaction, however, would restore production of luciferase, viewed quantitatively as luminescence in a plate reader. In the event, compounds exhibited strong dose–response behavior for restoration of luciferase activity. Compound **45**, used as a control, showed no activity.

While the performance of these compounds in cell lines was encouraging, could they work in a live animal? To address that question, we tested three compounds (**42**, **47**, and **49**) in the HSA^{LR} mouse model of DM1. Developed by the Thornton group [71], these transgenic mice carry approximately 250 CTG repeats inserted into a human skeletal actin (*HSA*) gene. HSA^{LR} mice are characterized by a DM1-like phenotype, and at the molecular level have splicing defects in transcripts targeted by MBNL1. Mice were treated with 40 mg/kg doses of compounds **42**, **49**, or **47** (as a control) via interperitoneal injection once daily for 5 days. When compared with age-matched controls, compounds **42** and **49** produced a statistically significant improvement in the MBNL1-dependent splicing of *Clcn1*, and *Atp2a1* in hind limb quadriceps muscle. No statistically significant change was observed on treatment with compound **47**. Two additional splicing events were examined for compound **49**: *Ttn*, and MBNL1-dependent splicing event, showed compound-dependent improvement, while *Capzb*, an MBNL1-independent splicing event, showed no change. These results confirmed that designed, high-affinity compounds targeting $(CUG)^{exp}$ RNA could have activity in an animal model of DM1.

11 Beyond Dimers: Accelerating Compound Development Through Orthogonal and Bidirectional Exchange

While the approach outlined above has produced compounds with exceptional selectivity, affinity, and target-relevant activity in biological systems, the process of dynamic combinatorial selection followed by iterative optimization via traditional medicinal chemistry is not as rapid as one might wish. Are there ways to accelerate RNA-targeted compound discovery via expansion of the chemical diversity space addressed by a DCL? Two possibilities examined to date are, first, introduction of additional, orthogonal exchange chemistries, and second, incorporation of bifunctional linkers into DCC. While still at an early stage, results from both methods suggest promise.

11.1 Orthogonal Exchange Chemistry

Successful use of two different exchange chemistries operating either simultaneously or orthogonally depending on conditions was first reported in the context of a dynamic combinatorial system by Rodriguez-Docampo and Otto in 2008 [72]. Here, hydrazone and disulfide exchange were implemented either separately, or together, depending on the pH of the solution. The Furlan group later reported extending this to thioester exchange [73]. Building on these initial reports, we sought to transfer the concept to a resin-bound mode, or "ternary resin-bound DCC" [74]. Here, monomers incorporating both an S-t-butyl disulfide and an acyl hydrazone were synthesized on resin (B1 and B2, Fig. 16), and allowed to equilibrate with either thiols (A1 and A2), acyl hydrazines (C1 and C2), or both. Thiopropanol was employed to accelerate disulfide exchange, while aniline was used as a hydrazone exchange catalyst [75, 76]. After a 1-week equilibration period followed by resin cleavage, mass spectral peaks corresponding to all possible AB and BC dimers were observed, as well as several ABC trimers. Surprisingly, no mass corresponding to trimer A1B1C1 was detected, suggesting that even in this simple system self-selection is at play. Recent work by the Kool group suggests that the hydrazone exchange process can be dramatically accelerated through the use of more effective catalysts [77], so further study of this system is warranted.

11.2 Dynamic Linker Incorporation

As discussed above, an intriguing result during work on triplet repeat RNA was the observation that the number of carbons separating the two halves of a modular $(CUG)^{exp}$ binding molecule had a dramatic effect on both its affinity and selectivity.

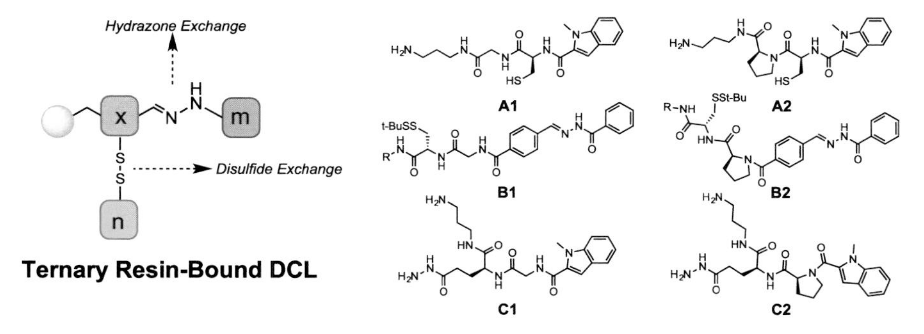

Fig. 16 Ternary resin-bound DCL

To test whether one could more rapidly access similar types of compounds in a DCL, we set out to implement the scheme shown in Fig. 17 [78]. Conceptually, the idea was to provide a "new option" to a previously studied solution-phase dynamic system through the addition of one or more linker dithiols (Fig. 17, green hexagon). When allowed to equilibrate, this library would generate new trimers incorporating the dithiol linker. One would also expect to see simple linear and (potentially) cyclic oligomers of dithiol derived through exchange and oxidative processes, by analogy to previous DCLs consisting of dithiols [79]. When presented with a previously screened RNA sequence, would this library select a new compound incorporating a linker module (for example, compound **C-D1-C**), or the originally identified structure?

Initial DCL experiments incorporating *trans*-1,2-cylohexanedithiol (**D1**) confirmed that in the absence of an RNA target, trimers B-D1-B, B-D1-C, and C-D1-C were formed in the library mixture, as well as all possible dimers with the exception of A-A. No trimers incorporating monomer A were observed. Both the latter results suggest that self-selection for A-X heterodimers occurs preferentially. When equilibrated in the presence of HIV-1 FSS RNA, however, trimer structures disappeared, leaving only dimers. Of these, homodimer C-C, the compound originally selected by RBDCC experiment (described earlier), displayed the largest degree of amplification. Thus, although these experiments did not yield new RNA-binding compounds, the experiment served both as proof-of-concept for incorporating linker variation in the form of dithiols as well as solution-phase confirmation of the earlier library selection experiment conducted on solid support (Fig. 18).

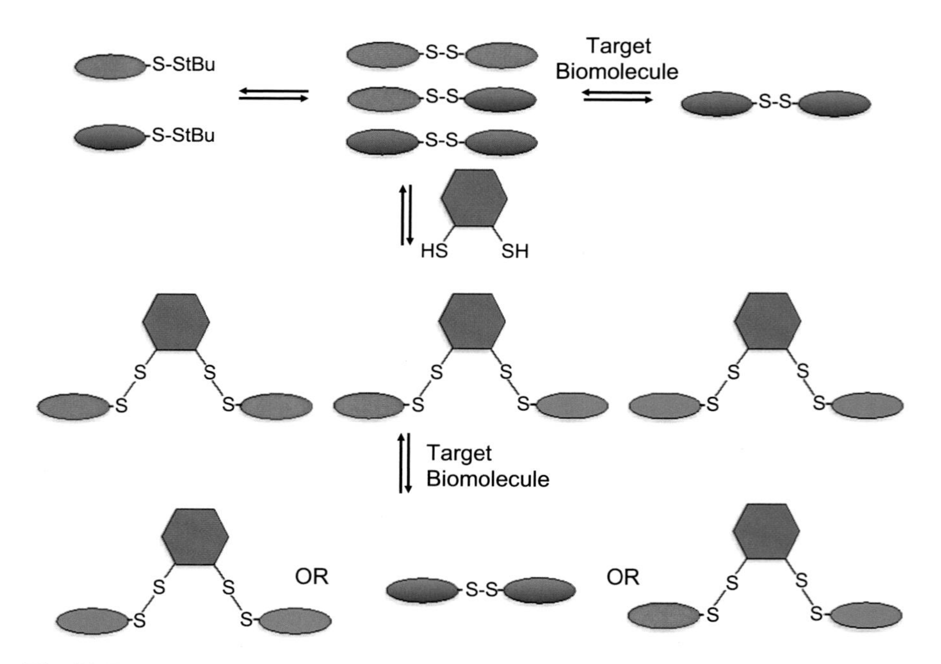

Fig. 17 Dynamic linker incorporation

12 Conclusions

While much work remains to be done, the experiments described in this chapter demonstrate that a resin-bound dynamic combinatorial library (RBDCL) approach to natural products-inspired compound discovery, followed by analog synthesis, can yield compounds able to bind biomedically relevant RNA sequence with high affinity and good specificity. For the two targets studied, compounds display target-relevant biological activity in human cells, and, for $(CUG)^{exp}$ targeted compounds, in an animal model of disease. Current efforts are focused on expansion of the methodology to other types of structures, and a broader range of exchange chemistries. We can anticipate this remaining an area of rapid growth for the foreseeable future.

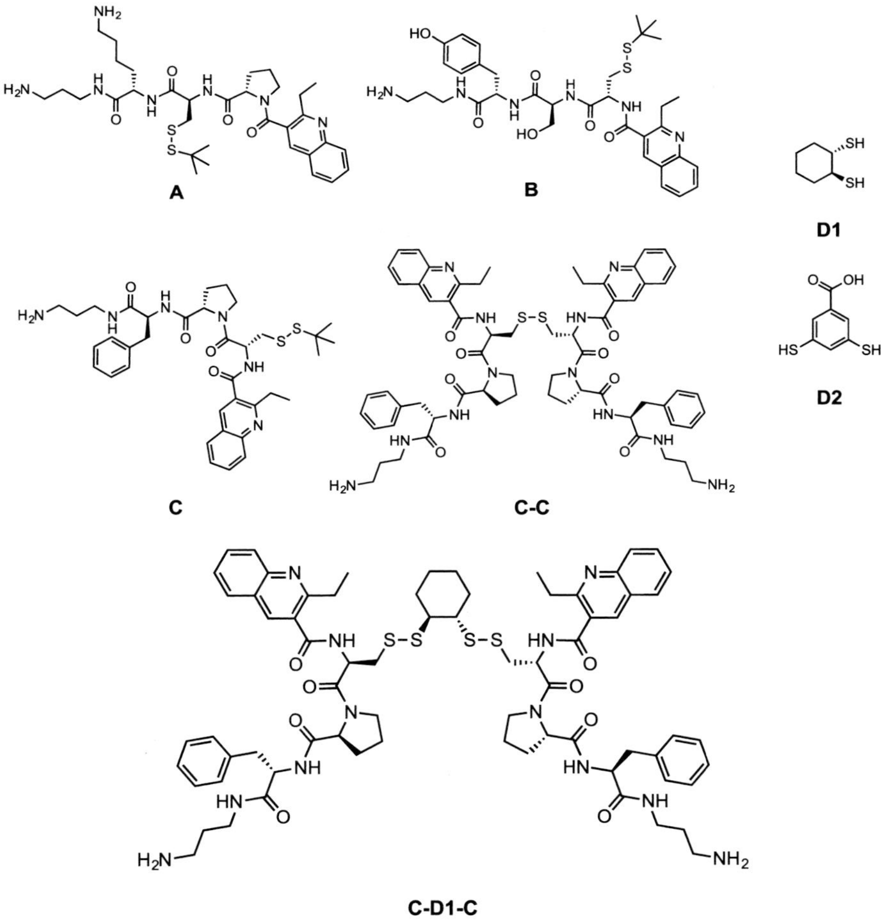

Fig. 18 Compounds employed in a proof-of-concept linker DCL

References

1. Dervan PB, Burli RW (1999) Sequence-specific DNA recognition by polyamides. Curr Opin Chem Biol 3:688–693
2. Thomas JR, Hergenrother PJ (2008) Targeting RNA with small molecules. Chem Rev 108: 1171–1224
3. Georgianna WE, Young DD (2011) Development and utilization of non-coding RNA-small molecule interactions. Org Biomol Chem 9:7969–7978
4. Guan L, Disney MD (2012) Recent advances in developing small molecules targeting RNA. ACS Chem Biol 7:73–86
5. Corbett PT, Leclaire J, Vial L, West KR, Wietor JL, Sanders JKM, Otto S (2006) Dynamic combinatorial chemistry. Chem Rev 106:3652–3711
6. Ramström O, Bunyapaiboonsri T, Lohmann S, Lehn JM (2002) Chemical biology of dynamic combinatorial libraries. Biochim Biophys Acta 1572:178–186
7. Miller BL (ed) (2009) Dynamic combinatorial chemistry. Wiley, New York, NY
8. Rideout D (1986) Self-assembling cytotoxins. Science 233:561–563
9. Feldman KS, Bobo JS, Ensel SM, Lee YB, Weinreb PH (1990) Template-controlled oligomerization support studies. Template synthesis and functionalization. J Org Chem 55:474–481
10. Miller BL, Bonner WA (1995) Enantioselective autocatalysis. III. Configurational and conformational studies on a 1,4-benzodiazepinooxazole derivative. Orig Life Evol Biosph 25:539–547
11. Krämer R, Lehn JM, Marquis-Rigault A (1993) Self-recognition in helicate self-assembly: spontaneous formation of helical metal complexes from mixtures of ligands and metal ions. Proc Natl Acad Sci U S A 90:5394–5398
12. Brady PA, Sanders JKM (1997) Thermodynamically-controlled cyclisation and interconversion of oligocholates: metal ion templated "living" macrolactonisation. J Chem Soc Perkin Trans 1 1997:3237–3253
13. Hioki H, Still WC (1998) Chemical evolution: a model system that selects and amplifies a receptor for the tripeptide (D)Pro(L)Val(D)Val. J Org Chem 63:904–905
14. Huc I, Lehn JM (1997) Virtual combinatorial libraries: dynamic generation of molecular and supramolecular diversity by self-assembly. Proc Natl Acad Sci U S A 94:2106–2110
15. Klug A (2010) The discovery of zinc fingers and their applications in gene regulation and genome manipulation. Annu Rev Biochem 79:213–231
16. Schepartz A, McDevitt JP (1989) Self-assembling ionophores. J Am Chem Soc 111:5976–5977
17. Klekota B, Hammond MH, Miller BL (1997) Generation of novel DNA-binding compounds by selection and amplification from self-assembled combinatorial libraries. Tetrahedron Lett 38:8639–8643
18. Klekota B, Miller BL (1999) Selection of DNA_binding compounds via multistage molecular evolution. Tetrahedron 55:11687–11697
19. Karan C, Miller BL (2001) RNA-selective coordination complexes identified via dynamic combinatorial chemistry. J Am Chem Soc 123:7455–7456
20. Ludlow RF, Otto S (2008) Two-vial, LC-MS identification of ephedrine receptors from a solution-phase dynamic combinatorial library of over 9000 components. J Am Chem Soc 130: 12218–12219
21. Bugaut A, Jantos K, Wietor JL, Rodriguez R, Sanders JKM, Balasubramanian S (2008) Exploring the differential recognition of DNA G-quadruplex targets by small molecules using dynamic combinatorial chemistry. Angew Chem Int Ed Engl 47:2677–2680
22. Buhler E, Sreenivasachary N, Candau SJ, Lehn JM (2007) Modulation of the supramolecular structure of G-quartet assemblies by dynamic covalent decoration. J Am Chem Soc 129:10058–10059
23. Harrington CA, Rosenow C, Retief J (2000) Monitoring gene expression using DNA microarrays. Curr Opin Microbiol 3:285–291
24. Tan DS, Burbaum JJ (2000) Ligand discovery using encoded combinatorial libraries. Curr Opin Drug Discov Devel 3:439–453

25. Hattori K, Koike K, Okuda K, Hirayama T, Ebihara M, Takenaka M, Nagasawa H (2016) Solution-phase synthesis and biological evaluation of triostin A and its analogues. Org Biomol Chem 14:2090–2111
26. Szajewski RP, Whitesides GM (1980) Rate constants and equilibrium-constants for thiol-disulfide interchange reactions involving oxidized glutathione. J Am Chem Soc 102:2011–2026
27. McNaughton BR, Miller BL (2006) Resin-bound dynamic combinatorial chemistry. Org Lett 8:1803–1806
28. Addess KJ, Sinsheimer JS, Feigon J (1993) Solution structure of a complex between [N-MeCys3, N-MeCys7]TANDEM and [d(GATATC)]2. Biochemistry 32:2498–2508
29. Addess KJ, Feigon J (1994) Sequence specificity of quinoxaline antibiotics. 2. NMR studies of the binding f [N-MeCys3, N-MCys7]TANDEM and triostin A to DNA containing a CpI step. Biochemistry 33:12397–12404
30. Mankodi A, Teng-Umnuay P, Krym M, Henderson D, Swanson M, Thornton CA (2003) Ribonuclear inclusions in skeletal muscle in myotonic dystrophy types 1 and 2. Ann Neurol 54:760–768
31. Ashizawa T, Dubel JR, Harati Y (1993) Somatic instability of CTG repeat in myotonic dystrophy. Neurology 43:2674–2678
32. Thornton CA, Johnson K, Moxley RT (1994) Myotonic dystrophy patients have larger CTG expansions in skeletal muscle than in leukocytes. Ann Neurol 35:104–107
33. Pushechnikov A, Lee MM, Childs-Disney JL, Sobczak K, French JM, Thornton CA, Disney MD (2009) Rational design of ligands targeting triplet repeating transcripts that cause RNA dominant disease: application to myotonic muscular dystrophy type 1 and spinocerebellar ataxia type 3. J Am Chem Soc 131:9767–9779
34. Arambula JF, Ramisetty SR, Baranger AM, Zimmerman SC (2009) A simple ligand that selectively targets CUG trinucleotide repeats and inhibits MBNL protein binding. Proc Natl Acad Sci U S A 106:16068–16073
35. Warf MB, Nakamori M, Matthys CM, Thornton CA, Berglund JA (2009) Pentamidine reverses the splicing defects associated with myotonic dystrophy. Proc Natl Acad Sci U S A 106:18551–118556
36. Liquori CL, Ricker K, Moseley ML, Jacobsen JF, Kress W, Naylor SL, Day JW, Ranum LPW (2001) Myotonic dystrophy type 2 caused by a CCTG expansion in intron 1 of ZNF9. Science 293:864–867
37. Tian B, White RJ, Xia T, Welle S, Turner DH, Mathews MB, Thornton CA (2000) Expanded CUG repeat RNAs form hairpins that activate the double-stranded RNA-dependent protein kinase PKR. RNA 6:79–87
38. Belew AT, Dinman JD (2015) Cell cycle control (and more) by programmed -1 ribosomal frameshifting: implications for disease and therapeutics. Cell Cycle 14:172–178
39. Atkins JF, Loughran G, Bhatt PR, Firth AE, Baranov PV (2016) Ribosomal frameshifting and transcriptional slippage: from genetic steganography and cryptography to adventitious use. Nucleic Acids Res 44:7007–7078
40. Biswas P, Jiang X, Pacchia AL, Dougherty JP, Peltz SW (2004) The human immunodeficiency virus type 1 ribosomal frameshifting site is an invariant sequence determinant and an important target for antiviral therapy. J Virol 78:2082–2087
41. Staple DW, Butcher SE (2005) Solution structure and thermodynamic investigation of the HIV-1 frameshift inducing element. J Mol Biol 349:1011–1023
42. Low JT, Garcia-Miranda P, Mouzakis KD, Gorelick RJ, Butcher SE, Weeks KM (2014) Structure and dynamics of the HIV-1 frameshift element RNA. Biochemistry 53:4282–4291
43. Shehu-Xhilaga M, Crowe SM, Mak J (2001) Maintenance of the Gag/Gag-Pol ratio is important for human immunodeficiency virus type 1 RNA dimerization and viral infectivity. J Virol 75:1834–1841
44. Dulude D, Berchiche YA, Gendron K, Brakier-Gingras L, Heveker N (2006) Decreasing the frameshift efficiency translates into an equivalent reduction of the replication of the human immunodeficiency virus type 1. Vaccine 345:127–136

45. Hung M, Patel P, Davis S, Green SR (1998) Importance of ribosomal frameshifting for human immunodeficiency virus type 1 particle assembly and replication. J Virol 72:4819–4824
46. Marcheschi RJ, Tonelli M, Kumar A, Butcher SE (2011) Structure of the HIV-1 frameshift site RNA bound to a small molecule inhibitor of viral replication. ACS Chem Biol 6:857–864
47. Tan DS, Foley MA, Stockwell BR, Shair MD, Schreiber SL (1999) Synthesis and preliminary evaluation of a library of polycyclic small molecules for use in chemical genetic assays. J Am Chem Soc 121:9073–9087
48. McNaughton BR, Gareiss PC, Miller BL (2007) Identification of a selective small-molecule frameshift-inducing stem-loop RNA from an 11,325 member resin bound dynamic combinatorial library. J Am Chem Soc 129:11306–11307
49. Gareiss PC, Sobczak K, McNaughton BR, Palde PB, Thornton CA, Miller BL (2008) Dynamic combinatorial selection of molecules capable of inhibiting the (CUG) repeat RNA-MBNL1 interaction in vitro: discovery of lead compounds targeting myotonic dystrophy (DM1). J Am Chem Soc 130:16254–16261
50. Olson KR, Eglen RM (2007) Beta galactosidase complementation: a cell-based luminescent assay platform for drug discovery. Assay Drug Dev Technol 5:137–144
51. McGovern SL, Helfand BT, Feng B, Shoichet BK (2003) A specific mechanism of nonspecific inhibition. J Med Chem 46:4265–4272
52. Feng BY, Toyama BH, Wille H, Colby DW, Collins SR, May BC, Prusiner SB, Weissman J, Shoichet BK (2008) Small-molecule aggregates inhibit amyloid polymerization. Nat Chem Biol 4:197–199
53. Schmidt MW, Baldridge KK, Boatz JA, Elbert ST, Gordon MS, Jensen JH, Koseki S, Matsunaga N, Nguyen KA, Su SJ, Windus TL, Dupuis M, Montgomery JA (1993) General atomic and molecular electronic structure system. J Comput Chem 14:1347–1363
54. Fotouhi N, Joshi P, Tilley JW, Rowan K, Schwinge V, Wolitzky B (2000) Cyclic thioether peptide mimetics as VCAM-VLA-4 antagonists. Bioorg Med Chem Lett 10:1167–1169
55. Stymiest JL, Mitchell BF, Wong S, Vederas JC (2003) Synthesis of biologically active dicarba analogues of the peptide hormone oxytocin using ring-closing metathesis. Org Lett 5:47–49
56. Berezowska I, Chung NN, Lemieux C, Wilkes BC, Schiller PW (2007) Dicarba analogues of the cyclic enkephalin peptides H-Tyr-c[D-Cys-Gly-Phe-D(or L)-Cys]NH(2) retain high opioid activity. J Med Chem 50:1414–1417
57. Mollica A, Guardiani G, Davis P, Ma S, Porreca F, Lai J, Mannina L, Sobolev AP, Hruby VJ (2007) Synthesis of stable and potent delta/mu opioid peptides: analogues of H-Tyr-c [D-Cys-Gly-Phe-D-Cys]-OH by ring-closing metathesis. J Med Chem 50:3138–3142
58. Nicolaou KC, Hughes R, Cho SY, Winssinger N, Smethurst C, Labischinski H, Endermann R (2000) Target-accelerated combinatorial synthesis and discovery of highly potent antibiotics effective against vancomycin-resistant bacteria. Angew Chem Int Ed Engl 39:3823–3828
59. Neto BAD, Lapis AAM (2009) Recent developments in the chemistry of deoxyribonucleic acid (DNA) intercalators: principles, design, synthesis, applications and trends. Molecules 14: 1725–1746
60. Chatterjee J, Gilon C, Hoffman A, Kessler H (2008) N-methylation of peptides: a new perspective in medicinal chemistry. Acc Chem Res 41:1331–1342
61. Luedtke N, Tor Y (2000) A novel solid-phase assembly for identifying potent and selective RNA ligands. Angew Chem Int Ed Engl 39:1788–1790
62. Ofori LO, Hilimire TA, Bennett RP, Brown NW, Smith HC, Miller BL (2014) High-affinity recognition of HIV-1 frameshift-stimulating RNA alters frameshifting in vitro and interferes with HIV-1 infectivity. J Med Chem 57:723–732
63. Buttke TMT, McCubrey JAJ, Owen TCT (1993) Use of an aqueous soluble tetrazolium/formazan assay to measure viability and proliferation of lymphokine-dependent cell lines. J Immunol Methods 157:233–240
64. Grentzmann G, Ingram JA, Kelly PJ, Gesteland RF, Atkins JF (1998) A dual-luciferase reporter system for studying recoding signals. RNA 4:479–486

65. Miller JH, Presnyak V, Smith HC (2007) The dimerization domain of HIV-1 viral infectivity factor Vif is required to block virion incorporation of APOBEC3G. Retrovirology 4:81
66. Finnegan CM, Rawat SS, Puri A, Wang JM, Ruscetti FW, Blumenthal R (2004) Ceramide, a target for antiretroviral therapy. Proc Natl Acad Sci U S A 101:15452–15457
67. Hilimire TA, Bennett RP, Stewart RA, Garcia-Miranda P, Blume A, Becker J, Sherer N, Helms ED, Butcher SE, Smith HC, Miller BL (2016) N-methylation as a strategy for enhancing the affinity and selectivity of RNA-binding peptides: application to the HIV-1 frameshift-stimulating RNA. ACS Chem Biol 11:88–94
68. Miller SC, Scanlan TS (1997) Site-selective N-methylation of peptides on solid support. J Am Chem Soc 119:2301–2302
69. Ofori LO, Hoskins J, Nakamori M, Thornton CA, Miller BL (2012) From dynamic combinatorial "hit" to lead: in vitro and in vivo activity of compounds targeting the pathogenic RNAs that cause myotonic dystrophy. Nucleic Acids Res 40:6380–6390
70. Crothers DM (1968) Calculation of binding isotherms for heterogeneous polymers. Biopolymers 6:575–584
71. Mankodi A, Logigian E, Callahan L, McClain C, White R, Henderson D, Krym M, Thornton CA (2000) Myotonic dystrophy in transgenic mice expressing an expanded CUG repeat. Science 298:1769–1773
72. Rodriguez-Docampo Z, Otto S (2008) Orthogonal or simultaneous use of disulfide and hydrazone exchange in dynamic covalent chemistry in aqueous solution. Chem Commun 2008:5301–5303
73. Escalante AM, Orrillo AG, Furlan RLE (2010) Simultaneous and orthogonal covalent exchange processes in dynamic combinatorial libraries. J Comb Chem 12:410–413
74. Gromova AV, Ciszewski JM, Miller BL (2012) Ternary resin-bound dynamic combinatorial chemistry. Chem Commun 2012:2131–2133
75. Dirksen A, Dirksen S, Hackeng TM, Dawson PE (2006) Nucleophilic catalysis of hydrazone formation and transimination: implications for dynamic covalent chemistry. J Am Chem Soc 128:15602–15603
76. Bhat VT, Caniard AM, Luksch T, Brenk R, Campopiano DJ, Greaney MF (2010) Nucleophilic catalysis of acylhydrazone equilibration for protein-directed dynamic covalent chemistry. Nat Chem 2:490–497
77. Larsen D, Pittelkow M, Karmakar S, Kool ET (2015) New organocatalyst scaffolds with high activity in promoting hydrazone and oxime formation at neutral pH. Org Lett 17:274–277
78. McAnany JD, Reichert JP, Miller BL (2016) Probing the geometric constraints of RNA binding via dynamic covalent chemistry. Bioorg Med Chem 24:3940–3946
79. Corbett PT, Tong LH, Sanders JKM, Otto S (2005) Diastereoselective amplification of an induced-fit receptor from a dynamic combinatorial library. J Am Chem Soc 127:8902–8903

Top Med Chem (2018) 27: 47–78
DOI: 10.1007/7355_2016_29

Published online: 17 March 2017

Structure-Based Discovery of Small Molecules Binding to RNA

Thomas Wehler and Ruth Brenk

Abstract Ribonucleic acids (RNAs) constitute attractive drug targets. The wealth of structural information about RNAs is steadily increasing making it possible to use this information for the design of new ligands. Two methods that make heavy use of structural knowledge for ligand discovery are molecular docking and fragment screening. In molecular docking the structure of the binding site is used as a template for the design of new ligands using computational methods whereas in fragment screening biophysical methods are used for the detection of weak binding ligands which are subsequently elaborated into tighter binding molecules. In this chapter, we give an overview of both methods in the context of ligand discovery for RNA targets and illustrate their applications for hit discovery.

Keywords Docking, Fragment screening, RNA, Structure-based design

Contents

1 Introduction 48
2 Molecular Docking for RNA Targets 50
  2.1 Conformational Search Algorithms 51
  2.2 Scoring Functions 52
  2.3 RNA-Ligand Docking 53
  2.4 New Ligands Discovered by RNA-Ligand Docking 58
  2.5 New Approaches for RNA-Ligand Docking 60
3 Fragment Screening 61
  3.1 RNA-Directed Fragment Libraries 62
  3.2 Fragment Screening Technologies to Identify RNA-Binding Ligands 64

T. Wehler
Johannes Gutenberg University Mainz, Institute for Pharmacy and Biochemistry, Staudinger Weg 5, 55128 Mainz, Germany

R. Brenk (✉)
Department of Biomedicine, University of Bergen, Jonas Lies vei 91, 5020 Bergen, Norway
e-mail: ruth.brenk@uib.no

3.3 Ligands Discovered by Fragment Screening on RNA Targets 66
3.4 Alternative Strategies to Discover RNA Ligands by FBDD: Surface Plasmon Resonance (SPR) 70
4 Conclusions 71
References 71

1 Introduction

Ribonucleic acids (RNAs) constitute attractive drug targets as they are involved in many essential steps of the cell cycle. For example, RNAs regulate transcription and translation, catalyze protein synthesis, and control gene expression [1]. In addition, RNAs can fold into complex 3D structures reflecting their diverse functions [2, 3]. RNA normally adopts an A-form helix. This structure is less attractive for binding small molecules due to the lack of a suitable binding site. However, perturbations in the A-from helix frequently occur through un- or mispaired bases resulting in hairpin loops, internal loops, or bulges. Thus, binding sites are created that can be addressed by small molecules [4].

Several drugs are known to act via RNA binding. It is well established that many antibiotics including aminoglycosides, macrolides, tetracyclines, and oxazolidinones target ribosomal RNA [5]. In recent years, also riboswitches were suggested as drug targets [6–8]. Riboswitches are noncoding RNAs that function as genetic switches in bacteria. They sense the concentration of their small molecule ligands whereas binding of the ligand directs folding of downstream elements in the expression platform that in turn influence expression. Thus, they contain a pocket that can be targeted for developing new antibiotics. Indeed, it was shown that compounds binding to the guanine riboswitch in vitro have activity in an in vivo infection model [9–11]. Further, it was demonstrated that the antibiotic compounds pyrithiamine and roseoflavin most likely act via the thiamine pyrophosphate (TPP) and flavin mononucleotide (FMN) riboswitches, resp. [12–16]. Other FMN analogs were shown to be active against *C. difficile* [17]. Only last year, the antibiotic compound ribocil which was discovered via a phenotypic screen was published. Ribocil is chemically not related to FMN but it was also found to exert its antibiotic activity via the FMN riboswitch (Fig. 1) [18].

While RNA molecules are suited to bind small molecules, special considerations have to be taken into account when designing ligands. RNA has polyanionic character and is surrounded by an envelope of well-ordered water molecules and positively charged metal ions. Therefore, ligands have to be of highly polar character to displace this shell and indeed RNA-binding ligands have the predisposition to be positively charged. This illustrates the particular challenges in drug design towards RNA, since charged compounds are less likely to penetrate membranes and often interact in a non-specific fashion with RNA requiring rigorous hit validation with unrelated RNA [19]. Furthermore, the small molecule ligands form different interactions with their RNA targets compared to small molecule interactions with proteins. Hydrophobic interactions with side chains similar to proteins

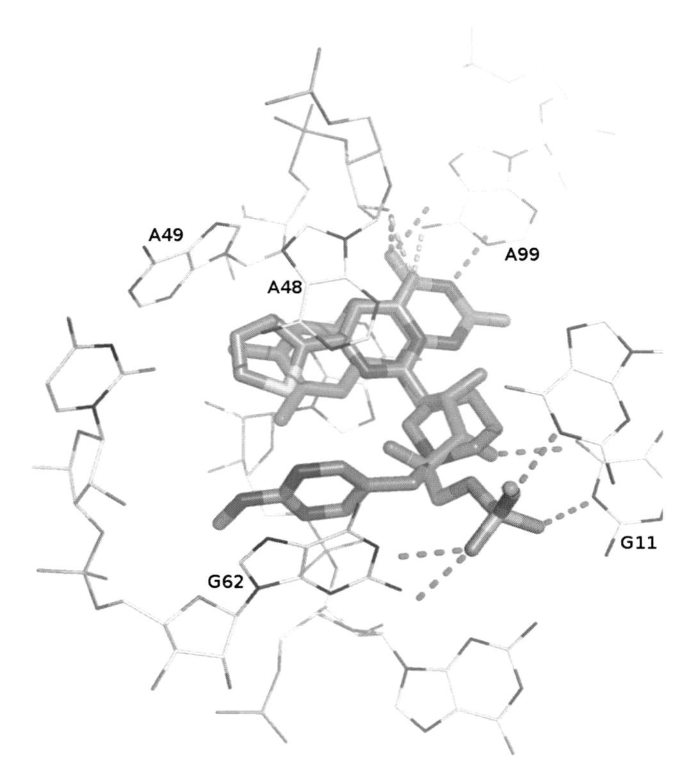

Fig. 1 The FMN-riboswitch binding site. The binding mode of FMN (*green* carbon atoms) was superimposed with the binding mode of ribocil (*blue* carbon atoms). Both ligands form stacking and face-to-face interactions with A48 and A49. FMN also forms a double hydrogen bond with A99, a hydrogen bond with G11 via a hydroxyl group and multiple hydrogen bonds via its terminal phosphate group (indicated as *green dashed lines*). In contrast, only the hydroxyl oxygen of ribocil is involved in hydrogen bonds (*gray dashed line*)

are lacking and the main interactions between small molecules and RNA turn out to be π-π-stacking, hydrogen bonding, and electrostatic interactions [20]. The different properties of small molecules binding to RNA targets compared to protein ligands can be rationalized by investigating the properties of antibacterial compounds. Antibacterials that modulate bacterial protein targets generally comply with the "rule-of-five" guidelines [21] for classical oral human drugs [22]. In contrast, antibacterials interfering with ribosomes often are larger than 500 Da and more polar. In the case of aminoglycosides *a*log*P*-values as low as -8.5 and polar surface areas (PSA) $>130\,\text{Å}^2$ are possible [22, 23].

The number of crystal structures of RNA molecules is steadily increasing. In September 2016, 303 crystal structures containing RNA and a ligand between 70 and 1000 Da were deposited in the Protein Data Bank (PDB) [24]. This structural wealth enables a structure-based approach for the design of new ligands. Two methods that make heavy use of structural knowledge for ligand discovery are molecular docking and fragment screening [25, 26]. We will expand on both methods in the following and give examples of their applications for the discovery of RNA ligands.

2 Molecular Docking for RNA Targets

The aim of molecular docking is to predict the binding mode of a small molecule in the binding site of its target and its binding affinity using computational methods (Fig. 2) [26]. Docking can also be used to screen large databases for ligands binding to a particular binding site (this is also referred to as structure-based virtual screening). In this context, all ligands in the database are sequentially placed into the binding site and scored for steric and chemical complementarity. Thus, a score-ranked database with ligands enriched among the top ranked molecules is obtained.

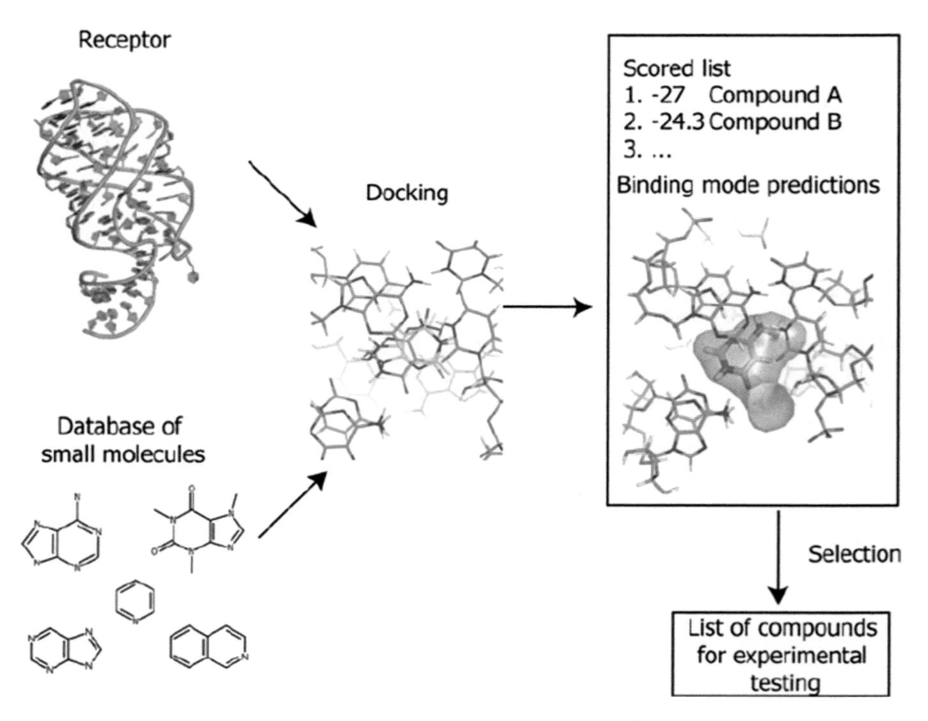

Fig. 2 Outline of molecular docking. Each entry of a small molecule database is sequentially placed in multiple orientations and conformations in the cavity and scored for steric and chemical complementarity. The result is a predicted binding mode for each database entry together with a score. High scoring ligands are visually inspected and selected for experimental testing. (Figure taken from [27])

Molecular docking consists of two parts: a conformational search to generate different binding modes and a scoring function to rank the generated binding modes and to estimate the binding energy. The main difference between the different conformational search algorithms is how the degrees of freedom of the ligand are treated while the scoring functions differ based on which underlying theory they are built on. Docking is routinely used for ligand discovery for protein targets [28] and there are several comprehensive textbooks and review articles about docking in general (e.g., [26, 29, 30]). Therefore, we will only give a concise overview and focus instead on what is specific for RNA-ligand docking compared to the more established protein-ligand docking.

2.1 Conformational Search Algorithms

There are three principle conformational search algorithms used in molecular docking to treat the degrees of freedom of a ligand: pre-generation of multi-conformer ligand libraries, incremental construction, and stochastic methods [26]. In the first approach, low energy conformations of the ligand are pre-calculated and subsequently docked rigidly into the binding site of the target. This approach is, for example, used in the earlier versions of DOCK [31]. In the branched-out version DOCK3.5 the generated ligand conformations are superimposed on a rigid substructure and the resulting ensemble is simultaneously placed into the binding site which goes along with a considerable increase in speed [32].

An alternative method is to incrementally build up the ligand in the binding site. For this approach, the ligand is first split into basic fragments. One of the fragments is then placed into the binding site. Subsequently, the ligand is reconstructed in the binding site by incrementally adding the remaining fragments. During the reconstruction of the ligand, different torsion angles are sampled. Examples of docking programs using this approach are FlexX [33], Surflex [34], and later versions of DOCK [35].

Ligand conformations and orientations can also be sampled on the fly using stochastic methods. The two main computational methods used are Monte Carlo simulations (MC) and genetic algorithms (GA). Both methods have in common that a random change in the binding mode (e.g., rotation or translation of the ligand or change of a torsion angle) is generated. The fit of the resulting pose is subsequently evaluated. If the fit is worse than the starting pose it is rejected, if not it is kept for the next cycle until a convergence criteria is reached. The docking programs ICM [36] and Glide [37] are based on MC simulations while GOLD [38] and AutoDock [39] use a GA.

2.2 Scoring Functions

Scoring functions fulfill two purposes: (1) They serve as fitness function in order to identify the binding mode with the lowest binding energy (i.e., the one found in the crystal structure) among the poses generated in the conformational search and (2) they rank different compounds relative to each other to identify the ones that bind best to a given target (Fig. 2). There is a wide variety of different techniques available which differ in speed and accuracy [26, 30]. In the context of virtual screening, typically only fast scoring functions are applied. They can be divided into three major classes: force-field based, empirical and knowledge-based scoring functions.

Force-field based scoring functions are based on the terms of intermolecular interactions of a classical molecular mechanics force field, i.e. they contain terms for van der Waals (vdW) and electrostatic interactions [26, 30]. They can also include terms for the intramolecular (strain) energy of the ligand. Their performance is improved when also the desolvation energy of the ligand is taken into account [40, 41]. The empirical parameters of the force field are either derived from physical measurements or ab initio calculations. Thus, there is no need for binding affinity data for parametrization. Examples of docking programs with force-field based scoring functions include DOCK and GOLD [31, 38].

Empirical scoring functions include several terms describing properties that are known to be important for forming a noncovalent complex whereas each term consists of a descriptor multiplied by a weighting coefficient [26, 30]. The descriptors typically take into account polar interactions such as hydrogen bonds and ionic interactions, apolar interactions such as hydrophobic and aromatic interactions and loss of ligand flexibility. The weighting coefficients are derived by multilinear regressions using a set of complexes for which both the binding affinity and the binding mode are known. As with force-field based scoring functions, considering the desolvation energy can increase their performance [42]. Among others, the docking programs FlexX [33], Glide [37], GOLD [38], AutoDock Vina [43], ICM [36], and Surflex [34] contain this type of scoring function.

Knowledge-based scoring functions attempt to capture the knowledge about receptor-ligand binding that is implicitly stored in large structural databases [26, 30]. The underlying assumption is that close intermolecular interactions between certain atom types or functional groups that occur more frequently than expected by random are likely to be energetically favourable and should therefore enhance the binding affinity. They are derived solely from structural data found, for example, in the PDB or the Cambridge Structural Database (CSD) [44] without including any affinity data. Examples of this type of scoring function include DrugScore [45], PMF [46], and the Astex statistical potential (ASP) [47] which is available in GOLD [38].

2.3 *RNA-Ligand Docking*

The physico-chemical principles for RNA-ligand binding are the same as for protein-ligand binding. Therefore, in general the methods developed for proteins can also be applied for RNA [20, 27, 48–50]. However, there are several challenges which have to be considered when transferring the methods. Firstly, RNA molecules are highly charged resulting in strong solvation and association of ionic molecules [4, 51, 52]. Water molecules and ions are often found at the binding site. These can either mediate ligand contacts or be displaced by the ligand which has to be considered during docking. Further, the long-range electrostatic interactions have to be modelled appropriately to avoid overly favouring of charged molecules, especially when using force-field based scoring functions. Secondly, RNA structures are highly flexible [53, 54]. Thus, ligand binding can be associated with a conformational change or induced fit movements which need to be considered when modelling RNA–ligand interactions [52]. Thirdly, depending on the chosen descriptors and atom type definition, scoring functions that were parameterized on protein-ligand complexes (i.e., empirical and knowledge based scoring functions) need to be re-parameterized on RNA-ligand complexes [51]. As there are nearly 100 times less crystal structures of RNA-ligand complexes than for protein-ligand complexes, less data can be used for parametrization which will in turn affect their performance [46, 47]. In the light of these challenges, researches have opted to either apply docking programs that were originally developed for protein-ligand docking to RNA, mostly with some adaptations, or to develop new methods and scoring functions (Tables 1 and 2).

2.3.1 Protein-Ligand Docking Programs Applied to RNA-Ligand Docking

All major docking programs developed for protein-ligand docking with the exception of FlexX have been applied to RNA-ligand docking (Table 1). The docking programs AutoDock Vina, GOLD, Glide, and Surflex were used as provided while AutoDock was used with and without adoptions. Barbault et al. [55] re-parameterized the AutoDock scoring function based on only eight RNA-ligand complexes while Detering and Varani [56] used solvation parameters for RNA atoms based on similar atom types found in amino acids. In addition, nitrogen atoms were treated differently, depending on them acting as hydrogen-bond donors or acceptors. Moitessier et al. extended AutoDock to account for RNA flexibility and changes in the first hydration shell [59]. When using different versions of DOCK below version 6, the atom parameters were taken from the AMBER force field [78], however, the phosphate backbone was neutralized by adjusting the partial charges of the involved atoms [56, 65]. DOCK 6 is an extension of DOCK 5 but was specifically optimized for RNA-ligand docking [68]. The incremental construction algorithm was modified to obtain better sampling and more sophisticated scoring schemes based on the generalized Born (GB) and Poisson–Boltzmann

Table 1 Overview of docking programs used for RNA-ligand docking

Docking program	Specifically developed for RNA-ligand docking	Retrospective validation	Prospective predictions
AutoDock [39]	In some cases, scoring function was re-parameterized [55] or solvation and atom parameters were adopted [56]	[56–60]	[61–63]
AutoDock Vina [43]	–	[64]	
DOCK 3.5 [32, 41]	Phosphate backbone was neutralized [65]	[65]	[65]
DOCK 4 and 5 [35, 66]	In some cases phosphate backbone was neutralized [56]	[56]	[61, 63, 67]
DOCK 6 [68]	Sampling method and scoring function was optimized for RNA [68]	[68, 69]	
GOLD [38]	–	[58, 70]	
Glide [37]	–	[58, 64, 70]	
ICM [36]	In some cases scoring function was tailored to RNA [67, 71]	[71, 72]	[67, 71, 72]
MODOR [73]	Yes	[73]	[74]
rDock [64]	Yes	[58, 64]	
RiboDock [75]	Yes	[75]	[76]
Surflex [34]	–	[58]	

Table 2 Overview of scoring functions developed for RNA-ligand docking

Scoring function	Retrospective validation
DrugScoreRNA [60]	[60, 69]
iMDLScore1 [58]	[58]
iMDLScore2 [58]	[58]
KScore [77]	[77]
LigandRNA [69]	[69]

(PB) implicit solvent models were added. Filikov et al. used ICM, but tailored the scoring function to RNA [71]. The new scoring function included terms for electrostatic, vdW, and hydrogen-bond interactions, desolvation energy, and the loss of conformational energy of the ligand. The weighting factors were fitted using affinity data of just five NMR structures of small ligands complexed with RNA.

2.3.2 Docking Programs and Scoring Functions Developed for RNA-Ligand Docking

Three docking programs were specifically developed for RNA-ligand docking (Table 1). MODOR generates first several orientations of the ligand in the binding site which are subsequently energy minimized whereas ligand and receptor are kept flexible to account for induced fit movements [73]. The resulting poses are scored with a force-field based scoring function which also considers the solvation energy based on an implicit solvent model. RiboDock treats the ligand as flexible and uses MC sampling to generate the binding modes [75]. It contains an empirical scoring function which includes terms for typical RNA–ligand interactions such as guanidinium–RNA interactions and π-π stacking of aromatic rings in addition to terms found in conventional functions such as hydrogen-bond and lipophilic interactions. The program rDock has evolved from RiboDock and is applicable for protein and RNA targets [64]. A GA is used to generate an initial docking pose which is refined using MC simulations. The ligand and terminal OH- and NH_3-groups of the receptor are treated flexible during docking. The terms for polar interactions in the original scoring function were reformulated and terms for vdW and solvation potentials were added. The scoring function was trained on a dataset composed of 43 protein-ligand and 15 RNA-ligand complexes together with their binding data.

Five standalone functions for scoring RNA-ligand complexes were published (Table 2). DrugScoreRNA, KScore, and LigandRNA are all knowledge-based scoring functions [60, 69, 77]. DrugScoreRNA uses distance-dependent pair potentials derived in a similar manner as the original DrugScore function [60]. Deriving the pair potentials was hindered by the small number of available crystal structures of RNA-ligand complexes. To overcome this hurdle the authors included also RNA-protein as well as DNA-ligand and -protein complexes in their knowledge base. KScore is built based on similar principles as DrugScoreRNA, but uses different atom types to describe the interactions in the complexes [77]. Further, a combination of protein-ligand, DNA-ligand, and RNA-ligand complexes was used to derive the scoring function making it applicable to not only score RNA–ligand but also DNA– and protein–ligand interactions. Due to the limited number of available crystal structures of RNA-ligand complexes also NMR structures were included in the training set. In contrast to DrugScoreRNA and KScore, LigandRNA uses distance- and angle-dependent potentials to describe the interactions between ligand and RNA atoms [77]. As knowledge base, NMR and crystal structures of RNA-protein and RNA-ligand complexes were used. iMDLScore1 and 2 are empirical scoring functions [58]. The basis of the iMDLScore functions is the AutoDock scoring function of which the terms for loss of torsional degrees of freedom upon binding, vdW, hydrogen bond and electrostatic interactions were optimized using a data set of NMR and crystal structures of RNA-ligand complexes together with their binding affinities. For deriving iMDLScore1 all structures were considered while for iMDLScore2 only complexes with positively charged ligands were included.

2.3.3 Performance of RNA-Ligand Docking

Only a few studies evaluating the performance of RNA-ligand docking were published (Tables 1 and 2). Most are retrospective studies comparing the performance based on previously published data. In general, docking programs are assessed based on their ability to (1) predict the binding mode correctly (mostly measured as root mean square deviation (rmsd) between docked and crystallographically determined binding mode), (2) to enrich known ligands out of a database of decoys (simulated virtual screening), and (3) to predict the binding affinity [30]. Comparison of the different methods for RNA-ligand docking is complicated by the fact that often different test data sets and evaluation criteria were used in the assessment studies.

Predicting the correct binding mode is generally considered to be the easiest among the three docking tasks [30]. Detering and Varani were among the firsts to validate the performance of DOCK and AutoDock for RNA-ligand docking in that respect [56]. They showed that for 60% of the test cases (16 complexes) a binding mode within 2.5 Å rmsd compared to the one found in the crystal structure was generated. Li et al. obtained similar results for Glide and GOLD using a test set of 60 RNA-ligand complexes (60 and 62%, respectively) [70]. On a test set composed of 56 RNA-ligand complexes, in 54% of the cases rDOCK was able to predict the correct binding mode while AutoDock Vina and Glide obtained success rates of 29 and 18%, respectively [64]. For MODOR, a success rate of 74% (based on a data set of 57 complexes), for RiboDock 50% (out of 10 complexes), and for ICM 53% was found (test set containing 96 complexes) [72, 73, 75]. With DOCK 6 the best results were obtained using a scoring scheme considering explicit water molecules and positively charged counterions in combination with the more advanced implicit solvent model PB/SA. In this scenario, for 80% of the complexes in the test set containing ligands with less than 7 rotatable bonds (in total 10 complexes) the top scoring binding mode had an rmsd of less than 2.0 Å [68]. This number dropped to 47% when also ligands with up to 13 rotatable bonds were considered (38 complexes in test set). Using the same measure, DrugScoreRNA predicted 42% of the binding modes correctly while the adopted AutoDock scoring function achieved a success rate of only 26% (31 complexes in test set) [60]. With a different test set (42 complexes). LigandRNA and DOCK 6 found the best solution in 36% and DrugScoreRNA in 31% of the cases [69]. Moitessier et al. were able to dock a set of 11 aminoglycosides with an average rmsd of 1.4 Å back into their receptors using an extended version of AutoDock accounting for structural water molecules and RNA flexibility [59]. A comprehensive assessment was carried out by Chen et al. [58]. They compared 5 docking programs (AutoDock, GOLD, Glide, MODOR, and RiboDock) and 11 scoring functions that come along with these programs. In terms of binding mode predictions GOLD with the GOLD fitness function and rDOCK with the rDOCK scoring function including the solvation term (see Sect. 2.3.2) worked best. Both methods were able to generate at least one pose with an rmsd $<$3 Å among the five top scoring poses for 73% of the cases (56 complexes in test set).

We are only aware of five studies in which the suitability of RNA-ligand docking for virtual screening was assessed [56, 58, 65, 71, 75]. Encouragingly, it was demonstrated that virtual screening performs better than random selection using either ICM, GOLD, Ribodock, Autodock, DOCK 3.5 or DOCK 4.0 as docking engine [56, 58, 65, 71, 75].

In four studies the ability to predict binding affinities was investigated [57, 60, 72, 77]. Barbault et al. used AutoDock together with their modified scoring function to establish structure-activity relationships for aminoglycoside derivatives [57]. When molecular dynamic (MD) simulations were used to optimize the binding modes, an R^2 of 0.84 between measured and calculated binding energies was obtained. On a data set of 15 diverse complexes, DrugScoreRNA achieved a Spearman rank correlation coefficient of $R_S = 0.61$ and KScore a linear correlation coefficient of $R = 0.81$ [60, 77]. Using ICM and a test set of 48 complexes, a linear correlation coefficient of $R = 0.71$ was obtained [72].

Collectively, RNA-ligand docking performance is promising but there is clearly room for improvements. Success rates for pose prediction in protein-ligand docking can be as high as 80% [30] which was only observed for one docking program/scoring function combination in the RNA field for a rather small and restricted data set (DOCK 6 with PB/SA, see above). The simulated virtual screening studies were carried out on very limited data sets, and with the exception of Detering and Varani, contained only one or two targets. Similar, the obtained correlation coefficients are in the area found for protein-ligand predictions, but the data sets used in the RNA field were mostly rather small. From protein-ligand docking, it is well known that the performance is target and software dependent [79, 80]. Therefore, extended evaluation studies are needed to obtain a better picture on the strengths and weaknesses of RNA-ligand docking with respect to virtual screening.

From the current studies, it is difficult to understand the underlying reasons for success and failure. Therefore, we recently introduced an experimentally and computationally tractable model system that allows entangling the different contributions to binding energy and such to guide the improvement of the computational methods [65]. The *Bacillus subtilis* xpt-pbuX guanine riboswitch carrying a C74U mutation (called GRA) was used to probe RNA-ligand docking [81]. The adenine binding site in this riboswitch is rather small (108 Å^3) and 98% shielded from bulk solvent making this pocket less complex. In retrospective tests based on previously published data we were able to correctly predict binding modes of known ligands, separate ligands from non-binders, and enrich ligands from a large database. Prospective predictions identified four new riboswitch ligands with binding affinities in the micromolar range; two of them were based upon molecular scaffolds not previously observed to bind to this riboswitch (Fig. 3a). High resolution crystal structures were determined to elucidate the binding modes of three of the new ligands (Fig. 3b). Interestingly, the difficulties that were encountered in this docking study were not specific to RNA-ligand docking but related to deficiencies in treating structural water molecules, predicting protonation states of ligands and anticipating multiple binding modes, all issues that are still challenging for protein-ligand docking as well. To the best of our knowledge, this is still the only RNA-ligand docking study which did not only evaluate RNA-ligand docking based on previously published data but also made predictions of new ligands which were subsequently tested using binding assays and X-ray crystallography.

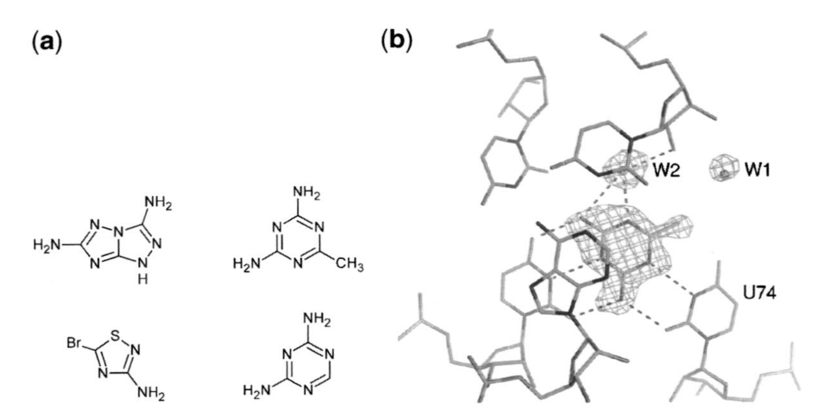

Fig. 3 (**a**) New riboswitch ligands identified by RNA-ligand docking. (**b**) Crystallographically determined binding mode of a new riboswitch ligand together with Fo-Fc omit map (contoured on 3.0σ) which was calculated by omitting the ligand from the final model. Putative hydrogen bonds are marked as *dashed lines*. (PDB code: 2nxz)

2.4 New Ligands Discovered by RNA-Ligand Docking

For several targets, RNA-ligand docking was successfully used for the discovery of new ligands, demonstrating that despite its deficiencies the method is a valuable tool for hit discovery for RNA (Table 1 and Fig. 4). In the following we only present studies for which interactions of the screening hits with the target were demonstrated in a binding assay.

In one of the first applications, Filikov et al. used adopted versions of DOCK and ICM for the discovery of ligands interrupting the formation of the transactivation response (TAR) element RNA-Tat complex [71]. Three compounds interrupting the complex were revealed. In a subsequent study, docking and evaluation were refined [67]. In this study, 11 ligands binding to TAR were found and for some of them cell activity could be demonstrated and binding to the intended binding site was confirmed using NMR. Also screening for TAR ligands, a different strategy was applied by Stelzer et al. [72]. They considered the flexibility of the target by not only docking into one structure of TAR, but by using an ensemble of 20 conformations, partially determined by NMR and partially derived from MD simulations. Thus, by using ICM as docking engine, six compounds that bind to TAR and inhibit complex formation with K_i values in the range of 10 nM–169 μM were found. One of the compounds was also shown to inhibit Tat-mediated activation of the HIV-1 long terminal repeat in T cell lines and HIV replication in an HIV-1 indicator cell line.

The RiboDock developers used their program for the discovery of ligands that target the bacterial ribosomal A-site [76]. Screening of a small molecule library revealed seven new ligands for which binding by NMR could be confirmed. These ligands had $K_{i,\mathrm{app}}$ values in the range of 17–419 μM.

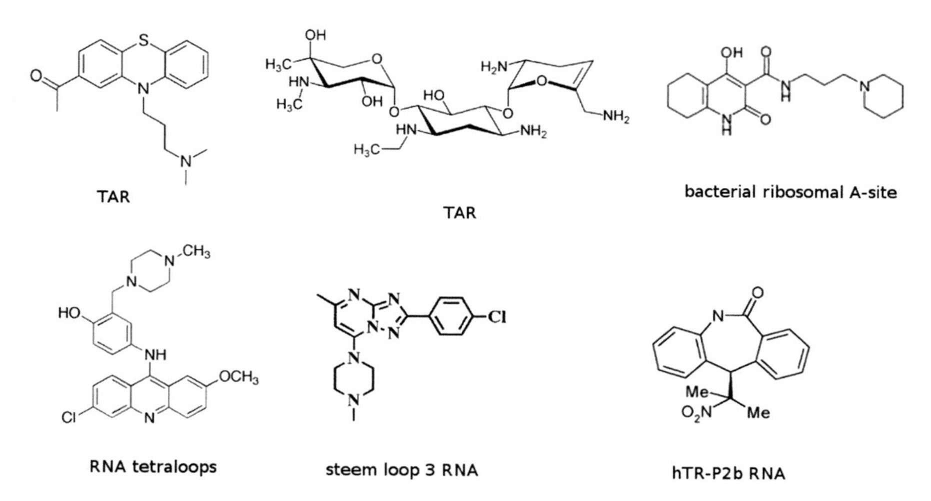

Fig. 4 Selected ligands discovered by RNA-ligand docking. The target name is given below the compound

Baranger et al. applied DOCK followed by AutoDock for finding ligands binding to RNA tetraloops and to stem loop 3 (SL3) RNA of the packaging element ψ of HIV-1. For the RNA tetraloop, they found one compound for which they could confirm binding to its target by NMR [63]. For the second target, two molecules were obtained that were selective for SL3 RNA over double- and single-stranded RNA, and one of them showed selectivity for SL3 RNA over other stem loops in ψ RNA [61]. In a third study, Warui and Baranger searched for compounds preventing binding of the nucleocapsid protein (NCp7) to the ψ-RNA. They started with a similarity search using one of the compounds from their previous study as starting point followed by docking using AutoDock [62]. As an alternative strategy, they conducted a high-throughput screen (HTS). Taken together with the previous report, in total five compounds that bind selectively to SL3 RNA compared to single- or double-stranded RNA were identified. Four of these compounds were identified using computational screening and one by HTS. Collectively, they also revealed four inhibitors of the SL3–NCp7 complex, all discovered by docking, however, these were not selective.

MODOR was used for the discovery of ligands binding to hTR-P2b which is part of the human telomerase RNA [74]. As MODOR is rather slow, an iterative strategy was employed. First, a small database of about 3000 molecules was docked. Promising hits were purchased and tested for binding by NMR. Subsequently, a similarity search in a much larger database was conducted based on the active compounds as starting structures. The hits from the similarity screen were then again docked using MODOR to obtain a ranking. Testing of promising hits resulted in 24 compounds which bound selectively to hTR-P2b RNA with K_D values between 84 μM and 1.5 mM. All of them bound to the U-rich region as it was intended in the docking approach.

2.5 New Approaches for RNA-Ligand Docking

The presented docking studies demonstrate the success of RNA-ligand docking. Nevertheless, there is scope for enhancements. One issue is how to handle RNA flexibility which was ignored in most of the studies. One option is to use predetermined RNA conformations for so-called ensemble docking [82] as done by Stelzer et al. and Moitessier et al. [59, 72]. Another option is to allow (parts of) the RNA to move during docking as, for example, done in MODOR and ICM [67, 73]. The required energy minimization steps in the latter two approaches makes them rather slow and not applicable to screen large databases. An alternative faster approach using elastic potentials grids was recently introduced by Krüger et al. [83]. The underlying idea is to adopt a 3D grid of potential field values, precalculated from an initial RNA conformation by DrugScoreRNA, to another conformation by moving grid intersection points in space, but keeping the potential field values constant. Thus, by using an apo- and a holo-structure as input, intermediate conformations can be considered during docking. It was demonstrated that this approach is twice as successful as docking into an apo structure and still half as successful as redocking to the holo structure.

A further difficulty in structure-based ligand design is to confirm the modelled binding mode experimentally in a timely manner. To this end, Frank demonstrated that measured and simulated chemical shifts of an RNA-ligand complex ("*holo* NMR chemical shifts") could be used to elucidate the binding mode [84]. First, a predictor was trained to predict the *holo* chemical shifts of a complex. In the second step, using docking, several potential binding modes of the ligand were generated and the chemical shifts of their complexes were predicted. Subsequently, the measured and predicted chemical shifts were compared in order to filter out the likely binding pose. Remarkably, for the models for which good agreement between measured and predicted *holo* chemical shifts was obtained, the binding modes were generally well predicted.

Finally, the holy grail in structure-based design is the accurate prediction of free-energies of binding or at least the difference in binding energy between two compounds. With this respect, free energy perturbation (FEP) methods have recently gained much attention in the protein field [85]. In this approach, MD or Monte Carlo simulations are carried out to determine the free-energy difference between two related ligands. For that purpose, the ligands are transformed from one to the other in silico by via either a chemical or alchemical path. In a recent study, the performance of FEP to predict binding energy differences for ligands binding either to the guanine riboswitch (GA) or the guanine riboswitch carrying a C74U mutation (GRA) was evaluated [86]. The calculations yielded an overall mean unsigned error (MUE) of 1.46 kcal/mol for the entire data set used. This value improved, if only ligands for which the binding data was published in the same paper were considered. These are promising results and it will be interesting to see the performance of FEP calculations for RNAs with more complex binding sites. However, the lack of suitable data sets currently prevents such studies.

3 Fragment Screening

In fragment-based drug discovery (FBDD) relatively small libraries (a few hundred to a few thousand) of low-molecular weight compounds (150–300 Da) are screened for weak-affinity ligands of the investigated target (Fig. 5). Despite the low potency of the discovered hits (K_D-values ≈10 μM-mM) this approach leads to a higher hit rate compared to HTS and offers the opportunity for a more efficient optimization process [25, 87–89]. The reason for this is that fragment libraries are able to sample larger proportions of chemical space in terms of scaffold diversity with a relatively low number of compounds compared to their HTS counterparts containing larger molecules and that less complex fragment molecules are more likely to bind to a target, albeit at the cost of potent binding affinity [90, 91]. The weakly binding hits

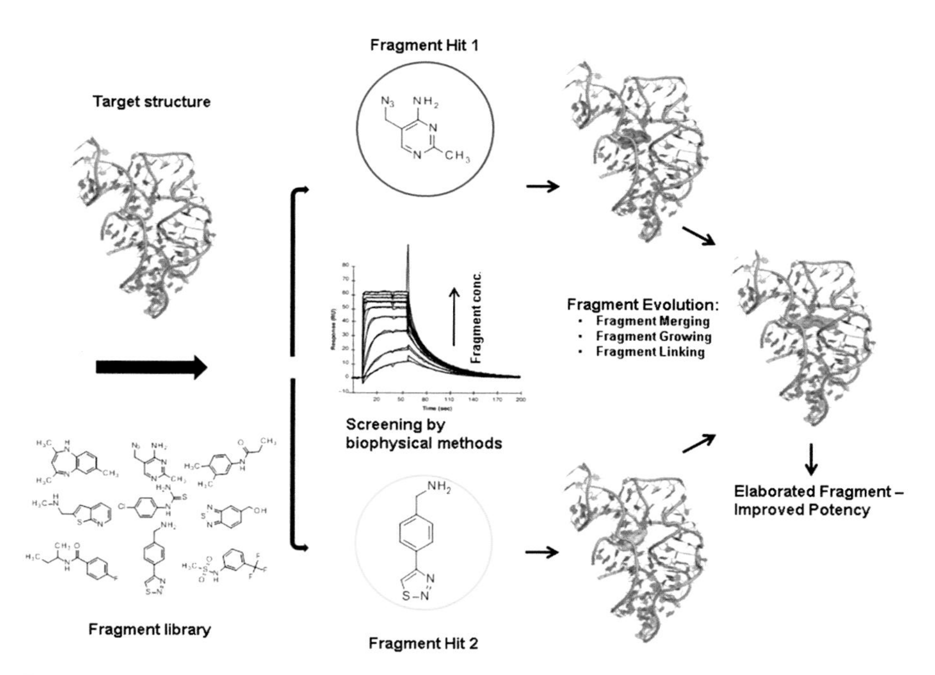

Fig. 5 Fragment screening. A library of fragments covering a large area of chemical space is screened for affinity to a certain target. Ligands are detected using biophysical methods with a high throughput and sufficient sensitivity. For affinity improvement the weakly binding hits are subsequently subjected to a fragment evolution process

are subsequently subjected to a fragment evolution process. The affinity of the fragments can be increased by either merging several fragments, linking them together or growing the fragments to explore additional areas in the binding site (Fig. 5).

3.1 *RNA-Directed Fragment Libraries*

Analysis of diverse sets of fragment hits against various targets revealed that these hits on average follow a "Rule of Three." This means that the hits usually have a molecular weight $\leq$300 Da, $\leq$3 h-bond donors, $\leq$3 h-bond acceptors and a ClogP $\leq$3. Further, the number of rotatable bonds is often $\leq$3 and the PSA $\leq$60 [92]. Therefore, these criteria are typically used for assembling fragment libraries. In addition, the chemical space covered should include key pharmacophores related to the target, and the fragments should have a balanced complexity and contain appropriate positions for fragment growing during lead optimization [93]. Based on these general guidelines, attempts were made to design fragment libraries tailored for RNA ligands.

3.1.1 Assembly of RNA-Binding Fragment Libraries by Literature Search

One possible approach to assemble a fragment library of RNA-binding compounds is to search published literature for small molecules binding to RNA and to use this information to derive a focused library. Bodoor et al. compiled a small molecule database of 121 published RNA-binding ligands with K_D-values of 50 μM or less [94]. Subsequently, the ligands were fragmented at suitable linker positions between rings and the obtained fragments were clustered according to physico-chemical descriptors and chemical fingerprints. Finally commercially available fragments representing each cluster were purchased resulting in a fragment library of 102 compounds. The performance of the library was validated by screening it against the bacterial ribosomal A site which resulted in two A site ligands that compete for the same binding site as gentamycin (see Sect. 3.3).

3.1.2 Kinase-Directed Libraries for RNA Fragment Screening

Foloppe et al. postulated that the pharmacophores for ligands binding to the bacterial ribosomal A-site (Fig. 6a) and the ATP binding site of kinases (Fig. 6b) overlap and therefore suggested using kinase-directed libraries for fragment screening against RNA targets [51]. Crystal structure analysis and NMR experiments demonstrated two main features of aminoglycosides binding to the A-site RNA. They all stack on a conserved guanine (G1491) and form two hydrogen bonds to the

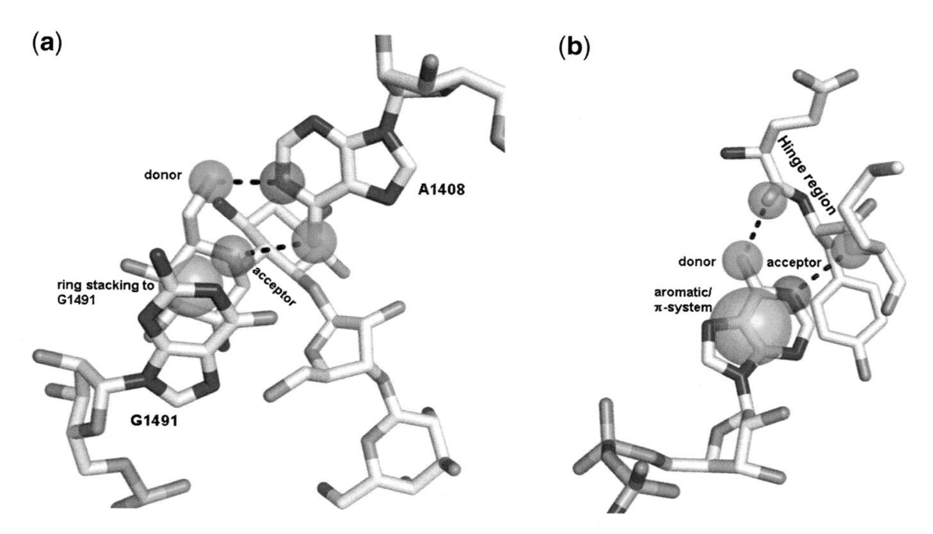

Fig. 6 (**a**) Binding mode of the aminoglycoside paromomycin in the ribosomal A-site (PDB code:1j7t). Potent ligands contain an aromatic ring stacking on G1491 (*orange sphere*), a hydrogen-bond donor interacting with N1 of A1408 (*green* and *red sphere*, resp.), and a hydrogen-bond acceptor interacting with N6 of A1408 (*red* and *green sphere*, resp.) (**b**) ATP analogue binding to the hinge region of lymphocyte-specific kinase (PDB code:1qpc). Typical kinase inhibitors contain an aromatic ring (*orange sphere*) and a hydrogen-bond donor and acceptor interacting with the hinge region (*green* and *red spheres*). Hydrogen bonds are marked as *black dashed lines*

Watson-Crick face of a conserved adenine (A1408) [95]. The pharmacophore of ligands binding to the ribosomal A-site RNA is therefore expected to include an aromatic ring stacking on G1491, a hydrogen-bond donor interacting with N1 of A1408 and a hydrogen-bond acceptor interacting with N6 of A1408 (Fig. 6a). This pharmacophore is shared by ligands binding to ATP binding sites of protein kinases [96]. Kinase inhibitors binding to the hinge region typically also contain an aromatic ring in addition to a hydrogen-bond acceptor and -donor in a similar spatial arrangement as required for A-site ligands (Fig. 6b). Docking of a kinase library against the A-site resulted in several plausible hit compounds, however, experimental verification of the results has not been reported [51].

There might be the opportunity to employ kinase-directed libraries also on other RNA targets, since due to the properties of the RNA bases, quite often a ligand will have to stack on a base while hydrogen bonding to the Watson-Crick face of another base in the binding pocket [51]. An example for another RNA target also requiring this pharmacophore are purine responsive riboswitches [97].

3.1.3 RNA-Focused Small Molecule Collection Based on Privileged Substructures

Benzimidazole containing compounds have proven to be a privileged scaffold in RNA drug design. Some derivatives compete with aminoglycosides for binding to the bacterial rRNA A-site, others bind to the hepatitis C internal ribosomal entry site (IRES) and *bis*-benzimidazoles interact with CUG repeats that cause myotonic dystrophy type 1 [98]. Based on this knowledge a similarity search for small, drug-like molecules containing the *bis*-benzimidazole or similar core structures was performed [99]. This resulted in a library of about 300 compounds with the potential to bind to RNA. The approach was validated by screening the library against expanded repeating RNA (r(CUG)^{exp}) that causes myotonic dystrophy type 1. The screening resulted in much higher hit rates compared to screening a conventional HTS library (9% vs. 0.01–3%) and in compounds with more suitable physico-chemical properties. It can be imagined that such an approach can also be applied to assemble a library for fragment screening by restricting the selected compounds to those that apply with the Rule of Three.

3.2 Fragment Screening Technologies to Identify RNA-Binding Ligands

The higher hit rates in fragment screening compared to conventional HTS come at the cost of binding affinity [91]. Therefore, methods are needed that are able to screen fragment libraries with the required sensitivity and throughput. In general, biophysical methods such as nuclear magnetic resonance (NMR) and surface plasmon resonance (SPR) are used [89]. However, alternative methods such as isothermal titration calorimetry (ITC), mass spectrometry (MS), thermal shift assays, X-ray crystallography, and computational screening have also been employed. In the following, we only concentrate on methods for which screening against RNA was reported.

3.2.1 NMR-Based Detection Methods

By far the most studies reported on fragment screening against RNA used NMR methods to detect ligand binding (see Sect. 3.3). NMR-based screening methods can be subdivided into two groups. On the one hand are ligand-detected methods and on the other hand are target-detected methods [100].

Ligand-detected methods trace the protons of the ligand involved in binding to the RNA. They rely on the intermolecular transfer of magnetization between ligand and target molecule and include saturation transfer difference (STD), waterLOGSY and nuclear Overhauser effect spectroscopy (NOESY). For more details, we refer to a recent review [100].

Target-detected NMR methods monitor the change of RNA chemical shifts upon ligand binding. The modified chemical environment of RNA atoms close to the binding site induces chemical shift variations in these atoms. Reporter atoms in RNA targets are either the imino protons of guanine and uridine or the aromatic H5/H6 atoms of uridine and cytosine. Three types of setups are typically used to detect small molecule binding to RNA targets: 1D-^{1}H spectra, TROSY, and TOCSY. For more details, we refer again to Moumne et al. [100]. All target-detected NMR methods for identification of small molecules binding to RNA have in common that more RNA is needed compared to the ligand-detected methods. Instead of using RNA-ligand ratios of $\approx$1:100 typically equimolar ratios are required for the former methods.

3.2.2 Mass Spectrometry (MS)

Using mass spectrometry (MS)-based methods, especially electrospray ionization (ESI-MS), rapid characterization of small molecules binding to RNA is achievable. Mixtures of compounds can be screened in parallel without the need for specific labeling. Identification of compounds is possible by using the exact molecular weight as label for each ligand [101, 102]. Native MS analysis of RNA-ligand complexes allows the simultaneous determination of stoichiometry and dissociation constants. The method is sensitive enough to detect ligands with K_D-values in the typical range of fragments. However, one of the drawbacks is the difficulty in identifying nonspecific binding [103].

There is an excellent review by Hofstadler and Griffey summarizing the adequacy of MS in analyzing non-covalent nucleic acid-ligand complexes. In particular RNA-ligand complexes were analyzed by ESI-MS. The most prominent examples are aminoglycosides binding either to the HIV-1 transactivation response element (TAR) or to the 16S A-site rRNA [102]. The group of Griffey developed an MS method to study the structure-activity relationships (SAR) for lead discovery against RNA targets. The so-called SAR by MS process starts by screening a set of compounds to identify hits of an RNA target of interest. SAR and specific binding of ligands are then probed through chemical elaboration of compounds and additional MS experiments. The output from SAR studies in turn is the key input for the elucidation of pharmacophoric features corresponding to the most important interactions [104].

3.2.3 Computational Approaches for Fragment Screening

In principle, the computational methods described above can also be used for virtual fragment screening. In fact, the overall performance of fragment docking and docking of larger molecules is comparable [105]. However, the reasons for failures are different. Sampling the binding modes becomes less of an issue in fragment docking while the need for accurate scoring functions increases. In principle our study about the discovery of GRA ligands using molecular docking described above can also be considered as fragment screening as all docked compounds were fragment-sized [65]. Admittedly, the binding site of GRA is rather small so that larger molecules are excluded from binding making this a rather easy target for FBDD. In a different study, Setny and Trylska established a workflow for searching for 16S RNA A site ligands using pharmacophore-based virtual fragment screening combined with 3D-QSAR. Twenty one potential ligands were shortlisted, however, binding measurements were not carried out [106].

3.3 Ligands Discovered by Fragment Screening on RNA Targets

Fragment-based drug design and lead discovery was successfully used to identify new ligands for various RNA targets using various screening approaches. The HIV-1 TAR element RNA was addressed by Zeiger et al. Ligand candidates contained basic molecular building blocks such as benzene rings, guanidine or amino groups. Screening was performed using a competitive fluorimetric assay with a fluorescently labelled Tat-peptide. The two most promising structural classes discovered were aminoquinolines and quinazolines. Their combination led to 2,4,6-triaminoquinazoline (Fig. 7) as the compound with the highest affinity to TAR RNA (IC_{50} = 40 μM) [107]. Further investigation of the interaction between 2,4,6-triaminoquinazoline and TAR RNA by ^{1}H-NMR titrations revealed that this fragment has two distinctive binding sites. A second fragment screening on HIV-1 TAR RNA was carried out by Davidson et al. [108]. They designed a chemical probe, the arginine derivative MV2003 (Fig. 7), to lock the RNA in a conformation capable of binding fragments at an adjacent site. At the same time, the bound probe served as reporter to detect fragment binding by ligand-based NMR during screening. After screening a 250 compound fragment library by STD-NMR 20 fragments were shortlisted out of which six were confirmed using NOESY experiments. In the presence of MV2003 and TAR RNA strong NMR signals were observed while in the absence of MV2003 ligand binding could not be detected indicating specific binding.

Another RNA target FBDD was successfully applied for is the bacterial 23S rRNA. This was done using the "SAR by MS" method developed by Swayze et al. (see Sect. 3.2.2) [104]. Screening of diverse compound libraries revealed two structurally different binding motifs for 23S rRNA-ligands: D-amino acids with a positively charged side chain and the quinoxaline-2,3-dione scaffold. Competition MS experiments showed that both classes were bound at different sites. Therefore, the compounds were linked together resulting in a compound (**10a**) with a K_D of 6.5 μM (Fig. 7) [104].

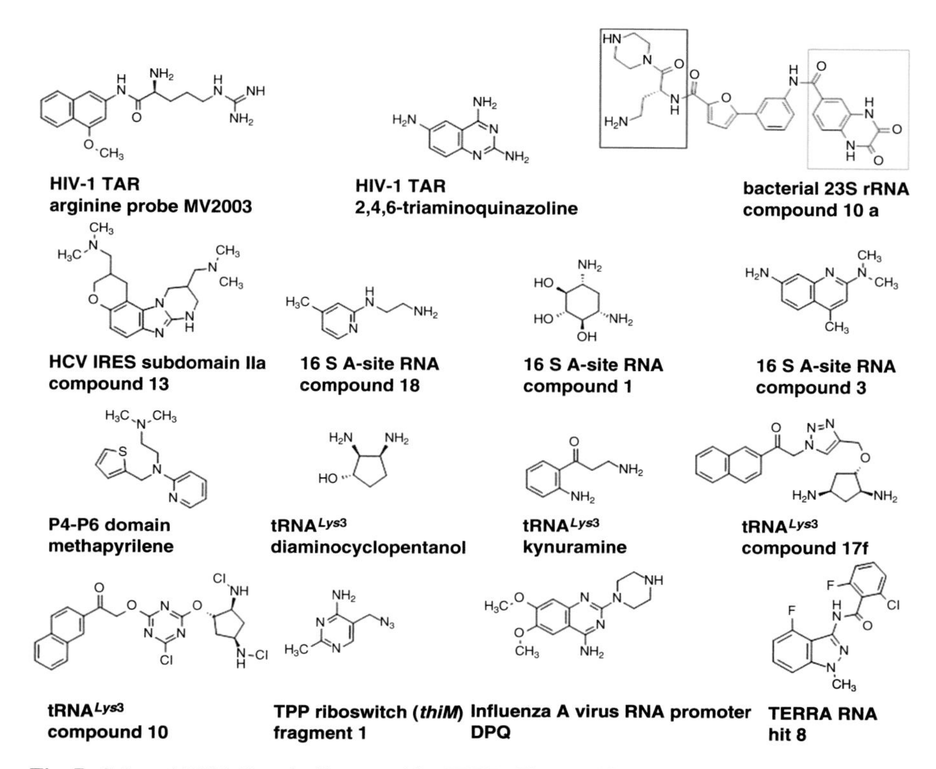

Fig. 7 Selected RNA-ligands discovered by FBDD. The specific target structure is given below the compound as well as the name of the compound in Sect. 3.3. For compound **10a** the originally different structural motifs linked together are marked by a *darkblue* and *yellow box*, respectively

The SAR by MS method was also used for the discovery of ligands binding to the internal ribosome entry site (IRES) subdomain IIa of HCV RNA. The benzimidazole screening hit displayed a K_D-value of 100 μM. Optimization led to the development of benzimidazoles with constrained side chains that were able to lower HCV RNA levels in replicon assays with the best hit (compound **13**) having a submicromolar dissociation constant (Fig. 7) [109]. It was subsequently shown that the compounds act by inducing conformational rearrangements in the IRES subdomain IIa, in particular in a widened interhelical angle which in turn facilitates the undocking of subdomain IIb from the ribosome and thus inhibits translation [110].

Another important antibiotic drug target is the ribosomal 16S A-site RNA. NMR-based fragment screening led to the discovery of aminoglycoside mimetics that are structurally very different from the known aminoglycosides [111]. The imino proton region of 1D NMR spectra was used to monitor spectral changes upon the addition of compounds for the identification of ligands. Screening a 10,000 compound library in such a fashion revealed several classes of ligands binding to the target. Optimization of a 2-aminopyridine to a 4-methyl analogue (compound

18) resulted in the tightest binding ligand (K_D of 3 μM, Fig. 7). Competition NMR experiments in the presence of an aminoglycoside confirmed that the optimized compound shared the same binding site. Further, the RNA-directed fragment library described above (Sect. 3.1.1) was evaluated by screening it against the A-site RNA using waterLOGSY NMR as main detection method. Five hits were identified out of which two (compounds **1** and **3**) showed competitive binding with gentamycin (Fig. 7) [94].

The P4-P6 domain of the *Tetrahymena thermophila* group I intron is another RNA target that possibly accommodates small molecules due to its stable and well-defined globular fold and the presence of several binding pockets [112]. For the fragment screen on this target the SHAPES approach was employed [113]. Here, NMR is used to detect binding of scaffolds derived from shapes most commonly found in known drug molecules [113]. Johnson et al. screened a SHAPES library of 112 compounds to find fragments binding to the P4-P6 domain of the group I intron. They evaluated three different NMR methods with waterLOGSY being the most suitable one. In summary, 23 of the SHAPES compounds interacted with the RNA. These primary hits were counterscreened against a double-stranded control RNA to exclude nonspecific binders which removed all compounds but methapyrilene (Fig. 7) [114].

Further, ligands for the antiviral target $tRNA^{Lys3}$ were developed using an FBDD approach [115, 116]. Similar as with the 16S A-Site RNA ([111], see above) the imino proton region of 1D NMR spectra was monitored to detect ligand binding. The primary screen of a focused library of 50 small organic compounds that were either synthesized in-house or commercially available revealed diaminocyclopentanol (K_D-value = 2.1 mM) and kynuramine (K_D-value = 0.6 mM) as ligands (Fig. 7). Subsequently, the binding sites of the compounds were determined using 2D HMQC NMR with ^{15}N-labelled RNA. Kynuramine interacted with the D stem of the tRNA and diaminocyclopentanol bound to at least two sites located in the T and D stems. Fragment elaboration and linking resulted in a compound (**17f**) with increased affinity (K_D-value of 1.8 μM, Fig. 7). However, this compound is not specific and also binds to other tRNAs with comparable affinity. Subsequently, it was shown that replacing the 1,4-triazole linker with a 1,3,5-triazine linker (compound **10**, Fig. 7) resulted in compounds that are selective for the D-stem over the T-stem (K_D-value of 1.4 μM) [117].

The thiamine pyrophosphate (TPP) riboswitch *thiM* of *E.coli* was also approached using FBDD [118, 119]. The initial screening tool was equilibrium dialysis with [^{3}H]thiamine as reporter ligand. Out of a library of ~1300 compounds 20 hits were identified. Binding was confirmed by recording proton spectra of compounds in the presence and absence of riboswitch RNA. The binding affinities of 17 of these compounds could be determined using ITC and were in the range of 22–280 μM [118]. Subsequently, the crystal structures of four of the fragments in complex with the riboswitch were determined [120]. The pyrimidine ring of fragment **1** (Fig. 7) with a K_D-value of 49 μM superimposes very well with the one of TPP (Fig. 8a). Crystal structure analysis of all the four co-crystallized fragments further revealed that these fragments induce an alternative conformation

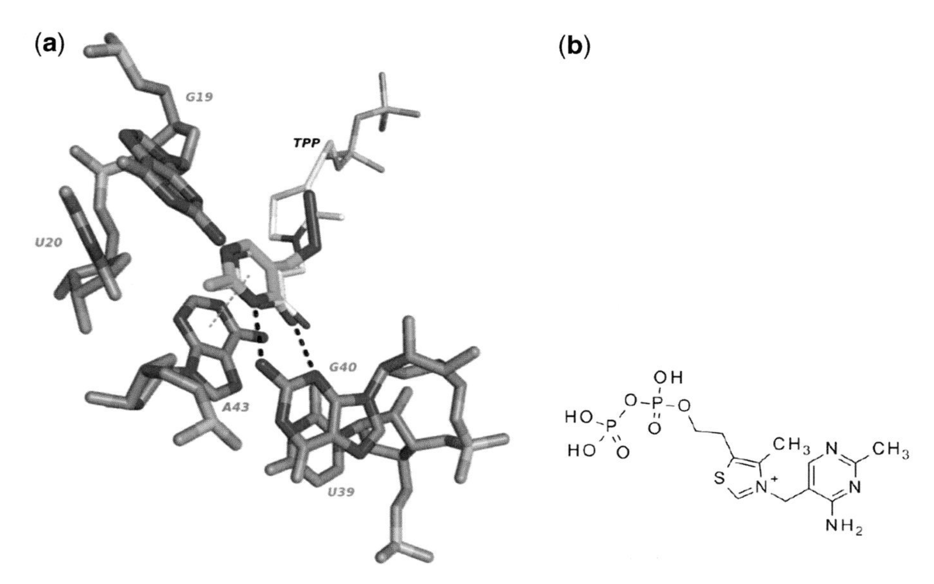

Fig. 8 (**a**) Overlay of the binding modes of fragment **1** (*green* carbon atoms) and TPP (*gray* carbon atoms) binding to the TPP riboswitch (*magenta* carbon atoms). *Black dashed lines* indicate the hydrogen bond network and the *green dashed line* π-π-stacking between the ligands and A43. (**b**) The chemical structure of the natural ligand TPP

of the TPP riboswitch which might be further exploited for the development of tightly binding ligands.

NMR spectroscopy was further successfully applied for the identification of fragment-like molecules binding to the influenza A virus RNA promoter [121]. Different influenza A virus subtypes carry viral RNAs in their genome with highly conserved nucleotides on their 5′- as well as their 3′-ends. These ends fold together and thereby promote the activity of the RNA polymerase making this partial duplex RNA a suitable antiviral drug target. 1D NMR spectra were used to screen small molecular weight compounds derived from a fragment library of 4279 compounds. Amongst seven hit substances 6,7-dimethoxy-2-(1-piperazinyl)-4-quinazolinamine (DPQ) (Fig. 7) had the most dramatic effect on imino protons of the RNA resulting in a K_D value of 50.5 μM. The structure of the complex between the RNA promoter and DPQ was also determined using NMR spectroscopy. In the complex structure, parts of the major grove are widened in order to accommodate the small molecule. Subsequent cell-based assays revealed that DPQ has antiviral activity in influenza A and B infected cell lines.

The suitability of FBDD for the discovery of telomeric RNA G-quadruplexes (TERRA) ligands was also demonstrated [122]. A library of 355 fluorinated fragments was screened by ^{19}F-NMR. The spectra of 20 compounds were affected by the presence of telomeric RNA. Seven primary hits were carried on and validated by ^{19}F-NMR and STD experiments. K_D-values in the range of 120–1900 μM were reported with hit **8** (Fig. 7) displaying the highest affinity [122].

3.4 Alternative Strategies to Discover RNA Ligands by FBDD: Surface Plasmon Resonance (SPR)

Surface plasmon resonance (SPR)-based biosensor assay methods are a useful tool to directly measure macromolecule–ligand interactions by determining binding affinities and kinetic parameters at the same time [123]. In most cases, the macromolecule is immobilized on a sensor chip surface coated with a thin gold film and a chemically modified (dextran-)matrix above it to keep the macromolecule bound to this functionalized surface. SPR is routinely used for fragment screening for protein targets [123]. The fragment molecules are applied to the sensor surface via a continuous flow in buffer and the binding event causes a change in the refractive index measured near the surface. This is mainly effected by a change of the size of the bound complex [124, 125]. An advantage of this method is its low consumption of macromolecule, since the immobilization theoretically enables screening of up to thousands of compounds without renewing the protein or nucleic acid [126]. However, proper experimental design is necessary to analyze the data and distinguish nonspecific binding from the interaction of interest.

To date there are only a few studies, where an SPR-based evaluation of small molecules binding to RNA was realized. Gonzalez-Fernandez et al. used SPR to test binding of tobramycin to anti-tobramycin aptamers with different levels of 2′ O-methylation in their aptamer sequence. In this case a reverse configuration compared to a typical SPR-experiment was chosen and the smaller binding partner tobramycin was immobilized to obtain a better signal [127]. For the detection of coenzyme B12 (AdoCbl) binding to the *btuB* riboswitch from *Escherichia coli* the conventional strategy was chosen. The expression platform of the *btuB* riboswitch is naturally not part of the interaction network to AdoCbl [128] and therefore was chosen to be immobilized. In detail, a biotinylated DNA oligonucleotide complementary to the 3′-end of the expression platform was immobilized to a streptavidin-coated sensor chip in order to attach the riboswitch RNA via Watson-Crick base pairing of the RNA-DNA-Hybrid. Determination of kinetic constants was achieved by testing an AdoCbl concentration series [129]. In case that the overall fold of the binding site will not be compromised, it is also possible to biotinylate the terminal residue of the RNA and immobilize it on a streptavidin-coated sensor chip. To study the preQ1 riboswitch of *Thermoanaerobacter tengcongensis* in the metabolite-bound and free states, the riboswitch was 5′-end biotinylated and the apparent K_D-values were determined [130]. A similar method with 5′-end biotinylated RNA was also described earlier for aminoglycoside antibiotics binding to RNA structures [131]. In principle, these approaches are also amenable to fragment screening, however the practical suitability still has to be demonstrated.

4 Conclusions

Structure-based methods clearly have an impact on the discovery of new RNA ligands. The ligands revealed using these methods are chemically diverse and many of them are below 500 Da and carry only one positive charge or are even neutral making them attractive starting points for further optimization (Figs. 3a, 4 and 7) As anticipated when using these approaches, the affinities of the hits were typically in the higher micromolar range. In the few studies in which potency optimization was attempted affinity could be improved to reach the single digit μM range. It can be expected that making use of structural information will help to improve their affinities even further. The drug-like compound ribocil which was recently disclosed together with its binding mode has an affinity in the low nanomolar range for the FMN riboswitch (Fig. 1, [18]) and can serve as an inspiration to intensify the hunt for drug-like and potent ligands for other RNA targets using structure-based methods.

Acknowledgements We are grateful for internal university research funding from the Johannes Gutenberg University Mainz.

References

1. Guan L, Disney MD (2012) Recent advances in developing small molecules targeting RNA. ACS Chem Biol 7(1):73–86
2. Batey RT, Rambo RP, Doudna JA (1999) Tertiary motifs in RNA structure and folding. Angew Chem Int Ed Engl 38(16):2326–2343
3. Reiter NJ, Chan CW, Mondragón A (2011) Emerging structural themes in large RNA molecules. Curr Opin Struct Biol 21(3):319–326
4. Thomas JR, Hergenrother PJ (2008) Targeting RNA with small molecules. Chem Rev 108 (4):1171–1224
5. Wilson DN (2013) Ribosome-targeting antibiotics and mechanisms of bacterial resistance. Nat Rev Microbiol 12(1):35–48
6. Blount KF, Breaker RR (2006) Riboswitches as antibacterial drug targets. Nat Biotechnol 24 (12):1558–1564
7. Deigan KE, Ferre-D'Amare AR (2011) Riboswitches: discovery of drugs that target bacterial gene-regulatory RNAs. Acc Chem Res 44(12):1329–1338
8. Matzner D, Mayer G (2015) (Dis)similar analogues of riboswitch metabolites as antibacterial lead compounds. J Med Chem 58(8):3275–3286
9. Kim JN, Blount KF, Puskarz I, Lim J, Link KH, Breaker RR (2009) Design and antimicrobial action of purine analogues that bind Guanine riboswitches. ACS Chem Biol 4(11):915–927
10. Mulhbacher J, Brouillette E, Allard M, Fortier LC, Malouin F, Lafontaine DA (2010) Novel riboswitch ligand analogs as selective inhibitors of guanine-related metabolic pathways. PLoS Pathog 6(4):e1000865
11. Ster C, Allard M, Boulanger S, Lamontagne Boulet M, Mulhbacher J, Lafontaine DA, Marsault E, Lacasse P, Malouin F (2013) Experimental treatment of *Staphylococcus aureus* bovine intramammary infection using a guanine riboswitch ligand analog. J Dairy Sci 96 (2):1000–1008

12. Lee ER, Blount KF, Breaker RR (2009) Roseoflavin is a natural antibacterial compound that binds to FMN riboswitches and regulates gene expression. RNA Biol 6(2):187–194
13. Mansjo M, Johansson J (2011) The riboflavin analog roseoflavin targets an FMN-riboswitch and blocks *Listeria monocytogenes* growth, but also stimulates virulence gene-expression and infection. RNA Biol 8(4):674–680
14. Ott E, Stolz J, Lehmann M, Mack M (2009) The RFN riboswitch of Bacillus subtilis is a target for the antibiotic roseoflavin produced by *Streptomyces davawensis*. RNA Biol 6 (3):276–280
15. Pedrolli DB, Matern A, Wang J, Ester M, Siedler K, Breaker R, Mack M (2012) A highly specialized flavin mononucleotide riboswitch responds differently to similar ligands and confers roseoflavin resistance to *Streptomyces davawensis*. Nucleic Acids Res 40 (17):8662–8673
16. Sudarsan N, Cohen-Chalamish S, Nakamura S, Emilsson GM, Breaker RR (2005) Thiamine pyrophosphate riboswitches are targets for the antimicrobial compound pyrithiamine. Chem Biol 12(12):1325–1335
17. Blount KF, Megyola C, Plummer M, Osterman D, O'Connell T, Aristoff P, Quinn C, Alan Chrusciel R, Poel TJ, Schostarez HJ, Stewart CA, Walker DP, Wuts PGM, Breaker RR (2015) Novel riboswitch-binding flavin analog that protects mice against *Clostridium difficile* infection without inhibiting cecal flora. Antimicrob Agents Chemother 59(9):5736–5746
18. Howe JA, Wang H, Fischmann TO, Balibar CJ, Xiao L, Galgoci AM, Malinverni JC, Mayhood T, Villafania A, Nahvi A, Murgolo N, Barbieri CM, Mann PA, Carr D, Xia E, Zuck P, Riley D, Painter RE, Walker SS, Sherborne B, de Jesus R, Pan W, Plotkin MA, Wu J, Rindgen D, Cummings J, Garlisi CG, Zhang R, Sheth PR, Gill CJ, Tang H, Roemer T (2015) Selective small-molecule inhibition of an RNA structural element. Nature 526 (7575):672–677
19. Aboul-ela F (2009) Strategies for the design of RNA-binding small molecules. Future Med Chem 2(1):93–119
20. Disney MD, Yildirim I, Childs-Disney JL (2014) Methods to enable the design of bioactive small molecules targeting RNA. Org Biomol Chem 12(7):1029–1039
21. Lipinski CA, Lombardo F, Dominy BW, Feeney PJ (1997) Experimental and computational approaches to estimate solubility and permeability in drug discovery and development settings. Adv Drug Deliv Rev 23(1–3):3–25
22. Mugumbate G, Overington JP (2015) The relationship between target-class and the physicochemical properties of antibacterial drugs. Bioorg Med Chem 23(16):5218–5224
23. O'Shea R, Moser HE (2008) Physicochemical properties of antibacterial compounds: implications for drug discovery. J Med Chem 51(10):2871–2878
24. Berman HM, Westbrook J, Feng Z, Gilliland G, Bhat TN, Weissig H, Shindyalov IN, Bourne PE (2000) The protein data bank. Nucleic Acids Res 28(1):235–242
25. Baker M (2012) Fragment-based lead discovery grows up. Nat Rev Drug Discov 12(1):5–7
26. Rognan D (2011) Docking methods for virtual screening: principles and recent advances. In: Sotriffer C (ed) Virtual screening. Wiley-VCH Verlag GmbH & Co. KGaA, Weinheim, pp 153–176
27. Daldrop P, Brenk R (2014) Structure-based virtual screening for the identification of RNA-binding ligands. Methods Mol Biol 1103:127–139
28. Irwin JJ, Shoichet BK (2016) Docking screens for novel ligands conferring new biology. J Med Chem 59(9):4103–4120
29. Kitchen DB, Decornez H, Furr JR, Bajorath J (2004) Docking and scoring in virtual screening for drug discovery: methods and applications. Nat Rev Drug Discov 3(11):935–949
30. Sotriffer C, Matter H (2011) The challenge of affinity prediction: scoring functions for structure-based virtual screening. In: Sotriffer C (ed) Virtual screening. Wiley-VCH Verlag GmbH & Co. KGaA, Weinheim, pp 177–221
31. Kuntz ID, Blaney JM, Oatley SJ, Langridge R, Ferrin TE (1982) A geometric approach to macromolecule-ligand interactions. J Mol Biol 161:269–288

32. Lorber DM, Shoichet BK (1998) Flexible ligand docking using conformational ensembles. Protein Sci 7(4):938–950
33. Rarey M, Kramer B, Lengauer T, Klebe G (1996) A fast flexible docking method using an incremental construction algorithm. J Mol Biol 261(3):470–489
34. Jain AN (2003) Surflex: fully automatic flexible molecular docking using a molecular similarity-based search engine. J Med Chem 46(4):499–511
35. Ewing TJ, Makino S, Skillman AG, Kuntz ID (2001) DOCK 4.0: search strategies for automated molecular docking of flexible molecule databases. J Comput Aided Mol Des 15 (5):411–428
36. Abagyan R, Totrov M, Kuznetsov D (1994) ICM – a new method for protein modeling and design – applications to docking and structure prediction from the distorted native conformation. J Comput Chem 15(5):488–506
37. Friesner RA, Banks JL, Murphy RB, Halgren TA, Klicic JJ, Mainz DT, Repasky MP, Knoll EH, Shelley M, Perry JK, Shaw DE, Francis P, Shenkin PS (2004) Glide: a new approach for rapid, accurate docking and scoring. 1. Method and assessment of docking accuracy. J Med Chem 47(7):1739–1749
38. Jones G, Willett P, Glen RC, Leach AR, Taylor R (1997) Development and validation of a genetic algorithm for flexible docking. J Mol Biol 267(3):727–748
39. Morris GM, Goodsell DS, Halliday RS, Huey R, Hart WE, Belew RK, Olson AJ (1998) Automated docking using a Lamarckian genetic algorithm and an empirical binding free energy function. J Comput Chem 19(14):1639–1662
40. Mysinger MM, Shoichet BK (2010) Rapid context-dependent ligand desolvation in molecular docking. J Chem Inf Model 50(9):1561–1573
41. Wei BQ, Baase WA, Weaver LH, Matthews BW, Shoichet BK (2002) A model binding site for testing scoring functions in molecular docking. J Mol Biol 322(2):339–355
42. Friesner RA, Murphy RB, Repasky MP, Frye LL, Greenwood JR, Halgren TA, Sanschagrin PC, Mainz DT (2006) Extra precision glide: docking and scoring incorporating a model of hydrophobic enclosure for protein-ligand complexes. J Med Chem 49(21):6177–6196
43. Trott O, Olson AJ (2010) AutoDock Vina: improving the speed and accuracy of docking with a new scoring function, efficient optimization, and multithreading. J Comput Chem 31 (2):455–461
44. Groom CR, Bruno IJ, Lightfoot MP, Ward SC (2016) The Cambridge Structural Database. Acta Crystallogr Sect B Struct Sci Cryst Eng Mater 72(2):171–179
45. Gohlke H, Hendlich M, Klebe G (2000) Predicting binding modes, binding affinities and "hot spots" for protein-ligand complexes using a knowledge-based scoring function. Persp Drug Discov Des 20:115–144
46. Muegge I (2006) PMF scoring revisited. J Med Chem 49(20):5895–5902
47. Mooij WT, Verdonk ML (2005) General and targeted statistical potentials for protein-ligand interactions. Proteins 61(2):272–287
48. Colizzi F, Lamontagne A-M, Lafontaine DA, Bussi G (2014) Probing riboswitch binding sites with molecular docking, focused libraries, and in-line probing assays. Methods Mol Biol 1103:141–151
49. Fulle S, Gohlke H (2010) Molecular recognition of RNA: challenges for modelling interactions and plasticity. J Mol Recognit 23(2):220–231
50. Stefaniak F, Chudyk EI, Bodkin M, Dawson WK, Bujnicki JM (2015) Modeling of ribonucleic acid-ligand interactions. Wiley Interdiscip Rev Comput Mol Sci 5(6):425–439
51. Foloppe N, Matassova N, Aboul-ela F (2006) Towards the discovery of drug-like RNA ligands? Drug Discov Today 11(21–22):1019–1027
52. Hermann T (2002) Rational ligand design for RNA: the role of static structure and conformational flexibility in target recognition. Biochimie 84(9):869–875
53. Cruz JA, Westhof E (2009) The dynamic landscapes of RNA architecture. Cell 136 (4):604–609

54. Dethoff EA, Chugh J, Mustoe AM, Al-Hashimi HM (2012) Functional complexity and regulation through RNA dynamics. Nature 482(7385):322–330
55. Barbault F, Zhang LR, Zhang LH, Fan BT (2006) Parametrization of a specific free energy function for automated docking against RNA targets using neural networks. Chemom Intel Lab Syst 82(1–2):269–275
56. Detering C, Varani G (2004) Validation of automated docking programs for docking and database screening against RNA drug targets. J Med Chem 47(17):4188–4201
57. Barbault F, Ren B, Rebehmed J, Teixeira C, Luo Y, Smila-Castro O, Maurel F, Fan B, Zhang L (2008) Flexible computational docking studies of new aminoglycosides targeting RNA 16S bacterial ribosome site. Eur J Med Chem 43(8):1648–1656
58. Lu C, Calin GA, Zhang S (2012) Novel insights of structure-based modeling for RNA-targeted drug discovery. J Chem Inf Model 52(10):2741–2753
59. Moitessier N, Westhof E, Hanessian S (2006) Docking of aminoglycosides to hydrated and flexible RNA. J Med Chem 49(3):1023–1033
60. Pfeffer P, Gohlke H (2007) DrugScoreRNA – knowledge-based scoring function to predict RNA-ligand interactions. J Chem Inf Model 47(5):1868–1876
61. Warui DM, Baranger AM (2009) Identification of specific small molecule ligands for stem loop 3 ribonucleic acid of the packaging signal Psi of human immunodeficiency virus-1. J Med Chem 52(17):5462–5473
62. Warui DM, Baranger AM (2012) Identification of small molecule inhibitors of the HIV-1 nucleocapsid–stem-loop 3 RNA complex. J Med Chem 55(9):4132–4141
63. Yan Z, Sikri S, Beveridge DL, Baranger AM (2007) Identification of an aminoacridine derivative that binds to RNA tetraloops. J Med Chem 50(17):4096–4104
64. Ruiz-Carmona S, Alvarez-Garcia D, Foloppe N, Garmendia-Doval AB, Juhos S, Schmidtke P, Barril X, Hubbard RE, Morley SD (2014) rDock: a fast, versatile and open source program for docking ligands to proteins and nucleic acids. PLoS Comput Biol 10(4): e1003571
65. Daldrop P, Reyes FE, Robinson DA, Hammond CM, Lilley DM, Batey RT, Brenk R (2011) Novel ligands for a purine riboswitch discovered by RNA-ligand docking. Chem Biol 18 (3):324–335
66. Moustakas DT, Lang PT, Pegg S, Pettersen E, Kuntz ID, Brooijmans N, Rizzo RC (2006) Development and validation of a modular, extensible docking program: DOCK 5. J Comput Aided Mol Des 20(10–11):601–619
67. Lind KE, Du Z, Fujinaga K, Peterlin BM, James TL (2002) Structure-based computational database screening, in vitro assay, and NMR assessment of compounds that target TAR RNA. Chem Biol 9(2):185–193
68. Lang PT, Brozell SR, Mukherjee S, Pettersen EF, Meng EC, Thomas V, Rizzo RC, Case DA, James TL, Kuntz ID (2009) DOCK 6: combining techniques to model RNA-small molecule complexes. RNA 15(6):1219–1230
69. Philips A, Milanowska K, Lach G, Bujnicki JM (2013) LigandRNA: computational predictor of RNA-ligand interactions. RNA 19(12):1605–1616
70. Li Y, Shen J, Sun X, Li W, Liu G, Tang Y (2010) Accuracy assessment of protein-based docking programs against RNA targets. J Chem Inf Model 50(6):1134–1146
71. Filikov AV, Mohan V, Vickers TA, Griffey RH, Cook PD, Abagyan RA, James TL (2000) Identification of ligands for RNA targets via structure-based virtual screening: HIV-1 TAR. J Comput Aided Mol Des 14(6):593–610
72. Stelzer AC, Frank AT, Kratz JD, Swanson MD, Gonzalez-Hernandez MJ, Lee J, Andricioaei I, Markovitz DM, Al-Hashimi HM (2011) Discovery of selective bioactive small molecules by targeting an RNA dynamic ensemble. Nat Chem Biol 7(8):553–559
73. Guilbert C, James TL (2008) Docking to RNA via root-mean-square-deviation-driven energy minimization with flexible ligands and flexible targets. J Chem Inf Model 48(6):1257–1268

74. Pinto IG, Guilbert C, Ulyanov NB, Stearns J, James TL (2008) Discovery of ligands for a novel target, the human telomerase RNA, based on flexible-target virtual screening and NMR. J Med Chem 51(22):7205–7215
75. Morley SD, Afshar M (2004) Validation of an empirical RNA-ligand scoring function for fast flexible docking using Ribodock. J Comput Aided Mol Des 18(3):189–208
76. Foloppe N, Chen IJ, Davis B, Hold A, Morley D, Howes R (2004) A structure-based strategy to identify new molecular scaffolds targeting the bacterial ribosomal A-site. Bioorg Med Chem 12(5):935–947
77. Zhao X, Liu X, Wang Y, Chen Z, Kang L, Zhang H, Luo X, Zhu W, Chen K, Li H, Wang X, Jiang H (2008) An improved PMF scoring function for universally predicting the interactions of a ligand with protein, DNA, and RNA. J Chem Inf Model 48(7):1438–1447
78. Pearlman DA, Case DA, Caldwell JW, Ross WS, Cheatham TE, DeBolt S, Ferguson D, Seibel G, Kollman P (1995) AMBER, a package of computer programs for applying molecular mechanics, normal mode analysis, molecular dynamics and free energy calculations to simulate the structural and energetic properties of molecules. Comput Phys Commun 91(1):1–41
79. Mysinger MM, Carchia M, Irwin JJ, Shoichet BK (2012) Directory of useful decoys, enhanced (DUD-E): better ligands and decoys for better benchmarking. J Med Chem 55 (14):6582–6594
80. Warren GL, Andrews CW, Capelli AM, Clarke B, LaLonde J, Lambert MH, Lindvall M, Nevins N, Semus SF, Senger S, Tedesco G, Wall ID, Woolven JM, Peishoff CE, Head MS (2006) A critical assessment of docking programs and scoring functions. J Med Chem 49 (20):5912–5931
81. Gilbert SD, Stoddard CD, Wise SJ, Batey RT (2006) Thermodynamic and kinetic characterization of ligand binding to the purine riboswitch aptamer domain. J Mol Biol 359 (3):754–768
82. Wong CF (2015) Flexible receptor docking for drug discovery. Expert Opin Drug Discovery 10(11):1189–1200
83. Krüger DM, Bergs J, Kazemi S, Gohlke H (2011) Target flexibility in RNA-ligand docking modeled by elastic potential grids. ACS Med Chem Lett 2(7):489–493
84. Frank AT (2016) Can holo NMR chemical shifts be directly used to resolve RNA-ligand poses? J Chem Inf Model 56(2):368–376
85. Wang L, Wu Y, Deng Y, Kim B, Pierce L, Krilov G, Lupyan D, Robinson S, Dahlgren MK, Greenwood J, Romero DL, Masse C, Knight JL, Steinbrecher T, Beuming T, Damm W, Harder E, Sherman W, Brewer M, Wester R, Murcko M, Frye L, Farid R, Lin T, Mobley DL, Jorgensen WL, Berne BJ, Friesner RA, Abel R (2015) Accurate and reliable prediction of relative ligand binding potency in prospective drug discovery by way of a modern free-energy calculation protocol and force field. J Am Chem Soc 137(7):2695–2703
86. Sund J, Lind C, Åqvist J (2015) Binding site preorganization and ligand discrimination in the purine riboswitch. J Phys Chem B 119(3):773–782
87. Carr RA, Congreve M, Murray CW, Rees DC (2005) Fragment-based lead discovery: leads by design. Drug Discov Today 10(14):987–992
88. Congreve M, Chessari G, Tisi D, Woodhead AJ (2008) Recent developments in fragment-based drug discovery. J Med Chem 51(13):3661–3680
89. Erlanson DA, McDowell RS, O'Brien T (2004) Fragment-based drug discovery. J Med Chem 47(14):3463–3482
90. Hall RJ, Mortenson PN, Murray CW (2014) Efficient exploration of chemical space by fragment-based screening. Prog Biophys Mol Biol 116(2–3):82–91
91. Hann MM, Leach AR, Harper G (2001) Molecular complexity and its impact on the probability of finding leads for drug discovery. J Chem Inf Comput Sci 41(3):856–864
92. Congreve M, Carr R, Murray C, Jhoti H (2003) A "rule of three" for fragment-based lead discovery? Drug Discov Today 8(19):876–877

93. Keserü GM, Erlanson DA, Ferenczy GG, Hann MM, Murray CW, Pickett SD (2016) Design principles for fragment libraries: maximizing the value of learnings from pharma fragment-based drug discovery (FBDD) programs for use in academia. J Med Chem 59(18):8189–8206
94. Bodoor K, Boyapati V, Gopu V, Boisdore M, Allam K, Miller J, Treleaven WD, Weldeghiorghis T, Aboul-Ela F (2009) Design and implementation of an ribonucleic acid (RNA) directed fragment library. J Med Chem 52(12):3753–3761
95. Francois B (2005) Crystal structures of complexes between aminoglycosides and decoding A site oligonucleotides: role of the number of rings and positive charges in the specific binding leading to miscoding. Nucleic Acids Res 33(17):5677–5690
96. Noble ME, Endicott JA, Johnson LN (2004) Protein kinase inhibitors: insights into drug design from structure. Science 303(5665):1800–1805
97. Batey RT, Gilbert SD, Montange RK (2004) Structure of a natural guanine-responsive riboswitch complexed with the metabolite hypoxanthine. Nature 432(7015):411–415
98. Velagapudi SP, Pushechnikov A, Labuda LP, French JM, Disney MD (2012) Probing a 2-aminobenzimidazole library for binding to RNA internal loops via two-dimensional combinatorial screening. ACS Chem Biol 7(11):1902–1909
99. Rzuczek SG, Southern MR, Disney MD (2015) Studying a drug-like, RNA-focused small molecule library identifies compounds that inhibit RNA toxicity in myotonic dystrophy. ACS Chem Biol 10(12):2706–2715
100. Moumne R, Catala M, Larue V, Micouin L, Tisne C (2012) Fragment-based design of small RNA binders: promising developments and contribution of NMR. Biochimie 94 (7):1607–1619
101. Griffey RH, Hofstadler SA, Sannes-Lowery KA, Ecker DJ, Crooke ST (1999) Determinants of aminoglycoside-binding specificity for rRNA by using mass spectrometry. Proc Natl Acad Sci U S A 96(18):10129–10133
102. Hofstadler SA, Griffey RH (2001) Analysis of noncovalent complexes of DNA and RNA by mass spectrometry. Chem Rev 101(2):377–390
103. Maple HJ, Garlish RA, Rigau-Roca L, Porter J, Whitcombe I, Prosser CE, Kennedy J, Henry AJ, Taylor RJ, Crump MP, Crosby J (2012) Automated protein-ligand interaction screening by mass spectrometry. J Med Chem 55(2):837–851
104. Swayze EE, Jefferson EA, Sannes-Lowery KA, Blyn LB, Risen LM, Arakawa S, Osgood SA, Hofstadler SA, Griffey RH (2002) SAR by MS: a ligand based technique for drug lead discovery against structured RNA targets. J Med Chem 45(18):3816–3819
105. Verdonk ML, Giangreco I, Hall RJ, Korb O, Mortenson PN, Murray CW (2011) Docking performance of fragments and druglike compounds. J Med Chem 54(15):5422–5431
106. Setny P, Trylska J (2009) Search for novel aminoglycosides by combining fragment-based virtual screening and 3D-QSAR scoring. J Chem Inf Model 49(2):390–400
107. Zeiger M, Stark S, Kalden E, Ackermann B, Ferner J, Scheffer U, Shoja-Bazargani F, Erdel V, Schwalbe H, Göbel MW (2014) Fragment based search for small molecule inhibitors of HIV-1 Tat-TAR. Bioorg Med Chem Lett 24(24):5576–5580
108. Davidson A, Begley DW, Lau C, Varani G (2011) A small-molecule probe induces a conformation in HIV TAR RNA capable of binding drug-like fragments. J Mol Biol 410 (5):984–996
109. Seth PP, Miyaji A, Jefferson EA, Sannes-Lowery KA, Osgood SA, Propp SS, Ranken R, Massire C, Sampath R, Ecker DJ, Swayze EE, Griffey RH (2005) SAR by MS: discovery of a new class of RNA-binding small molecules for the hepatitis C virus: internal ribosome entry site IIA subdomain. J Med Chem 48(23):7099–7102
110. Jerod Parsons M, Castaldi P, Dutta S, Dibrov SM, Wyles DL, Hermann T (2009) Conformational inhibition of the hepatitis C virus internal ribosome entry site RNA. Nat Chem Biol 5(11):823–825
111. Yu L, Oost TK, Schkeryantz JM, Yang J, Janowick D, Fesik SW (2003) Discovery of aminoglycoside mimetics by NMR-based screening of *Escherichia coli* A-site RNA. J Am Chem Soc 125(15):4444–4450

112. Cate JH, Gooding AR, Podell E, Zhou K, Golden BL, Kundrot CE, Cech TR, Doudna JA (1996) Crystal structure of a group I ribozyme domain: principles of RNA packing. Science 273(5282):1678–1685
113. Fejzo J, Lepre CA, Peng JW, Bemis GW, Ajay M, Murcko A, Moore JM (1999) The SHAPES strategy: an NMR-based approach for lead generation in drug discovery. Chem Biol 6(10):755–769
114. Johnson EC, Feher VA, Peng JW, Moore JM, Williamson JR (2003) Application of NMR SHAPES screening to an RNA target. J Am Chem Soc 125(51):15724–15725
115. Chung F, Tisné C, Lecourt T, Dardel F, Micouin L (2007) NMR-guided fragment-based approach for the design of tRNALys3 ligands. Angew Chem Int Ed 46(24):4489–4491
116. Chung F, Tisné C, Lecourt T, Seijo B, Dardel F, Micouin L (2009) Design of tRNA(Lys)3 ligands: fragment evolution and linker selection guided by NMR spectroscopy. Chemistry 15 (29):7109–7116
117. Moumné R, Larue V, Seijo B, Lecourt T, Micouin L, Tisné C (2010) Tether influence on the binding properties of tRNALys3 ligands designed by a fragment-based approach. Org Biomol Chem 8(5):1154–1159
118. Chen L, Cressina E, Leeper FJ, Smith AG, Abell C (2010) A fragment-based approach to identifying ligands for riboswitches. ACS Chem Biol 5(4):355–358
119. Cressina E, Chen L, Moulin M, Leeper FJ, Abell C, Smith AG (2011) Identification of novel ligands for thiamine pyrophosphate (TPP) riboswitches. Biochem Soc Trans 39(2):652–657
120. Warner KD, Homan P, Weeks KM, Smith AG, Abell C, Ferre-D'Amare AR (2014) Validating fragment-based drug discovery for biological RNAs: lead fragments bind and remodel the TPP riboswitch specifically. Chem Biol 21(5):591–595
121. Lee M-K, Bottini A, Kim M, Bardaro MF, Zhang Z, Pellecchia M, Choi B-S, Varani G (2014) A novel small-molecule binds to the influenza A virus RNA promoter and inhibits viral replication. Chem Commun 50(3):368–370
122. Garavis M, Lopez-Mendez B, Somoza A, Oyarzabal J, Dalvit C, Villasante A, Campos-Olivas R, Gonzalez C (2014) Discovery of selective ligands for telomeric RNA G-quadruplexes (TERRA) through 19F-NMR based fragment screening. ACS Chem Biol 9 (7):1559–1566
123. Shepherd CA, Hopkins AL, Navratilova I (2014) Fragment screening by SPR and advanced application to GPCRs. Prog Biophys Mol Biol 116(2–3):113–123
124. Cooper MA (2002) Optical biosensors in drug discovery. Nat Rev Drug Discov 1(7):515–528
125. David Wilson W (2002) Analyzing biomolecular interactions. Science 295(5562):2103–2105
126. Dalvit C (2009) NMR methods in fragment screening: theory and a comparison with other biophysical techniques. Drug Discov Today 14(21–22):1051–1057
127. González-Fernández E, Santos-Álvarez N d-l, Miranda-Ordieres AJ, Lobo-Castañón MJ (2012) SPR evaluation of binding kinetics and affinity study of modified RNA aptamers towards small molecules. Talanta 99:767–773
128. Peselis A, Serganov A (2012) Structural insights into ligand binding and gene expression control by an adenosylcobalamin riboswitch. Nat Struct Mol Biol 19(11):1182–1184
129. Schaffer MF, Choudhary PK, Sigel RKO (2014) The AdoCbl-riboswitch interaction investigated by in-line probing and surface plasmon resonance spectroscopy (SPR). Methods Enzymol 549:467–488
130. Jenkins JL, Krucinska J, McCarty RM, Bandarian V, Wedekind JE (2011) Comparison of a preQ1 riboswitch aptamer in metabolite-bound and free states with implications for gene regulation. J Biol Chem 286(28):24626–24637
131. Hendrix M, Priestley ES, Joyce GF, Wong CH (1997) Direct observation of aminoglycoside-RNA interactions by surface plasmon resonance. J Am Chem Soc 119(16):3641–3648

Top Med Chem (2018) 27: 79–110
DOI: 10.1007/7355_2017_3

Published online: 4 May 2017

Approaches for the Discovery of Small Molecule Ligands Targeting microRNAs

Daniel A. Lorenz and Amanda L. Garner

Abstract RNA is essential for life, serving as the intermediate between genomic storage and protein function. As our knowledge of biological systems has grown, so has our understanding of RNA, revealing additional functions of this critical class of biomolecules. One class of RNA, microRNAs (miRNA), highlights the fundamental role of non-coding RNA in higher organisms. miRNAs regulate nearly every biological pathway through targeted translational suppression, and dysregulation of miRNA expression has been implicated in many human disease states. Thus, therapeutically targeting miRNAs with small molecule ligands is of growing importance. Herein we focus on methods employed to discover small molecule miRNA ligands, their successes thus far, and future directions for the field.

Keywords High-throughput screening, microRNAs, Small molecules, Therapeutics

Contents

1 Introduction 80
2 microRNA Biogenesis and Function 81
3 Therapeutic Strategies 82
4 Screening Assays 83
4.1 Cellular 83
4.2 Biochemical 87
4.3 Computational 98
5 Conclusions 100
References 102

D.A. Lorenz and A.L. Garner (✉)
Department of Medicinal Chemistry, College of Pharmacy, and Program in Chemical Biology, University of Michigan, Ann Arbor, MI 48109, USA
e-mail: algarner@umich.edu

1 Introduction

MicroRNA-mediated gene silencing is a key regulatory mechanism of nearly all biological pathways and represents a new horizon for therapeutic intervention. First discovered in 1993, microRNAs (miRNA or miR) are a family of small RNAs typically ~22 nucleotides (nt) in length that function in the fine-tuning of gene expression [1, 2]. This regulatory function is exerted without the translation of the miRNA, and instead uses sequence complementarity to regulate protein synthesis. Thus, a single miRNA can regulate the expression of hundreds of mRNA transcripts. The extent of the physiological importance for miRNAs remained a mystery until 10 years after their discovery when a conserved miRNA was identified across multiple organisms [3, 4]. The let-7 miRNA family was found to play a fundamental role in embryonic development across species from *C. elegans* to humans, serving as an example of the tremendous regulatory power of these small RNAs. This excitement has continued, and 2,588 mature miRNAs have been identified for *Homo sapiens* alone, as according to the miRBase (http://www.mirbase.org). Consequently, greater than 60% of all mRNA transcripts are thought to be regulated by miRNAs [5].

Regulation of mRNA transcripts through miRNAs is not universal, and varied expression levels are found across cell types and states. Imbalance of miRNA expression has been linked to most human diseases, including cancer, diabetes, heart disease, neurodegeneration and viral infections [6]. Table 1 summarizes the

Table 1 Clinically relevant miRNAs discussed in this chapter

miRNA	Target gene	Expression level	Disease	Reference
let-7	*RAS*	Down	Cancer	[7]
1	*MET*	Down	Cancer	[8]
10b	*HOXD10*	Up	Cancer	[9]
21	*PTEN*	Up	Cancer	[10]
27a	*ZBTB10*	Up	Cancer	[11]
29a	*LOXL2*	Down	Cancer	[12]
34a	*BCL2*	Down	Cancer	[13]
96	*FOXO1*	Up	Cancer	[14]
122	*HCV, CCNG1*	Up, down	HCV, cancer	[15, 16]
125	*EIF4EBP1*	Down	Cancer	[17]
133a	*CDC42*	Down	Heart disease	[18]
142	*APC*	Up	Cancer	[19]
155	*TP53INP1*	Up	Cancer	[20]
206	*HDAC4*	Up	ALS	[21]
335	*RB1*	Up	Cancer	[22]
372	*LATS2*	Up	Cancer	[23]
373	*LATS2*	Up	Cancer	[23]
504	*P53*	Up	Cancer	[24]
525	*ZNF395*	Up	Cancer	[25]
544	*MTOR*	Up	Cancer	[26]

miRNA targets discussed within this chapter, which represent only a small portion of the known clinically relevant miRNAs. Biological details regarding these miRNAs will be discussed throughout the chapter. Due to their vast therapeutic potential, many researchers have invested effort in the manipulation of these important biomolecules. In this chapter, we will review the current state-of-the-art in the discovery of small molecule ligands for targeting miRNAs, in addition to compounds identified using these approaches.

2 microRNA Biogenesis and Function

As introduction to many of the approaches used to target miRNAs, the biogenesis of a miRNA will be briefly discussed (Fig. 1); this topic has been reviewed elsewhere in detail [27, 28]. miRNAs are primarily transcribed by RNA polymerase II as long transcripts called pri-miRNAs. These pri-miRNAs are recruited and processed in the nucleus by the Microprocessor complex, which at its core contains a RNase III enzyme, Drosha, and its accessory protein, DiGeorge Syndrome Critical Region 8 (DGCR8). The Microprocessor complex recognizes a hairpin loop structure in the pri-miRNA, resulting in its cleavage to generate a pre-miRNA stem loop of ~60–80 nt containing a distinctive 2 nt overhang at its 3′ end. Pre-miRNAs are then transported to the cytoplasm through Exportin 5 and are subsequently identified by another RNase III enzyme, Dicer. Dicer subsequently cleaves the loop region from the hairpin creating a ~22 nt mature miRNA duplex. After cleavage, Dicer forms a complex with Argonaute (AGO), and the miRNA strand with less thermodynamic stability at the 5′ end is transferred into the AGO protein. This, along with other accessory proteins, forms the RNA-Induced Silencing Complex (RISC). RISC uses nt 2–8 from the 5′ end of the mature single-stranded miRNA as a guide to find complementary sequences in the 3′ untranslated regions (UTR) of target mRNA transcripts. Upon finding a target, the RISC complex sequesters the

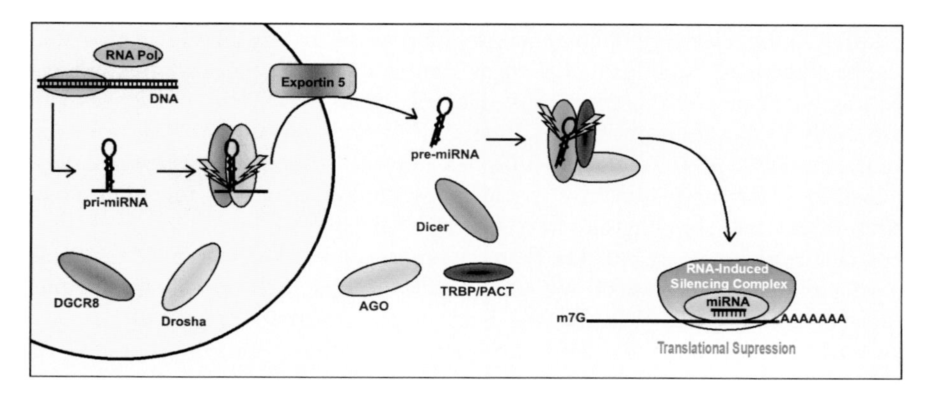

Fig. 1 miRNA biogenesis pathway

transcript from the translational machinery and recruits decapping and deadenylation proteins, thereby silencing the expression of that gene.

3 Therapeutic Strategies

Because each miRNA can regulate the expression of hundreds of mRNAs, targeting a single miRNA will result in a potentially dramatic shift in phenotype due to alteration of the global cellular transcriptome [29]. As a result, drug discovery efforts have focused on diseases for which this would be beneficial, such as cancer, or liver-centered disorders (e.g. hepatitis, hepatocellular carcinoma) due to the bioavailability of the most common method for targeting nucleic acids, including miRNAs. The current state-of-the-art for targeting miRNAs relies on antisense oligonucleotide (ASO)-based approaches [6, 30]. Anti-miRs have been the focus of most studies, and several candidates are in preclinical and clinical trials [6]. These molecules function by binding to miRNAs with high sequence complementarity to block their ability to regulate gene expression. For those miRNAs downregulated in disease, which is common for tumor suppressor miRNAs [31–33], miRNA mimetics are currently undergoing preclinical and clinical evaluation [6, 30]. Yet, despite the tremendous resources dedicated toward developing anti-miRs, miRNA mimetics and other RNA interference-based drugs, many liabilities exist with oligonucleotide-based approaches, including the requirement for chemical modifications to enhance in vivo pharmacokinetic and pharmacodynamic properties, hybridization-based off-target effects [34], toxicity (e.g. immunostimulation, inhibition of coagulation and liver toxicity), drug delivery problems and high cost [6, 35]. As such, no miRNA-targeted drugs have yet been approved.

Inviting comparison to protein drug targets, the use of small molecules to regulate miRNA biogenesis and function represents a promising, yet to date, unrealized therapeutic strategy. The challenges associated with the discovery of small molecules able to selectively target a specific RNA have been well-documented [36–38]. One major issue is RNA's hugely net negative charge, resulting in the discovery of compounds with poor selectivity due to electrostatic-based interactions. Additional challenges include its highly dynamic structure and the low abundance of therapeutically relevant RNAs [39]. These challenges will also apply to the discovery of small molecule ligands for miRNAs. However, in contrast to DNA, RNA, including miRNAs and their precursors, possesses unique secondary and tertiary structures, creating potential binding sites for small molecules. In fact, these binding sites have already been exploited for the design of small molecule RNA aptamers [40–42]. Furthermore, several endogenous RNAs are able to selectively recognize small molecule ligands, for example, metabolite-binding riboswitches [43]. This has recently been demonstrated for a pre-miRNA, and pre-miR-125 was found to exhibit low micromolar affinity for folic acid [44]. As additional support for this possibility, ribosomal RNA-binding antibiotics spanning several molecular scaffolds, including aminoglycosides, tetracyclines and macrolides, are widely used drugs. Thus, a significant hurdle that remains for targeting miRNAs, and RNAs generally, with small molecules is our lack of knowledge regarding drug-like RNA-binding scaffolds that are not susceptible to the types of non-specific interactions afforded by intercalators and positively-charged molecules. Throughout this chapter, we will focus on screening assays used to identify small molecule regulators of miRNAs and a selection of compounds identified using these approaches. Through these efforts, it is hoped that a path forward to overcome the challenges of RNA-targeted drug discovery will be illuminated.

4 Screening Assays

4.1 Cellular

The earliest efforts toward the goal of miRNA-targeted small molecule discovery employed cellular assays of miRNA activity. In this approach, miRNA pathway-based assays were developed, which utilize reporter genes designed to encode a sequence in the 3′ UTR that is complementary to the target protein or miRNA being investigated (Fig. 2) [45–47]. Thus, as miRNA levels are increased or decreased, the readout gene, typically luciferase or green fluorescent protein (GFP), will either decrease or increase, respectively, allowing for quantification of miRNA biogenesis and activity. While cellular pathway-based screens may be more biologically accurate, they suffer from a few drawbacks. The largest of these is the requirement for significant efforts in target identification following compound screening. In fact,

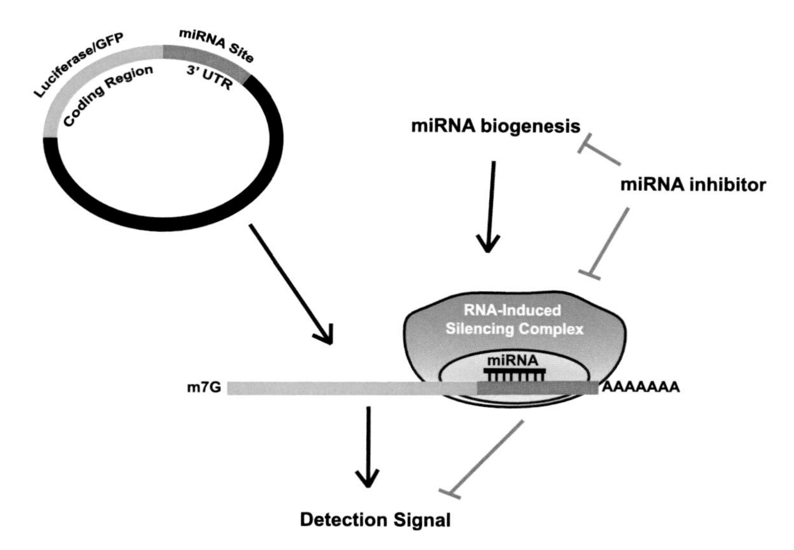

Fig. 2 Cellular assay for the identification of small molecule miRNA pathway modulators

for the compounds discovered using this methodology to date, specific targets have largely not yet been defined.

Several compounds that modulate miRNA expression have been discovered through pathway-based approaches. The first of these screens was reported by the Deiters group, who discovered the first small molecule miRNA modulator, compound **1** (Fig. 3), for miR-21 through a small screen of ~1,000 compounds [48]. miR-21 is one of the first miRNAs found to be overexpressed in nearly all tumor types [49–52]. The status of miR-21 as a bona fide oncomiR [53] was established in a mouse model of B-cell lymphoma, where ASO-mediated

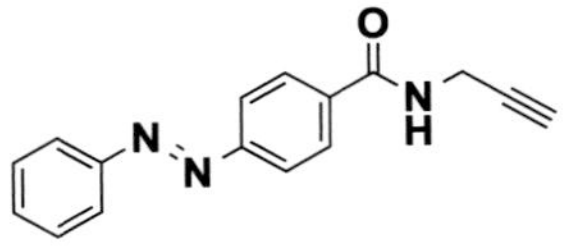

1
miR-21
Ref [48]

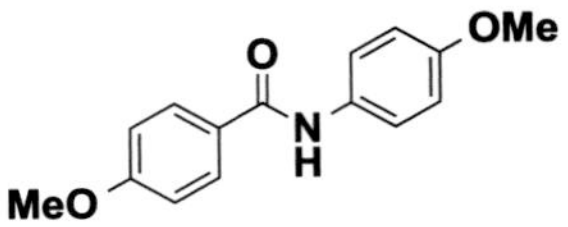

2
miR-21
Ref [57]

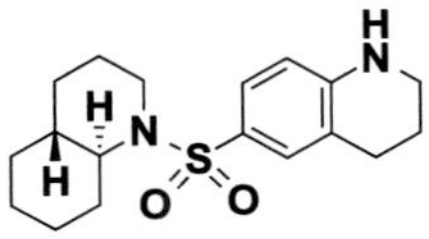

3
miR-122
Ref [58]

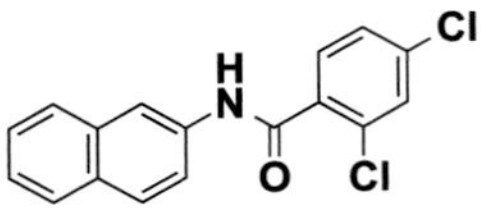

4
miR-122
Ref [58]

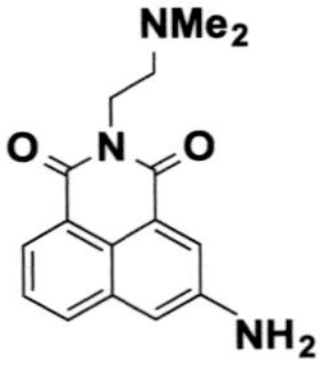

5
miR-122
Ref [58]

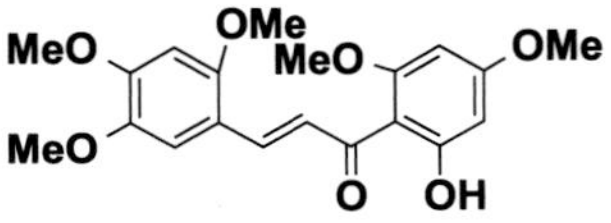

6
miR-34a
Ref [68]

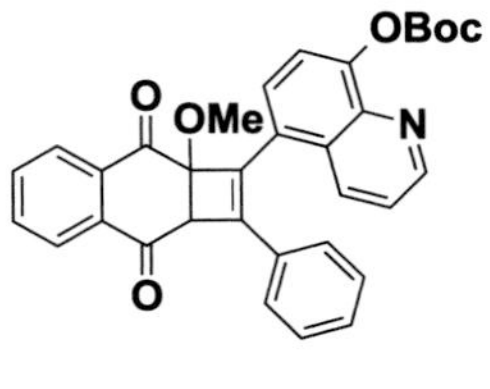

7
miR-1, -133a, -206
Ref [69]

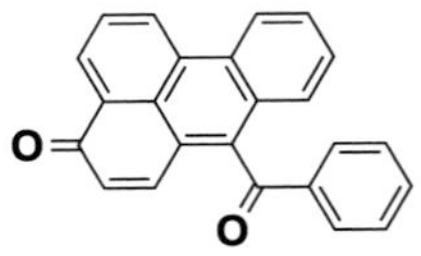

8
universal activator
Ref [71]

Fig. 3 Small molecules discovered using a cellular pathway-based approach inactivation of miR-21 alone led to rapid tumor regression [54]. This pioneering study demonstrated the concept of oncomiR addiction in cancer [55], and since then, miR-21 addiction has been reported in many cancers. By regulating its target mRNAs, including *PTEN*, miR-21 promotes cellular proliferation, migration and invasion, and prevents apoptosis [52]. Compound **1** was found to exhibit an EC_{50} of 2 μM, in addition to high specificity for miR-21 over miR-30 and miR-93, as determined via RT-PCR [48]. Derivatives of **1** have been synthesized by another group via scaffold hopping with several exhibiting improved cellular inhibition of miR-21 activity over the parent compound [56]. The Deiters group subsequently performed a much larger screen of over 300,000 compounds against their miR-21 reporter assay [57]. From this screen and structure-activity relationship (SAR) studies, the group identified aryl amide **2** (Fig. 3) with an EC_{50} of 0.86 μM and good selectivity for miR-21 over miRs-125b, -17, and -222. Preliminary mechanistic studies revealed no change in pri-miR-21 levels, suggesting that **2** does not act at the level of transcription; however, the exact mechanism-of-action remains unknown.

The Deiters group used the same assay format to identify scaffolds with the ability to regulate miR-122 expression (**3** and **4**; EC_{50} values of 0.6 μM and 3 μM, respectively; Fig. 3) [58]. miR-122 is the most abundant miRNA in the liver and is of critical importance in stimulating the stability, replication and translation of hepatitis C virus [59, 60]. ASOs targeting miR-122 are likely to be the first FDA approved anti-miR-targeting drugs on the market, and two are currently undergoing clinical trials, Miravirsen from Santaris Pharma and RG-101 from Regulus Therapeutics [6]. Miravirsen, in particular, has shown promising clinical data in phase I and phase II trials [61–64]. Interestingly, during their studies, the Deiters group also uncovered a small molecule that increased the maturation of miR-122 at 3 μM (**5**, Fig. 3) [58]. This activity is relevant to miR-122 in the context of its role in hepatocellular carcinoma where it has been found to be downregulated [16]. In general, compounds that are capable of increasing miRNA levels are also useful, as several miRNAs are downregulated in disease, in particular, tumor suppressors in cancer [31–33]. miR-34a is an example of such a miRNA [65, 66]. In fact, a miR-34a mimic, MRX34 from Mirna Therapeutics, was recently investigated in phase I clinical trials for several cancers; [67] however, the studies were recently halted due to the development of immune-related severe adverse effects. A small molecule, compound **6** (Rubone), has been discovered with the ability to increase mature miR-34a levels and reduce tumor volumes in vivo in a mouse model of hepatocellular carcinoma (Fig. 3) [68]. Thus, this compound or others targeting miR-34a may be valuable leads for future drug development campaigns.

Other groups have also used a pathway-based approach to discover miRNA biogenesis inhibitors. Zhang and co-workers used a similar luciferase assay to screen a focused photoadduct library of acetylenes and naphthalenequinones [69]. From this library, the group discovered compound **7** (Fig. 3), which was found to downregulate several miRNAs, including the myogenic miRs-1, -133a, and -206, which are enriched in heart and skeletal muscle, with selectivity over five other miRNAs. This molecule was later used as a chemical tool to elucidate a new pathway regarding the muscle differentiation protein, MyoD [70]. The same group used a similar photoadduct library to identify a universal activator of miRNA biogenesis (**8**; Fig. 3), which promoted the maturation of pre-miRNAs to mature miRNAs across several cell lines [71]. An FDA-approved drug, enoxacin, a fluoroquinolone antibiotic, has also been shown to enhance miRNA processing in cells by binding to a Dicer accessory protein, TAR RNA-binding protein 2 (TRBP) [72, 73]. In a slightly different approach, using GFP as a read-out, Shum and co-workers conducted a screen of ~7,000 compounds targeting miR-21 expression [74]. The group identified several molecules, including 6-hydroxy-DL-DOPA, deoxycorticosterone, pachyrrhizin and flutamide; however, the activity of these molecules has yet to be validated.

It is important to note that all of the compounds discussed thus far act either at the level of transcription or a target has yet to be defined; three studies are contrary to this. In the first, the Jeang group discovered that poly-L-lysine (PLL) hydrobromide (**9**; Fig. 4) can act as a general Dicer inhibitor through a luciferase reporter assay [75]. PLL, however, likely acts through non-specific, electrostatic interactions and will be subject to significant off-target effects. In the second, the Maiti group used a luciferase reporter assay for miR-21 and screened a small library

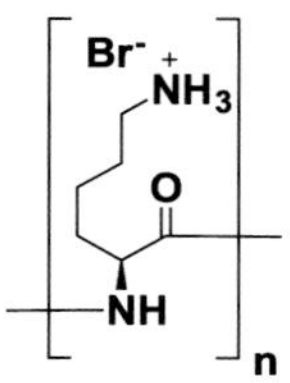

9
General Dicer inhibitor
Ref [75]

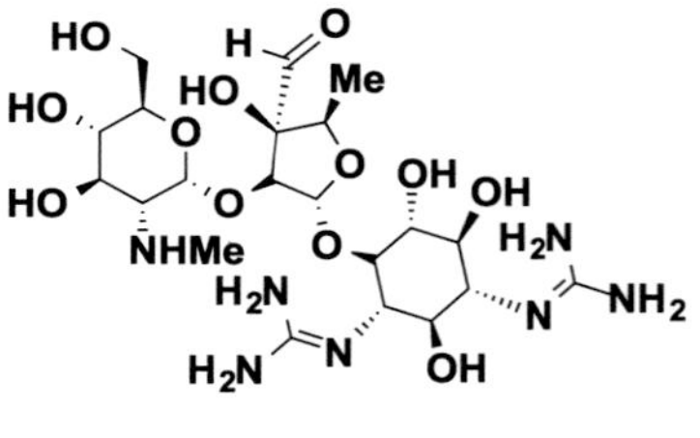

10
pre-miR-21
Ref [76]

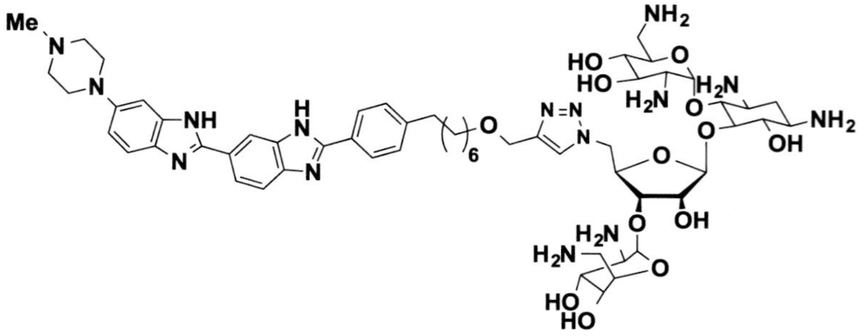

11
pre-miR-27a
Ref [79]

Fig. 4 Additional small molecules discovered using a cellular pathway-based approach of 15 aminoglycosides [76]. From these efforts and additional characterization, they determined that streptomycin (**10**; Fig. 4) specifically inhibited Dicer-mediated maturation of pre-miR-21; however, this finding has recently been called into question [77, 78]. Finally, Maiti, Arya and coworkers examined neomycin-bisbenzimidazole conjugates, represented by compound **11** (Fig. 4) [79]. This work was carried out in order to improve the activity of neomycin, which was previously found by the group to modulate the processing of pre-miR-27a [80]. miR-27a is part of a miRNA cluster that is upregulated in a number of human cancers, in addition to other diseases [11]. Conjugate **11** was discovered using a luciferase assay of miR-27a activity, and was found to decrease miR-27a levels by ~65% after treatment with 5 μM compound; specificity was not determined. Using AutoDock, compound **11** was found to bind at the minor groove of the pre-miRNA just below the loop region. Further characterization also showed inhibition of cell cycle progression in MCF-7 cells through flow cytometry, which was not observed with the parent neomycin due to poor cellular uptake.

4.2 Biochemical

4.2.1 Small Molecule Microarray

Schreiber and co-workers were the first to report using immobilized small molecule libraries, or small molecule microarrays, for high-throughput screening (HTS) [81]. This approach has since been applied to many biological targets, including RNAs and miRNAs [82–84]. Briefly, compounds are immobilized onto a slide, typically glass or agarose, via covalent conjugation chemistry (Fig. 5). Subsequently, fluorescently-labeled biomolecules, in this case, a target or control RNA, are incubated with the immobilized molecules, followed by washing. Those molecules that exhibit binding affinity for RNA are then identified via fluorescence imaging. Selectivity can be determined by comparing binders of the target RNA to a control RNA. Advantages of small molecule microarrays include their high-throughput potential and potential ease of identifying specific binders. A drawback, however, is the requirement for compound immobilization, which greatly limits the diversity of ligands that can be assayed. In fact, as will be seen, the majority of recent studies have focused solely on previously known chemical space for RNAs, including positively-charged peptides, aminoglycosides and benzimidazoles and related analogues. This restriction is due to the fact that library members must contain chemical handles for covalent conjugation to the slide, limiting library size and structure. Another disadvantage of the microarray approach is that the readout does not report function. More specifically, a compound may bind a target with high affinity; however, it may not elicit the desired biological effect (e.g. inhibition of miRNA processing). Nonetheless, small molecule microarray is a heavily used assay in the field and many interesting molecule have been discovered using this method for targeting miRNAs.

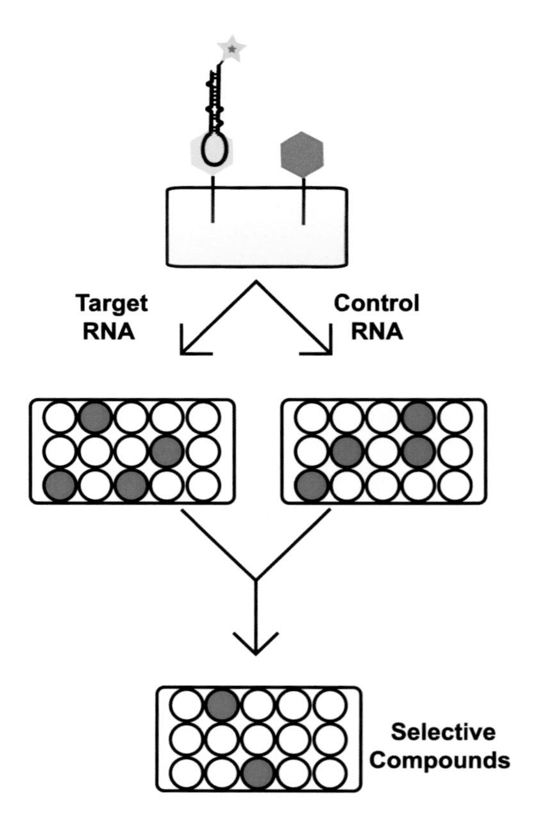

Fig. 5 Small molecule microarray

The Luebke group reported the first microarray screen targeting Drosha-catalyzed pri-miRNA maturation using an immobilized peptoid array consisting of 7,680 compounds [85]. The library was screened twice: once against their target, the hairpin loop of oncogenic pri-miR-21, and again against a control loop sequence. Through this screen, the authors identified peptoid **12** (Fig. 6) with a K_d of 1.8 μM and 20-fold specificity over the control. Of note, the control varied only in the loop region, demonstrating that the loop is targetable with specificity despite its dynamics [86]. To further this point, a Mg^{2+} cleavage assay was conducted in the absence and presence of **12**, which showed a change in cleavage localized to the loop region. Electrostatics was shown to play a key role in the mode of RNA binding; and when the arginine residue was removed, binding affinity was lost. Subsequently, it was reported that these compounds were not able to bind in the presence of Mg^{2+} concentrations necessary for Drosha cleavage (Fig. 1). To identify compounds that would inhibit Drosha processing in the presence of Mg^{2+}, a follow up screen of >14,000 peptoids yielded compound **13** (Fig. 6) with a K_d of 12 μM and selectivity over pri-miR-16 [87]. This new peptoid maintained the ability to block Drosha processing, albeit only at 250 μM with 3.3 nM of pri-miR-21. These findings highlight two points: The first is that these molecules

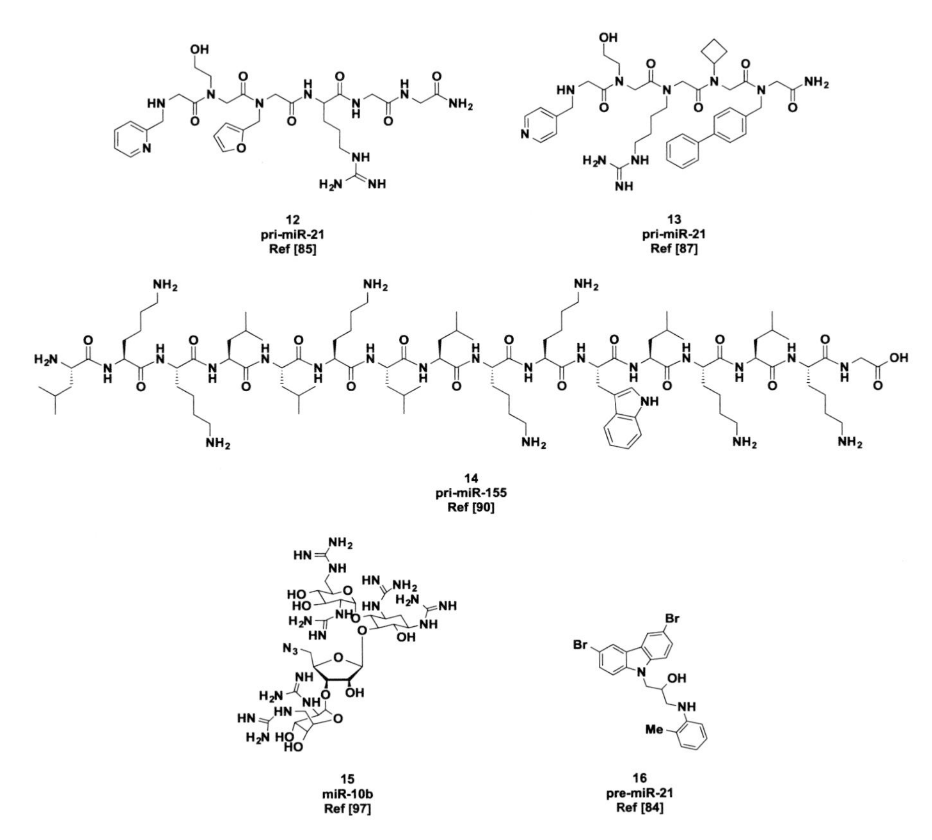

Fig. 6 miRNA ligands discovered via microarray assays

are likely binding to the RNA and not the protein components of the Microprocessor complex. The second is that a more complex mechanism of inhibition is likely occurring due to the vast excess of peptoid required to inhibit Drosha processing. Within the same library, the group also discovered three molecules, which while selective in the microarray, were not found to be selective in follow-up homogenous assays. This phenomenon has been reported before and may be due to linker effects or surface immobilization associated with the microarray format [88, 89].

The Shin group has also reported the use of a microarray screen, where they identified pre-miR-155-binding peptides from a library of 185 amphiphilic peptides with helical propensity [90]. The activity of miR-155, like many of the other miRNAs that have been targeted, is implicated in cancer development, and overexpression of this miRNA is signature in many solid and hematological tumors [20, 49, 91]. After systematic deletions and mutations, a 16-amino acid sequence, compound **14** (Fig. 6), was found to be the most promising, leading to inhibition of in vitro Dicer processing with an IC_{50} value of 770 nM. Peptide **14** was also shown to reduce mature miR-155 levels in cells by Northern blot. Further analysis showed the **14** was able to induce apoptosis in MCF-7 cells, as determined through flow cytometry; however, it displayed less than two-fold selectivity based on its K_d values for other pre-miRNAs tested. The authors also mentioned that while several peptides exhibited binding to pre-miR-155, they did not inhibit Dicer-mediated maturation. Thus, although a molecule exhibits specific binding to a RNA, it may not elicit a biological effect, and compounds discovered using assays that only monitor binding must be carefully followed up on in functional secondary assays. In addition to this work, another group has also screened a library of tryptophan-rich amphiphilic peptides to identify compounds that were able to enhance the processing of let-7a [92], a well-known tumor suppressor miRNA that is downregulated in cancer [7]. Additional peptide binders for miRNAs have also been discovered via phage display [93], cellular miRNA profiling [94] and a library of macrocyclic helix-threading peptides [95]. RNA-binding proteins have also been evolved for the targeted inhibition of miR-21 [96].

Using immobilized guanidinylated aminoglycoside derivatives based on kanamycin and neomycin, the Disney group targeted miRNAs using a bottom-up approach [97]. Instead of screening against a specific RNA target, a RNA internal loop library was screened against the two compounds using their two-dimensional combinatorial screening (2DCS) approach [98, 99]. Once a selective interaction was identified, the authors mined a miRNA database for structures containing the target site, and the identified motif was found within the Drosha processing site of pri-miR-10b. miR-10b is an oncogenic miRNA that is highly expressed in metastatic breast cancer cells and functions in promoting tumor cell invasion and metastasis by targeting the expression of homeobox D10 (*HOXD10*), which in turn leads to the enhanced expression of *RHOC*, a pro-metastatic gene [9, 100, 101]. Using guanidinylated neomycin B (**15**; Fig. 6) as a probe, the authors first validated the binding interaction of this molecule with the identified internal loop, and it exhibited a K_d of 417 nM. The group next demonstrated that **15** was able to inhibit the biogenesis of mature miR-10b and increase levels of pri-miR-10b in HeLa cells, demonstrating targeting of Drosha processing. Finally, the authors used a luciferase assay with the 3′ UTR of *HOXD10*, a known target of miR-10b, and were able to detect an increase of luciferase activity in the presence of 100 uM **15**. While not a direct approach to targeting, these types of screens can be deposited into databases for later discovery; this approach will be discussed in **4.3**.

More recently, the Schneekloth group used small molecule microarray to screen a pre-miR-21 hairpin fragment against a library of low molecular weight compounds unbiased for RNA binding [84]. From these efforts, compound **16** was discovered as a selective binder over a control HIV TAR hairpin RNA. This compound also was found to exhibit good binding affinity for the pre-miR-21 hairpin (K_d values of 3.2 μM and 2.3 μM in fluorescence intensity and 2-aminopurine assays, respectively) and inhibit Dicer processing due to its ability to bind proximal to the Dicer cleavage site. This study is important in that it highlights the potential for more drug-like molecules to bind to miRNA targets.

4.2.2 FRET

Förster, or fluorescence, resonance energy transfer (FRET) is a common technique used to assay intra- and intermolecular systems both in vitro and in vivo. The basis of FRET assays centers around the change in fluorescence upon excitation of an energetically-matched fluorophore (donor) and fluorescence quencher (acceptor) that are in close proximity, typically 1–10 nm. Arenz and co-workers reported the first FRET assay for pre-miRNA maturation, which they referred to as a beacon assay [102, 103]. In this approach, both ends of the pre-miRNA hairpin loop are first labeled with a FRET pair with the fluorophore at the 5′ end and quencher at the 3′ end, creating a quenched pre-miRNA beacon through base-pairing (Fig. 7). Upon treatment of the pre-miRNA substrate with Dicer, the enzyme then breaks the FRET pair through pre-miRNA processing to generate fluorescence signal. Of note, other researchers have also designed FRET-based pre-miRNA maturation assays [80, 104]. Unlike small molecule and peptide microarray discussed in Sect. 4.2.1, pre-miRNA beacon assays enable functional readout providing the benefit of identifying small molecules that block maturation and not simply bind to the pre-miRNA. A disadvantage, however, as with all fluorescence-based assays, is that FRET approaches often suffer from compound interference [105, 106]. Additionally, incorporation of the fluorophores is also critical as some beacon designs suggest that Dicer cleaves the substrate twice, once at the location of the fluorophore/quencher and again at the loop region, leading to potentially inaccurate data. Dicer, in fact, acts as a "molecular ruler" and counts from the 3′ end to digest the pre-miRNA into a mature miRNA duplex [107, 108].

Currently, most beacon assays have looked at aminoglycosides and their derivatives. Using neamine and 2-deoxystreptamine scaffolds armed with click chemistry handles, the Arenz group searched for inhibitors of Dicer-mediated pre-let-7 maturation [109]. The let-7 family regulates cancer development and progression by targeting cellular oncogenes including RAS and its mutant isoforms and Myc [7]. The most potent compound identified was neamine analogue **17** (Fig. 8), which was found to inhibit pre-let-7 maturation with an IC_{50} value (0.63 μM) two-orders of magnitude better than that of kanamycin A. The Duca group has also examined click chemistry-functionalized aminoglycosides, in this case targeting pre-miRs-

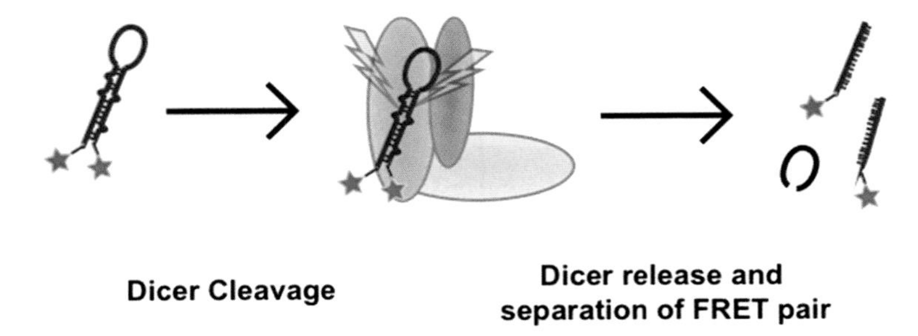

Fig. 7 FRET assay for Dicer-mediated pre-miRNA processing

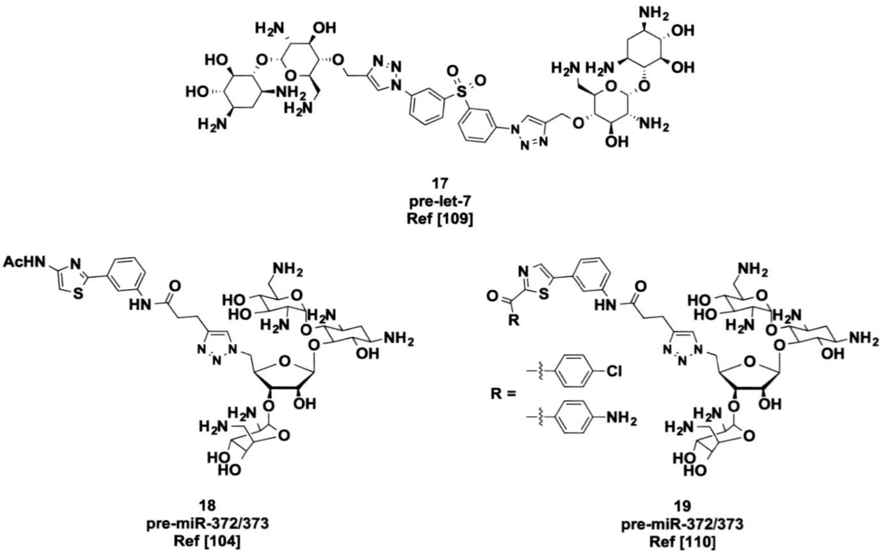

17
pre-let-7
Ref [109]

18
pre-miR-372/373
Ref [104]

19
pre-miR-372/373
Ref [110]

Fig. 8 Aminoglycoside inhibitors discovered using FRET assays of Dicer-mediated pre-miRNA processing

372 and -373 [104, 110]. These miRNAs are part of the miR-371–373 gene cluster that was originally linked to stem cell pluripotency and cell cycle regulation [111, 112]. miRs-373 and -373 have since been characterized as oncomiRs overexpressed in several cancers, including gastric carcinoma, and function by targeting tumor suppressor genes such as *Large Tumor Suppressor 2* (*LATS2*) [23]. In their first report, the group synthesized a library of neomycin-nucleobase conjugates containing natural and non-natural nucleobases and identified compound **18** (Fig. 8), which inhibited Dicer processing of pre-miRs-373 and -373 with IC_{50} values of 2.42 μM and 5.13 μM, respectively [104]. This compound also inhibited cellular pre-miR-372 and-373 processing and activity and slightly inhibited the growth of AGS gastric cancer cells at 50 μM. In a second report, SAR studies were performed on **18**, which yielded analogues **19** (Fig. 8) with enhanced IC_{50} values of 0.985/2.19 μM and 1.08/1.10 μM for inhibition of pre-miR-372 and -373 maturation, respectively [110]. These derivatives were also active in AGS cells; however, both had questionable specificity for the pre-miRNAs examined based on RT-qPCR analyses of treated cells. In addition to aminoglycosides, the Duca group has also explored other ribosome-targeted antibiotic scaffolds [78]. This investigation revealed neomycin B and select tetracycline derivatives as general inhibitors of Dicer-mediated pre-miRNA maturation.

Although FRET assays have yet to be used in large scale screening efforts to identify pre-miRNA maturation inhibitors, this type of assay has been employed to screen for general Dicer inhibitors using a duplexed siRNA substrate [113]. In this screen, 2,816 small molecules from the LOPAC, Prestwick and NIH collections were tested in 1,536-well format yielding an initial hit rate of 1.4%; however, upon analysis of the hits, the authors cited false hit identification as a significant limitation of their screen. This is likely due to both the assay format, which is subject to compound interference, particularly by fluorescence quenchers [105, 106], and the composition of the libraries tested.

4.2.3 Fluorescent Indicator Displacement

The simplest assay to set up for the discovery of RNA binding scaffolds is a fluorescence indicator displacement (FID) assay. The premise of FID assays is based upon the fact that many fluorogenic dyes (e.g. TO-PRO-1, SYBR green, ethidium bromide, X2S) change their emission properties upon binding to nucleic acids [114]. Thus, if a compound is able to bind to the RNA and compete with the dye, then a change in fluorescence will be detected (Fig. 9). This type of RNA assay format has been explored by several groups for both assay optimization [115] and HTS purposes [114]. Generally, FID assays are straightforward to implement and easily adaptable to any RNA target, allowing for expedited data collection. Yet, similar to FRET assays, dependence upon fluorescent signal generation may result in a large number of false positives and negatives, thereby complicating follow-up studies. Additionally, as mentioned before, identified compounds that displace the dye may not inhibit miRNA maturation or activity due to the lack of functional read-out.

Herdewijn and colleagues performed an early FID screen against a collection of aminoglycosides and antitumor drugs to identify binders of pre-miR-155 [116]. Compounds were screened for displacement of ethidium bromide. From this analysis, the aminoglycosides framycetin, kanamycin B and neomycin B were revealed to be the strongest binders (K_d values of 6 μM, 6 μM and 11 μM, respectively); however, these compounds only weakly ($\leq$10%) inhibited Dicer processing. This result reiterates the limitations of binding-based assays for miRNAs.

Nakatani and co-workers were the first to use xanthone dyes (**X2S**, **20**; Fig. 10) as fluorescent indicators for RNA [115]. Following up on this prior work, the group identified that substitution of the xanthone to a thioxanthone (**X2SS**, **21**; Fig. 10) showed enhanced binding to a pre-miRNA, pre-miR-29a [117]. This miRNA

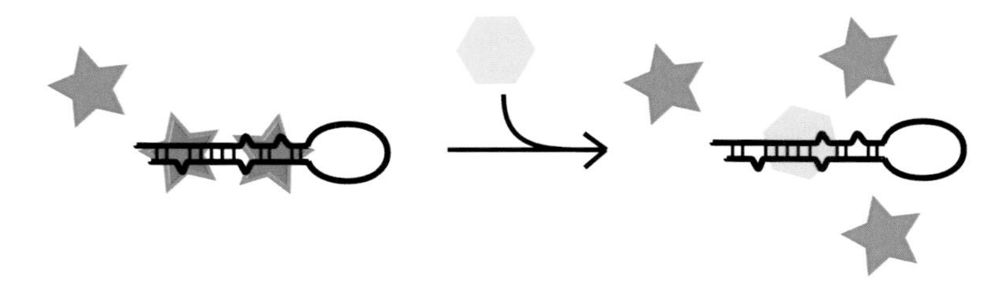

Fig. 9 Fluorescent indicator displacement assay

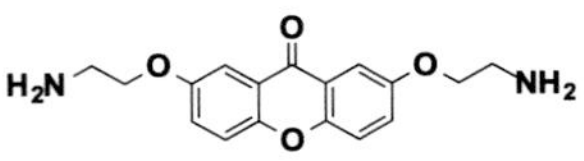

20
X2S
Ref [115]

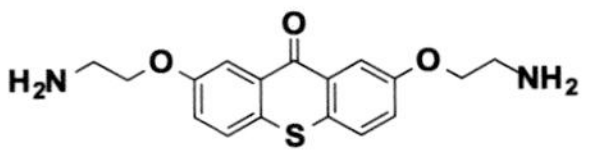

21
X2SS
Ref [117]

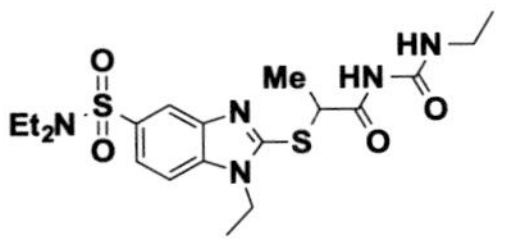

22
Ref [118]

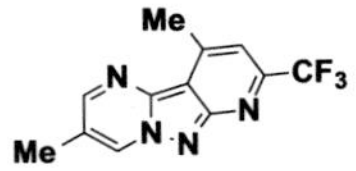

23
Ref [118]

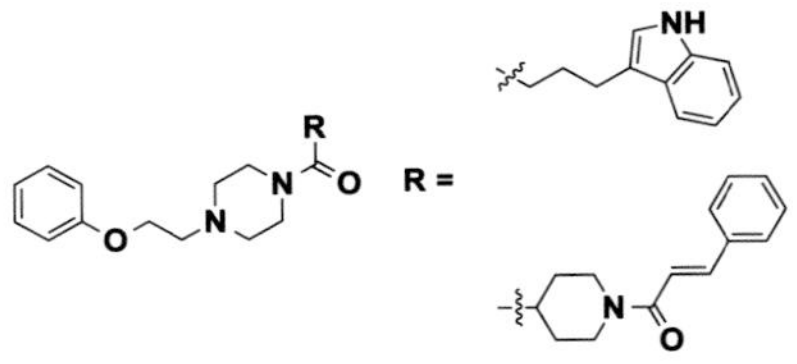

24
Ref [118]

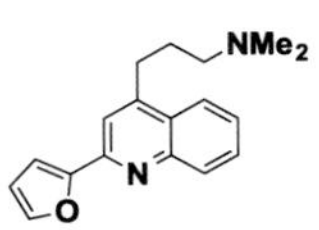

25
Ref [119]

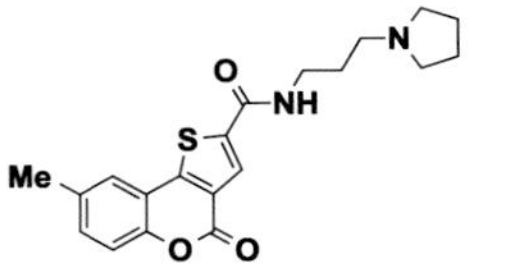

26
Ref [119]

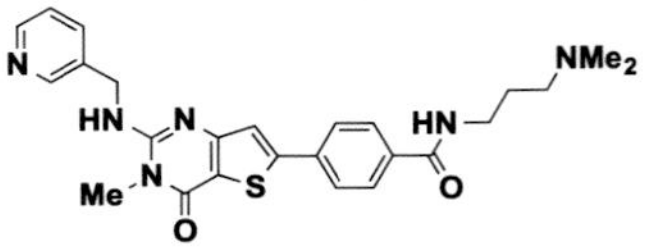

27
Ref [119]

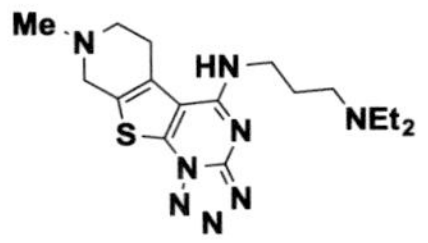

28
Ref [119]

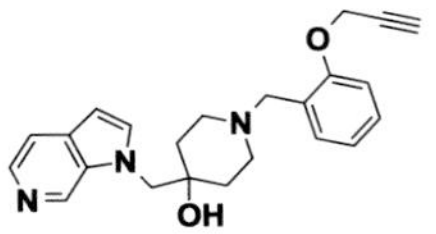

29
Ref [119]

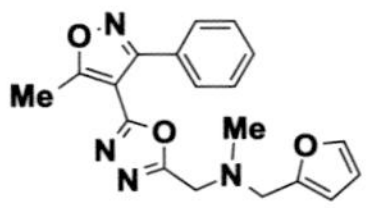

30
Ref [119]

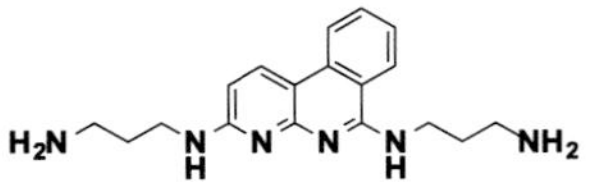

31
Ref [120]

Fig. 10 Compounds discovered for pre-miR-29a using xanthone dye-based FID assays

belongs to a family of three miRNAs (miR-29a, b, c) that function as tumor suppressors and immune modulators [12]. Using **21** and a structurally-related molecule, the Nakatani group performed two FID-based HTSs. The first study was a pilot screen of 9,600 compounds against ten pre-miRNA targets, including pre-miRs-21, -29a and -122 [118]. From this campaign, hit rates of 0.46–1.8% were obtained dependent upon the pre-miRNA; however, many fluorescent compounds were found to interfere with the assay. Focusing on pre-miR-29a, seven selective compounds were identified in the first FID screen with four showing reproducible binding in a second round of assay (**22–24**; Fig. 10). After demonstrating that xanthone-based FID assays are amenable to HTS, a follow-up screen of ~41,000 basic nitrogen-containing compounds was performed [119]. In this campaign, two pre-miRNA targets were chosen, pre-miR-29a and -122, and Rev Response Element (RRE) RNA was used as a control. For each RNA, 600–650 hits were identified, and the authors state that there was significant overlap between the RNA targets. 1,075 of the hit compounds were then rescreened using several rounds of surface plasmon resonance (SPR) to determine binding to pre-miR-29a. From this secondary screening, two series of binders were identified: compounds with rapid association and dissociation kinetics likely to act as intercalators (**25–27**; Fig. 10), and compounds with slow association and/or dissociation likely to function as groove binders (**28–30**; Fig. 10). No further biological characterization has been reported on these compounds. The Nakatani group has also taken a rational approach to the identification of pre-miR-29a binders, and compound **31** was designed to bind to a cytosine bulge in pre-miR-29a adjacent to the Dicer cleavage site [120]. Modest inhibition of Dicer processing was observed in the presence of 200 μM of **31**.

Key to the implementation of FID assays is the development of high quality and targetable fluorescent probes. Ethidium bromide and X2S are nucleic acid intercalators; however, RNA is capable of participating in many different types of binding interactions. Thus, an ideal FID probe should capture these diverse binding modes. As a new approach for FID assays, Arya and colleagues designed a fluorescein-conjugated neomycin (**F-Neo, 32**; Fig. 11) to prevent the identification of simple intercalators as screening hits [121]. Using their newly developed assay, the group designed and screened a library of dipeptide-neomycin conjugates represented by general scaffold **33** (Fig. 11) against three mature miRNAs, miR-142, -335, -504, and one pre-miRNA, pre-miR-504. From this study, several conjugates with higher affinity for the mature and pre-miRNA than neomycin alone were identified. These results translated to qPCR studies in cells, and a ~50% decrease in target miRNA expression was observed.

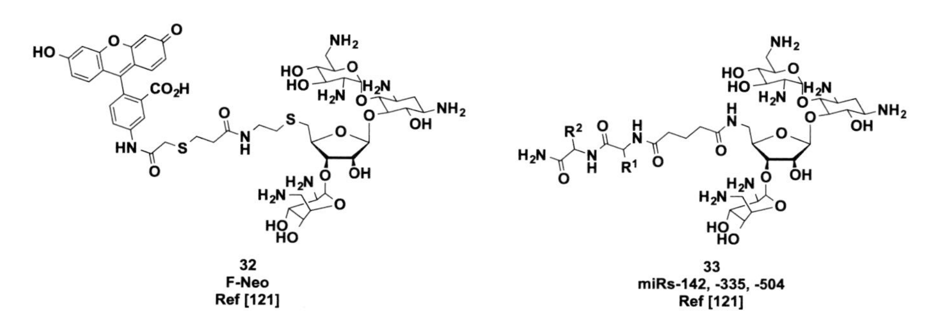

Fig. 11 Neomycin probe for FID and neomycin-peptide conjugate scaffold for miRNAs

4.2.4 cat-ELCCA

With the goal of discovering new chemical space for targeting RNAs, in particular pre-miRNAs, our laboratory has designed and developed a conceptually unique pre-miRNA maturation assay based on our group's novel assay technology, cat-ELCCA, or **cat**alytic **E**nzyme-**L**inked **C**lick **C**hemistry **A**ssay [77, 122–124]. In this approach, opposite of small molecule microarray, a labeled pre-miRNA substrate is immobilized onto a microtiter plate through a biotin-streptavidin interaction (Fig. 12). Biotin is incorporated at the 5′ end through a linker to maintain the 3′ 2 nt overhang that is recognized by Dicer. The pre-miRNA, which also contains a click chemistry handle (**X**) within the loop, is then subjected to Dicer-mediated maturation, which enzymatically cleaves the terminal loop bearing the click chemistry tag; however, in the presence of a small molecule inhibitor, the tag would remain. Subsequently, horseradish peroxidase (HRP) modified with a matching click chemistry handle (**Y**) is added to the wells, which covalently reacts with the remaining immobilized pre-miRNA substrate. Conjugation to HRP affords the assay catalytic signal amplification following addition of a pro-absorbent, -fluorogenic or -chemiluminescent HRP substrate similar to that observed in ELISA, which employs HRP-conjugated secondary antibodies. With respect to this click chemistry detection step, both copper(I)-catalyzed alkyne-azide cycloaddition chemistry [77] and inverse electron-demand Diels-Alder chemistry utilizing trans-cyclooctene and methyltetrazine [122] have been employed, with the later being far superior for detection and HTS.

Although more labor intensive than mix-and-measure assays such as FRET and FID, there are several important benefits to this new assay design over these traditional assay formats. First, the read-out is the result of catalytic signal

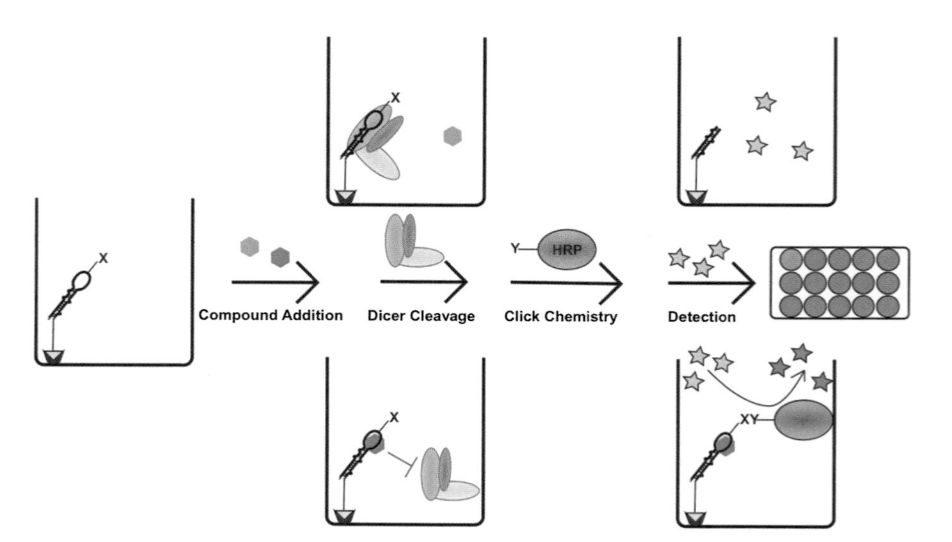

Fig. 12 cat-ELCCA where **X** and **Y** represent click chemistry handles

amplification, generating a robust and highly sensitive assay with large dynamic range. Assays armed with catalytic signal amplification like ELISA are known to be superior for bioanalyte detection; [123] however, antibodies for RNA targets are difficult to generate making this class of biomolecule incompatible with such an approach. By using click chemistry-mediated catalytic signal amplification, cat-ELCCA finally enables access to this approach for RNA assays. Additionally, despite the extra steps that washing adds to the overall protocol, their inclusion significantly reduces the chances of compound interference. In fact, fluorescent molecules and quenchers were not found to interfere with cat-ELCCA in both fluorescence and chemiluminescence detection [77].

cat-ELCCA also has advantages over the other types of assays used to discover miRNA modulators. Contrary to small molecule microarray (Sect. 4.2.1), by immobilizing the RNA instead of potential ligands, the assay allows for screening of any compound library to discover new RNA-binding motifs. Additionally, cat-ELCCA is dependent upon a functional read-out; thus, hits will be identified only for compounds that functionally inhibit Dicer processing of the pre-miRNA. This is similar to the FRET assays in Sect. 4.2.2, but unlike these beacon assays, a more biologically relevant pre-miRNA substrate is used in cat-ELCCA. Finally, also similar to FRET, because the assay monitors only the Dicer maturation step, target validation and subsequent medicinal chemistry will be expedited, unlike the cellular assays described in Sect. 4.1.

Pilot screening for assay development purposes using cat-ELCCA was performed using the LOPAC library of 1,280 pharmacologically active compounds [122]. From this campaign, the assay yielded an excellent Z' factor [125] of 0.69 with a signal-to-background ratio of 11.5. Representative hits are shown in Fig. 13 (1.64% overall hit rate); however, the activity of all compounds identified was very weak at 25 μM (9.9%, 7.5% and 11.3% inhibition, respectively, for **34–36**) and no follow-up studies were performed on these molecules (unpublished results). Nonetheless, encouraged by our assay statistics, to date, we have pushed forward with our efforts and ~85,000 small molecules and natural product extracts have been screened. The results of these efforts are forthcoming.

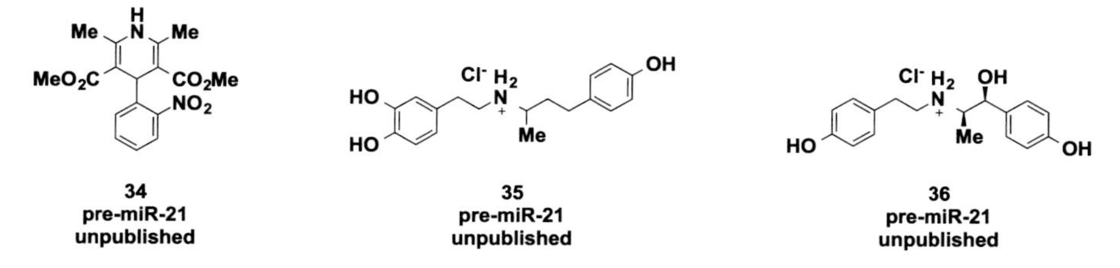

Fig. 13 Pilot screening hits using cat-ELCCA for Dicer-mediated pre-miRNA maturation

4.3 Computational

Drug discovery efforts have expanded in the last decade to include using powerful computational software to predict and analyze ligand binding. Most of these efforts have focused on proteins and are aided by the availability of high-resolution crystal structures. RNA, on the other hand, due to its highly dynamic nature, has proven to be much more challenging to gain accurate structural details of potential druggable sites, such as the loop regions of pre-miRNAs [86, 126]. To overcome these challenges, programs have been developed to model the structure of RNA [127, 128]. Additionally, methods have been developed to make use of more predictable regions of RNA, such as bulges and internal loops, to generate a large database of molecules that can recognize specific structures [129]. As will be described, this approach has been applied to miRNAs to enable ligand discovery efforts [130, 131]. Additionally, researchers have used published screening data to perform predictive modeling of potential small molecule miRNA-binding scaffolds [132]. One of the largest advantages of computational methods is the ability to screen enormous compound libraries at the relatively low cost of computer time. More details regarding structure-based drug design for RNA can be found in a following chapter of this volume [133].

With respect to computational drug discovery for miRNAs, the first report was from the Kang group [134]. Using the MC-Fold/MC-Sym pipeline in conjunction with AutoDock, the researchers first generated a three-dimensional model of pre-miR-21, and then docked 1,990 compounds from the NCI database into the predicted structure. These efforts resulted in the discovery of AC1MMYR2 (**37**; Fig. 14), which was demonstrated to inhibit miR-21 maturation across several cancer cell lines, as well as increase PTEN and PDCD4 expression, known miR-21 targets, after 30 μM treatment. RT-PCR experiments showed that **37** regulated miR-21 levels in addition to those of several other miRNAs. In cancer cells, AC1MMYR2 exhibited anti-proliferative activity, promoted apoptosis and suppressed tumor cell migration and invasion. Compound **37** was similarly found to be efficacious in mouse models of glioma and triple negative breast cancer. In a subsequent study, **37** was shown to inhibit breast cancer lymph node metastasis by targeting carcinoma-associated fibroblasts via NF-κB pathway attenuation [135].

Drawing together several technologies that enable the design of RNA-targeted small molecule ligands from only sequence, the Disney group has developed a platform for targeting oncogenic miRNAs [98, 99, 129, 136–138]. Inforna, which integrates the group's 2DCS and structure-activity relationships through sequencing (StARTS) methodologies with RNA secondary structure prediction, was piloted to identify small molecules able to bind selectively to precursor miRNAs [130]. Additional information regarding this approach can be found in a preceding chapter of this volume [139]. Inforna was used to match targetable motifs within the Drosha or Dicer processing sites of disesase-relevant miRNAs, which were predicted from miRBase, with small molecule ligands for that structural element. From these efforts, benzimidazole **38** (Fig. 14) was identified as a selective binder

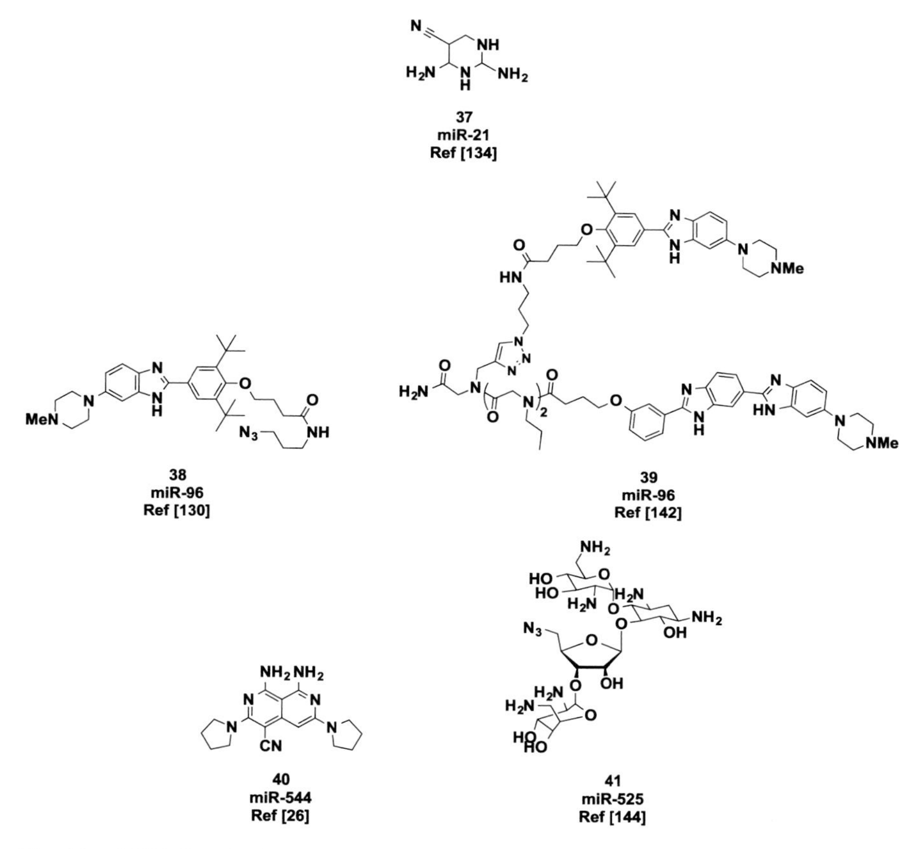

Fig. 14 miRNA ligands discovered via computational approaches

of the miR-96 precursor with binding affinities of 1.3 and 3.4 μM for the 5'UUU/3'AUA and 5'CGAUUU/3'GGUAUA motifs within the Drosha processing site of this miRNA. miR-96 is a member of the miR-183 cluster (miRs-183, -96 and -182), which are frequently overexpressed in cancer, neurological and autoimmune diseases [140]. In cancer, the miR-183 cluster functions to downregulate the expression of tumor suppressors, including *FOXO1* and *PDCD4* [140]. Through biological characterization, **38** was shown to reduce mature miR-96 levels by 90% at 40 μM compound in MCF-7 cells via qRT-PCR. Importantly, no inhibition was observed for miR-182 and -183 at the same dose. The selectivity of **38** was further assayed through miRNA profiling against 149 other disease-causing and abundant miRNAs. From this analysis, **38** affected miR-96 by ~19-fold, while a miR-96-targeted antagomir affected 12 miRNAs by ~2.5-fold. These findings suggest that small molecule RNA ligands may exhibit enhanced selectivity for a target over ASO-based reagents [141]. As designed, compound **38** was found to inhibit Drosha processing of pri-miR-96, in addition to promoting the expression of miR-96 target, *FOXO1*, and inducing apoptosis.

In a follow-up report, the Disney group again used Inforna, this time to optimize the activity of **38** [142]. An additional targetable site adjacent to the Drosha processing site, a 1 × 1 GG internal loop, was identified and found to bind to a bis-benzimidazole ligand. The researchers then synthesized heterodimeric compound **39** (Fig. 14), which they termed Targaprimir-96, to simultaneously bind both sites with one molecule. This bidentate molecule was shown to exhibit significantly enhanced binding affinity (K_d of 85 nM) and cellular inhibition of miR-96 maturation (IC_{50} value of ~50 nM). In triple negative breast cancer cell (TNBC) lines, **39** was found to induce apoptosis through the enhanced expression of miR-96 target genes [14]. Of particular importance, **39** was shown to be bioactive in vivo and inhibited tumor growth in a mouse model of TNCB through targeted downregulation of miR-96 without toxicity.

In addition to miR-96, small molecules have been designed for two other miRNAs using Inforna. Like several other miRNAs discussed, miR-544 is part of a cluster of miRNAs (miRs-300, -382, -494, -495, -539, -543 and -544) that are upregulated in cancer and target tumor suppressor genes [143]. The Phinney group discovered that one target of miR-544 is mammalian target of rapamycin (mTOR), which it regulates in response to hypoxia to ensure tumor cell adaptation to this cellular stress. Thus, they hypothesized that miR-544 inhibitors may be beneficial for inhibiting tumorigenesis by inhibiting this adaptive response. In collaboration with the Disney group, compound **40** (Fig. 14) was discovered to target the 1 × 1 UU loop within the Dicer processing site of pre-miR-544 with midnanomolar affinity [26]. In breast cancer cells, **40** (20 nM) was found to inhibit Dicer mediated cleavage of pre-miR-544 leading to reduced mature miR-544 levels, in addition to altering cellular growth under hypoxic conditions. To further evaluate the efficacy of **40**, MDA-MB-231 cells pretreated with compound were implanted into mice, leading to inhibition of tumor growth in comparison to an untreated control.

Finally, as part of their original report, neomycin azide (**41**; Fig. 14) was identified as a binder of the Drosha processing site of miR-525 [130], which is overexpressed in liver cancer leading to enhanced migration and invasion [25]. Of note, **41** was also found to inhibit the biogenesis of miRs-517c, -518e, and -519d [130]. In a follow-up study, the Disney group found that **41** bound to miR-525 with a K_d of 355 nM, inhibited its Drosha-mediated biogenesis, upregulated miR-525 target genes and inhibited cell invasion in HepG2 hepatocellular carcinoma cells from 6.25 to 25 μM [144].

5 Conclusions

The discovery of selective small molecule ligands for RNA targets remains an important challenge in both basic science and drug discovery. Focusing on work within the sphere of miRNAs, here we have highlighted several methodologies used to identify small molecule binders and modulators of these cellular micromanagers. From our analysis, two important conclusions can be drawn: the first regards the assays used, while the second relates to the molecules.

As described in the text, several assay types and approaches have been used for the identification of small molecule miRNA ligands and pathway modulators, and each methodology has both advantages and disadvantages for discovery purposes. By comparing these technologies, what is clear is that using methods that only enable the detection of molecules that bind to RNA may not yield compounds that modulate the function of that RNA. An exception to this is Inforna, which allows one to target molecules to Drosha or Dicer sites for inhibition of miRNA maturation. Other binding-based assays, like small molecule microarray and FID, rely on more general patterns of binding and have resulted in the identification of molecules without functional inhibition of the miRNA due to engagement in these potentially non-specific interactions. Newer assays, such as pre-miRNA beacon assays and cat-ELCCA, overcome some of these limitations. Cat-ELCCA, in particular, was developed with HTS in mind and offers additional advantages for screening with the goal of expanding the known chemical space for targeting RNAs and miRNAs. The need for new assays for targeting RNA was cited as a future goal in a previous comprehensive review of RNA-targeted drug discovery, and continues to be a growing area in the field [36]. An issue that remains to be addressed with respect to biochemical RNA assays is the fact that different assay types can yield different RNA ligands highlighting the potential importance of RNA dynamics in the discovery of RNA-targeted small molecules [145, 146]. This is an area that has been gaining significant ground in recent years and will be one to watch in the future, particularly its integration with computational drug discovery efforts [147].

Within this review, neomycin and other aminoglycosides, in addition to other previously known RNA-binding scaffolds, have been cited as ligands for various miRNAs several times. This reliance on known scaffolds for RNAs is primarily due to the fact that we have only scratched the surface in terms of our knowledge of privileged molecules for targeting RNA. Disney and co-workers have greatly enabled our understanding of the types of RNA motifs that are ideal for binding small molecules, such as internal loops and RNA hairpins; [114, 129] however, much remains to be learned with respect to RNA-targeted chemical space. A major challenge associated with such discovery is its reliance on HTS campaigns, which brings two issues: one is cost and the second is the composition of available screening libraries. The later is quite problematic, as commercially available compound libraries, which are most accessible, are not composed of molecules that are likely to bind avidly to RNA and the chemical diversity of these sets is limited. We, ourselves, have experienced this in our screening campaign (see Sect. 4.2.4); however, expansion to new areas of chemical space, including natural products and much larger and diverse pharmaceutical industry libraries may help to overcome this challenge. It is surely a worthy pursuit and is beginning to show some successes for RNA targets [148, 149].

Although not discussed within this review, several proteins interact with miRNAs to either promote or inhibit their processing and activity. Thus, we would be remiss not to mention the work done in targeting these proteins involved in various steps of miRNA regulation, which include shared components of the pathway, such as AGO [150–153], and miRNA-family specific modulators, like Lin28 that regulates let-7 maturation [154–157]. The targeting of RNA-protein interactions has been explored for viral RNA targets [158] and several RNA-binding antibiotics interact at the interface of RNA and protein [159]. This is an emerging area for targeting miRNAs and may yield new druggable paths for affecting the biology of these small RNAs. Going forward, the discovery of new chemical space and targets to regulate miRNAs will aid in our quest to manipulate these important biomolecules and ignite a new field of therapeutics. Using the technologies discussed herein and the initial compounds identified, the vision of regulating miRNAs with small molecules is becoming ever closer and will be exciting to watch.

References

1. Lee RC, Feinbaum RL, Ambros V (1993) The *C. elegans* heterochronic gene *lin-4* encodes small RNAs with antisense complementarity to *lin-14*. Cell 75:843–854
2. Wightman B, Ha I, Ruvkun G (1993) Posttranscriptional regulation of the heterochronic gene *lin-14* by *lin-4* mediates temporal pattern formation in *C. elegans*. Cell 75:855–862
3. Ruvkun G, Reinhart BJ, Slack FJ, Basson M, Pasquinelli AE, Bettinger JC, Rougvie AE, Horvitz HR (2000) The 21-nucleotide *let-7* RNA regulates developmental timing in *Caenorhabditis elegans*. Nature 403:901–906
4. Pasquinelli AE, Reinhart BJ, Slack F, Martindale MQ, Kuroda MI, Maller B, Hayward DC, Ball EE, Degnan B, Muller P, Spring J, Srinivasan A, Fishman M, Finnerty J, Corbo J, Levine M, Leahy P, Davidson E, Ruvkun G (2000) Conservation of the sequence and temporal expression of *let-7* heterochronic regulatory RNA. Nature 408:86–89
5. Friedman RC, Kai-How Farh K, Burge CB, Bartel DP (2009) Most mammalian mRNAs are conserved targets of microRNAs. Genome Res 19:92–105
6. Li Z, Rana TM (2014) Therapeutic targeting of microRNAs: current status and future challenges. Nat Rev Drug Discov 13:622–638
7. Roush S, Slack FJ (2008) The *let-7* family of microRNAs. Trends Cell Biol 18:505–516
8. Han C, Yu Z, Duan Z, Kan Q (2014) Role of microRNA-1 in human cancer and its therapeutic potentials. Biomed Res Int 2014:428371
9. Ma L (2010) Role of miR-10b in breast cancer metastasis. Breast Cancer Res 12:210
10. Pfeffer S, Yang CH, Pfeffer LM (2015) The role of miR-21 in cancer. Drug Dev Res 76:270–277
11. Chhabra R, Dubey R, Saini N (2010) Cooperative and individualistic functions of the microRNAs in the miR-23a-27a-24-2 cluster and its implications in human diseases. Mol Cancer 9:232
12. Wang Y, Zhang X, Li H, Yu J, Ren X (2013) The role of miRNA-29 family in cancer. Eur J Cell Biol 92:123–128
13. Hermeking H (2010) The miR-34 family in cancer and apoptosis. Cell Death Differ 17:193–199
14. Guttilla IK, White BA (2009) Coordinate regulation of FOXO1 by miR-27a, miR-96, and miR-182 in breast cancer cells. J Biol Chem 284:23204–23216

15. Henke JI, Goergen D, Zheng J, Song Y, Schuttler CG, Fehr C, Junemann C, Niepmann M (2008) microRNA-122 stimulates translation of hepatitis C virus RNA. EMBO J 27:3300–3310
16. Gramantieri L, Ferracin M, Fornari F, Veronese A, Sabbioni S, Liu C-G, Calin GA, Giovannini C, Ferrazzi M, Grazi GL, Croce CM, Bolondi L, Negrini M (2007) Cyclin G1 is a target of miR-122, a microRNA frequently down-regulated in human hepatocellular carcinoma. Cancer Res 67:6092–6099
17. Ozen M, Creighton CJ, Ozdemir M, Ittmann M (2008) Widespread deregulation of microRNA expression in human prostate cancer. Oncogene 27:1788–1793
18. Thum T, Catalucci D, Bauersachs J (2008) MicroRNAs: novel regulators in cardiac development and disease. Cardiovasc Res 79:562–570
19. Isobe T, Hisamori S, Hogan DJ, Zabala M, Hendrickson DG, Dalerba P, Cai S, Scheeren F, Kuo AH, Sikandar SS, Lam JS, Qian D, Dirbas FM, Somlo G, Lao K, Brown PO, Clarke MF, Shimono Y (2014) miR-142 regulates the tumorigenicity of human breast cancer stem cells through the canonical WNT signaling pathway. Elife 3:e01977
20. Faraoni I, Antonetti FR, Cardone J, Bonmassar E (2009) miR-155 gene: a typical multifunctional microRNA. Biochim Biophys Acta 1792:497–505
21. Williams AH, Valdez G, Moresi V, Qi X, McAnally J, Elliott JL, Bassel-Duby R, Sanes JR, Olson EN (2009) MicroRNA-206 delays ALS progression and promotes regeneration of neuromuscular synapses in mice. Science 326:1549–1554
22. Scarola M, Schoeftner S, Schneider C, Benetti R (2010) miR-335 directly targets Rb1 (pRb/P105) in a proximal connection to P53-dependent stress response. Cancer Res 70:6925–6933
23. Staedel C, Varon C, Nguyen PH, Vialet B, Chambonnier L, Rousseau B, Soubeyran I, Evrary S, Couillaud F, Darfeuille F (2015) Inhibition of gastric tumor cell growth using seed-targeting LNA as specific, long-lasting microRNA inhibitors. Mol Ther Nucleic Acids 4:e246
24. Hu W, Chan CS, Wu R, Zhang C, Sun Y, Song JS, Tang LH, Levine AJ, Feng Z (2010) Negative regulation of tumor suppressor P53 by microRNA miR-504. Mol Cell 38:689–699
25. Pang F, Zha R, Zhao Y, Wang Q, Chen D, Zhang Z, Chen T, Yao M, Gu J, He X (2014) MiR-525-3p enhances the migration and invasion of liver cancer cells by downregulating ZNF395. PLoS One 9:e90867
26. Haga CL, Velagapudi SP, Strivelli JR, Yang W-Y, Disney MD, Phinney DG (2015) Small molecule inhibition of miR-544 biogenesis disrupts adaptive responses to hypoxia by modulating ATM-mTOR signaling. ACS Chem Biol 10:2267–2276
27. Lin S, Gregory RI (2015) MicroRNA biogenesis pathways in cancer. Nat Rev Cancer 15:321–333
28. Ha M, Kim VN (2014) Regulation of microRNA biogenesis. Nat Rev Mol Cell Biol 15:509–524
29. Mack GS (2007) MicroRNA gets down to business. Nat Biotechnol 25:631–638
30. Wang V, Wu W (2009) MicroRNA-based therapeutics for cancer. BioDrugs 23:15–23
31. Calin GA, Sevignani C, Dumitru CD, Hyslop T, Noch E, Yendamuri S, Shimuzu M, Rattan S, Bullrich F, Negrini M, Croce CM (2004) Human microRNA genes are frequently located at fragile sites and genomic regions involved in cancers. Proc Natl Acad Sci U S A 101:2999–3004
32. Lu J, Getz G, Miska EA, Alvarez-Saavedra E, Lamb J, Peck D, Sweet-Cordero A, Ebert BL, Mak RH, Ferrando AA, Downing JR, Jacks T, Horvitz HR, Golub TR (2005) MicroRNA expression profiles classify human cancers. Nature 435:834–838
33. Ling H, Fabbri M, Calin GA (2013) MicroRNAs and other non-coding RNAs as targets for anticancer drug development. Nat Rev Drug Discov 12:847–865
34. Franceschini A, Meier R, Casanova A, Kreibich S, Daga N, Andritschke D, Dilling S, Ramo P, Emmenlauer M, Kaufmann A, Conde-Alvarez R, Low SH, Pelkmans L, Helenius A, Hardt W-D, Dehio C, von Mering C (2014) Specific inhibition of diverse

pathogens in human cells by synthetic microRNA-like oligonucleotides inferred from RNAi screens. Proc Natl Acad Sci U S A 111:4548–4553
35. White PJ, Anastasopoulos F, Pouton CW, Boyd BJ (2009) Overcoming biological barriers to *in vivo* efficacy of antisense oligonucleotides. Expert Rev Mol Med 11:e10
36. Thomas JR, Hergenrother PJ (2008) Targeting RNA with small molecules. Chem Rev 108:1171–1224
37. Guan L, Disney MD (2012) Recent advances in developing small molecules targeting RNA. ACS Chem Biol 7:73–86
38. Connelly CM, Moon MH, Schneekloth Jr JS (2016) The emerging role of RNA as a therapeutic target for small molecules. Cell Chem Biol 23:1077–1090
39. Disney MD (2013) Rational design of chemical genetic probes of RNA function and lead therapeutics targeting repeating transcripts. Drug Discov Today 18:1228–1236
40. Werstuck G, Green MR (1998) Controlling gene expression in living cells through small molecule-RNA interactions. Science 282:296–298
41. Sparano BA, Koide K (2007) Fluorescent sensors for specific RNA: a general paradigm using chemistry and combinatorial biology. J Am Chem Soc 129:4785–4794
42. Paige JS, Wu KY, Jaffrey SR (2011) RNA mimics of green fluorescent protein. Science 333:642–646
43. Serganov A, Nudler E (2013) A decade of riboswitches. Cell 152:17–24
44. Terasaka N, Futai K, Katoh T, Suga H (2016) A human microRNA precursor binding to folic acid discovered by small RNA transcriptomic SELEX. RNA 22:1918–1928
45. Connelly CM, Thomas M, Deiters A (2012) High-throughput luciferase reporter assay for small-molecule inhibitors of microRNA function. J Biomol Screen 17:822–828
46. Connelly CM, Deiters A (2014) Cellular microRNA sensors based on luciferase reporters. Methods Mol Biol 1095:135–146
47. Connelly CM, Deiters A (2014) Identification of inhibitors of microRNA function from small molecule screens. Methods Mol Biol 1095:147–156
48. Gumireddy K, Young DD, Xiong X, Hogenesch JB, Huang Q, Deiters A (2008) Small-molecule inhibitors of microRNA miR-21 function. Angew Chem Int Ed 47:7482–7484
49. Volinia S, Calin GA, Liu C-G, Ambs S, Cimmino A, Petrocca F, Visone R, Iorio M, Roldo C, Ferracin M, Prueitt RL, Yanaihara N, Lanza G, Scarpa A, Vecchione A, Negrini M, Harris CC, Croce CM (2006) A microRNA expression signature of human solid tumors defines cancer gene targets. Proc Natl Acad Sci U S A 103:2257–2261
50. Selcuklu SD, Donoghue MTA, Spillane C (2009) miR-21 as a key regulator of oncogenic processes. Biochem Soc Trans 37:918–925
51. Krichevsky AM, Gabriely G (2009) miR-21: a small multi-faceted RNA. J Cell Mol Med 13:39–53
52. Jazbutyte V, Thum T (2010) MicroRNA-21: from cancer to cardiovascular disease. Curr Drug Targets 11:926–935
53. Esquela-Kerscher A, Slack FJ (2006) Oncomirs – microRNAs with a role in cancer. Nat Rev Cancer 6:259–269
54. Medina PP, Nolde M, Slack FJ (2010) OncomiR addiction in an *in vivo* model of microRNA-21-induced pre-B-cell lymphoma. Nature 467:86–90
55. Cheng CJ, Slack FJ (2012) The duality of oncomiR addiction in the maintenance and treatment of cancer. Cancer J 18:232–237
56. Jiang C-S, Wang X-M, Zhang S-Q, Meng L-S, Zhu W-H, Xu J, Lu S-M (2015) Discovery of 4-benzoylamino-N-(prop-2-yn-1-yl)benzamides as novel microRNA-21 inhibitors. Bioorg Med Chem 23:6510–6519
57. Naro Y, Thomas M, Stephens MD, Connelly CM, Deiters A (2015) Aryl amide small-molecule inhibitors of microRNA miR-21 function. Bioorg Med Chem Lett 25:4793–4796
58. Young DD, Connelly CM, Grohmann C, Deiters A (2010) Small molecule modifiers of microRNA miR-122 function for the treatment of hepatitis C virus infection and hepatocellular carcinoma. J Am Chem Soc 132:7976–7981

59. Jopling CL, Yi M-K, Lancaster AM, Lemon SM, Sarnow P (2005) Modulation of hepatits C virus RNA abundance by a liver-specific microRNA. Science 309:1577–1581
60. Lanford RE, Hildebrandt-Eriksen ES, Petri A, Persson R, Lindow M, Munk ME, Kauppinen S, Orum H (2010) Therapeutic silencing of microRNA-122 in primates with chronic hepatits C virus infection. Science 327:198–201
61. Janssen HLA, Reesink HW, Lawitz EJ, Zeuzem S, Rodriguez-Torres M, Patel K, van der Meer AJ, Patrick AK, Chen A, Zhou Y, Persson R, King BD, Kauppinen S, Levin AA, Hodges MR (2013) Treatment of HCV infection by targeting microRNA. N Engl J Med 368:1685–1694
62. van der Ree MH, van der Meer AJ, de Bruijne J, Maan R, van Vliet A, Welzel TM, Zeuzem S, Lawitz EJ, Rodriguez-Torres M, Kupcova V, Wiercinska-Drapalo A, Hodges MR, Janssen HLA, Reesink HW (2014) Long-term safety and efficacy of microRNA-targeted therapy in chronic hepatitis C patients. Antivir Res 111:53–59
63. Ottosen S, Parsley TB, Yang L, Zeh K, van Doom L-J, van der Veer E, Raney AK, Hodges MR, Patrick AK (2014) In vitro antiviral activity and preclinical and clincal resistance profile of miravirsen, a novel anti-hepatitis C virus therapeutic targeting the human factor miR-122. Antimicrob Agents Chemother 59:599–608
64. van der Ree MH, van der Meer AJ, van Nuenen AC, de Bruijne J, Ottosen S, Janssen HLA, Kootstra NA, Reesink HW (2015) Miravirsen dosing in chronic hepatits C patients results in decreased microRNA-122 levels without affecting other microRNAs in plasma. Aliment Pharmacol Ther 43:102–113
65. Wiggins JF, Ruffino L, Kelnar K, Omotola M, Patrawala L, Brown D, Bader AG (2010) Development of a lung cancer therapeutic based on the tumor suppressor microRNA-34. Cancer Res 70:5923–5930
66. Adams BD, Parsons C, Slack FJ (2016) The tumor-suppressive and potential therapeutic functions of miR-34a in epithelial carcinomas. Expert Opin Ther Targets 20:737–753
67. Daige CL, Wiggins JF, Priddy L, Nelligan-Davis T, Zhao J, Brown D (2014) Systemic delivery of a miR-34a mimic as a potential therapeutic for liver cancer. Mol Cancer Ther 13:2352–2360
68. Xiao Z, Li CH, Chan SL, Xu F, Feng L, Wang Y, Jiang JD, Sung JJY, Cheng CHK, Chen Y (2014) A small-molecule modulator of the tumor-suppressor miR34a inhibits the growth of hepatocellular carcinoma. Cancer Res 74:6236–6247
69. Tan S-B, Huang C, Chen X, Wu Y, Zhou M, Zhang C, Zhang Y (2013) Small molecular inhibitors of miR-1 identified from photocycloadducts of acetylenes with 2-methoxy-1,4-napthalenequinone. Bioorg Med Chem 21:6124–6131
70. Tan S-B, Li J, Chen X, Zhang W, Zhang D, Zhang C, Li D, Zhang Y (2014) Small molecule inhibitor of myogenic microRNAs leads to a discovery of miR-221/222-myoD-myomiRs regulatory pathway. Chem Biol 21:1265–1270
71. Chen X, Huang C, Zhang W, Wu Y, Chen X, Zhang C-Y, Zhang Y (2012) A universal activator of microRNAs identified from photoreaction products. Chem Commun 48:6432–6433
72. Shan G, Li Y, Zhang J, Li W, Szulwach KE, Duan R, Faghihi MA, Khalil AM, Lu L, Paroo Z, Chan AWS, Shi Z, Liu Q, Wahlestedt C, He C, Jin P (2008) A small molecule enhances RNA interference and promotes microRNA processing. Nat Biotechnol 26:933–940
73. Melo S, Villanueva A, Moutinho C (2011) Small molecule enoxacin is a cancer-specific growth inhibitor that acts by enhancing TAR RNA-binding protein 2-mediated microRNA processing. Proc Natl Acad Sci U S A 108:4394–4300
74. Shum D, Bhinder B, Radu C, Farazi T, Landthaler M, Tuschl T, Calder P, Ramirez CN, Djaballah H (2012) An image-based biosensor assay strategy to screen for modulators of the microRNA 21 biogenesis pathway. Comb Chem High Throughput Screen 15:529–541
75. Watashi K, Yeung ML, Starost MF, Hosmane RS, Jeang K-T (2010) Identification of small molecules that suppress microRNA function and reverse tumorigenesis. J Biol Chem 285:24707–24716

76. Bose D, Jayaraj G, Suryawanshi H, Agarwala P, Pore SK, Banerjee R, Maiti S (2012) The tuberculosis drug streptomycin as a potential cancer therapeutic: inhibition of miR-21 function by directly targeting its precursor. Angew Chem Int Ed 51:1019–1023
77. Lorenz DA, Song JM, Garner AL (2015) High-throughput platform assay technology for the discovery of pre-microRNA-selective small molecule probes. Bioconjug Chem 26:19–23
78. Tran TPA, Vo DD, Di Giorgio A, Duca M (2015) Ribosome-targeting antibiotics as inhibitors of oncogenic microRNAs biogenesis: old scaffolds for new perspectives in RNA targeting. Bioorg Med Chem 23:5334–5344
79. Nahar S, Ranjan N, Ray A, Arya DP, Maiti S (2015) Potent inhibition of miR-27a by neomycin-benzimidazole conjugates. Chem Sci 6:5837–5846
80. Bose D, Jayaraj GG, Kumar S, Maiti S (2013) A molecule-beacon-based screen for small molecule inhibitors of miRNA maturation. ACS Chem Biol 8:930–938
81. MacBeath G, Koehler AN, Schreiber SL (1999) Printing small molecule as microarrays and detecting protein-ligand interactions en mass. J Am Chem Soc 121:7967–7968
82. Hong JA, Neel DV, Wassaf D, Caballero F, Koehler AN (2014) Recent discoveries and applications involving small-molecule microarrays. Curr Opin Chem Biol 18:21–28
83. Abulwerdi FA, Schneekloth Jr JS (2016) Microarray-based technologies for the discovery of selective, RNA-binding molecules. Methods 103:188–195
84. Connelly CM, Boer RE, Moon MH, Gareiss P, Schneekloth JS Jr (2017) Discovery of inhibitors of microRNA-21 processing using small molecule microarrays. ACS Chem Biol 12:435–443
85. Chirayil S, Chirayil R, Luebke KJ (2009) Discovering ligands for a microRNA precursor with peptoid microarrays. Nucleic Acids Res 37:5486–5497
86. Chirayil R, Wu Q, Amezcua C, Luebke KJ (2014) NMR characterization of an oligonucleotide model of the MiR-21 pre-element. PLoS One 9:e108231
87. Diaz JP, Chirayil R, Chirayil S, Tom M, Head KJ, Luebke KJ (2014) Association of a peptoid ligand with the apical loop of pri-miR-21 inhibits cleavage by Drosha. RNA 20:528–539
88. Carlson CB, Beal PA (2002) Points of attachment and sequence of immobilized peptide-acridine conjugates control affinity for nucleic acids. J Am Chem Soc 124:8510–8511
89. Disney MD, Seeberger PH (2004) Aminoglycoside microarray to explore interactions of antibiotics with RNAs and proteins. Chem Eur J 10:3308–3314
90. Pai J, Hyun S, Hyun JY, Park S-H, Kim W-J, Bae S-H, Kim N-K, Yu J, Shin I (2016) Screening of pre-miRNA-155 binding peptides for apoptosis inducing activity using peptide microarrays. J Am Chem Soc 138:857–867
91. Eis PS, Tam W, Sun L, Chadburn A, Li Z, Gomez MF, Lund E, Dahlberg JE (2005) Accumulation of miR-155 and BIC RNA in human B cell lymphomas. Proc Natl Acad Sci U S A 102:3627–3632
92. Hyun S, Han A, Jo MH, Hohng S, Yu J (2014) Dicer nuclease-promoted production of Let7a-1 microRNA is enhanced in the presence of tryptophan-containing amphiphilic peptides. Chembiochem 15:1651–1659
93. Bose D, Nahar S, Rai MK, Ray A, Chakraborty K, Maiti S (2010) Selective inhibition of miR-21 by phage display screened peptide. Nucleic Acids Res 43:4342–4352
94. Nahar S, Bose D, Pal S, Chakraborty TK, Maiti S (2015) Cyclic cationic peptides containing sugar amino acids selectively distinguishes and inhibits maturation of pre-miRNAs of the same family. Nucleic Acid Ther 25:323–329
95. Krishnamurthy M, Simon K, Orendt AM, Beal PA (2007) Macrocyclic helix-threading peptides for targeting RNA. Angew Chem Int Ed 46:7044–7047
96. Chen Y, Yang F, Zubovic L, Pavelitz T, Yang W, Godin K, Walker M, Zheng S, Macchi P, Varani G (2016) Targeting inhibition of oncogenic miR-21 maturation with designed RNA-binding proteins. Nat Chem Biol 12:717–723
97. Velagapudi SP, Disney MD (2014) Two-dimensional combinatorial screening enables the bottom-up design of a microRNA-10b inhibitor. Chem Commun 50:3027–3029

98. Disney MD, Labuda LP, Paul DJ, Poplawski SG, Pushechnikov A, Tran T, Velagapudi SP, Wu M, Childs-Disney JL (2008) Two-dimensional combinatorial screening identifies specific aminoglycoside-RNA internal loop partners. J Am Chem Soc 130:11185–11194
99. Velagapudi SP, Seedhouse SJ, Disney MD (2010) Structure-activity relationships through sequencing (StARTS) defines optimal and suboptimal RNA motif targets for small molecules. Angew Chem Int Ed 49:3816–3818
100. Ma L, Teruya-Feldstein J, Weinberg RA (2007) Tumour invasion and metastasis initiated by microRNA-10b in breast cancer. Nature 449:682–687
101. Ma L, Reinhardt F, Pan E, Soutschek J, Bhat B, Marcusson EG, Teruya-Feldstein J, Bell GW, Weinberg RA (2010) Therapeutic silencing of miR-10b inhibits metastasis in a mouse mammary tumor model. Nat Biotechnol 28:341–347
102. Davies BP, Arenz C (2006) A homogeneous assay for microRNA maturation. Angew Chem Int Ed 45:5550–5552
103. Davies BP, Arenz C (2008) A fluorescence probe for assaying microRNA maturation. Bioorg Med Chem 16:49–55
104. Vo DD, Staedel C, Zehnacker L, Benhida R, Darfeuille F, Duca M (2014) Targeting the production of oncogenic microRNAs with multimodal synthetic small molecules. ACS Chem Biol 9:711–721
105. Comley J (2003) Assay interference a limiting factor in HTS? Drug Discov World 4:91–98
106. Imbert P-E, Unterreiner V, Siebert D, Gubler H, Parker C, Gabriel D (2007) Recommendations for the reduction of compound artifacts in time-resolved fluorescence resonance energy transfer assays. Assay Drug Dev Technol 5:363–372
107. MacRae IJ, Zhou Z, Doudna JA (2007) Structural determinants of RNA recognition and cleavage by Dicer. Nat Struct Mol Biol 14:934–940
108. Feng Y, Zhang X, Graves P, Zeng Y (2012) A comprehensive analysis of precursor microRNA cleavage by human Dicer. RNA 18:2083–2092
109. Klemm CM, Berthelmann A, Neubacher S, Arenz C (2009) Short and efficient synthesis of alkyne-modified amino glycoside building blocks. Eur J Org Chem 17:2788–2794
110. Vo DD, Tran TPA, Staedel C, Benhida R, Darfeuille F, Di Giorgio A, Duca M (2016) Oncogenic microRNAs biogenesis as a drug target: structure-activity relationship studies on new aminoglycoside conjugates. Chemistry 22:5350–5362
111. Judson RL, Babiarz JE, Venere M, Blelloch R (2009) Embryonic stem cell-specific microRNAs promote induced pluripotency. Nat Biotechnol 27:459–461
112. Wang Y, Blelloch R (2009) Cell cycle regulation by microRNAs in embryonic stem cells. Cancer Res 69:4093–4096
113. Podolska K, Sedlak D, Bartunek P, Svoboda P (2014) Fluorescence-based high-throughput screening of Dicer cleavage activity. J Biomol Screen 19:417–426
114. Tran T, Disney MD (2012) Identifying the preferred RNA motifs and chemotypes that interact by probing millions of combinations. Nat Commun 3:1125
115. Zhang J, Umemoto S, Nakatani K (2010) Fluorescent indicator displacement assay for ligand-RNA interactions. J Am Chem Soc 132:3660–3661
116. Maiti M, Nauwelaerts K, Herdewijn P (2012) Pre-microRNA binding aminoglycosides and antitumor drugs as inhibitors of Dicer catalyzed microRNA processing. Bioorg Med Chem Lett 22:1709–1711
117. Murata A, Fukuzumi T, Umemoto S, Nakatani K (2013) Xanthone derivatives as potential inhibitors of miRNA processing by human Dicer: targeting secondary structures of pre-miRNA by small molecules. Bioorg Med Chem Lett 23:252–255
118. Murata A, Harada Y, Fukuzumi T, Nakatani K (2013) Fluorescent indicator displacement assay of ligands targeting 10 microRNA precursors. Bioorg Med Chem 21:7101–7106
119. Fukuzumi T, Murata A, Aikawa H, Harada Y, Nakatani K (2015) Exploratory study on the RNA-binding structural motifs by library screening targeting pre-miR-29a. Chemistry 21:16859–16867

120. Murata A, Otabe T, Zhang J, Nakatani K (2016) BZDANP, a small-molecule modulator of pre-miR-29a maturation by Dicer. ACS Chem Biol 11:2790–2796
121. Watkins D, Jiang L, Nahar S, Maiti S, Arya DP (2015) A pH sensitive high-throughput assay for miRNA binding of a peptide-aminoglycoside (PA) library. PLoS One 10:e0144251
122. Lorenz DA, Garner AL (2016) A click chemistry-based microRNA maturation assay optimized for high-throughput screening. Chem Commun 52:8267–8270
123. Garner AL, Janda KD (2010) cat-ELCCA: a robust method to monitor the fatty acid acyltransferase activity of ghrelin O-acyltransferase (GOAT). Angew Chem Int Ed 49:9630–9634
124. Garner AL, Janda KD (2011) A small molecule antagonist of ghrelin O-acyltransferase (GOAT). Chem Commun 47:7512–7514
125. Zhang J-H, Chung TDY, Oldenburg KR (1999) A simple statistical parameter for use in evaluation and validation of high throughput screening assays. J Biomol Screen 4:67–73
126. Mustoe AM, Brooks III CL, Al-Hashimi HM (2014) Hierarchy of RNA functional dynamics. Annu Rev Biochem 83:441–466
127. Turner DH, Sugimoto N, Freier SM (1988) RNA structure prediction. Annu Rev Biophys Biophys Chem 17:167–192
128. Parisien M, Major F (2008) The MC-Fold and MC-Sym pipeline infers RNA structure from sequence data. Nature 452:51–55
129. Disney MD, Winkelsas AM, Velagapudi SP, Southern M, Fallahi M, Childs-Disney JL (2016) Inforna 2.0: a platform for the sequence-based design of small molecules targeting structured RNAs. ACS Chem Biol 11:1720–1728
130. Velagapudi SP, Gallo SM, Disney MD (2014) Sequence-based design of bioactive small molecules that target precursor microRNAs. Nat Chem Biol 10:291–297
131. Liu B, Childs-Disney JL, Znosko BM, Wang D, Fallahi M, Gallo SM, Disney MD (2016) Analysis of secondary structural elements in human microRNA hairpin precursors. BMC Bioinformatics 17:112
132. Jamal S, Periwal V, Consortium OSDD, Scaria V (2012) Computational analysis and predictive modeling of small molecule modulators of microRNA. J Cheminform 4:16
133. Wehler T, Brenk R (2017) Structure-based discovery of small molecules binding to RNA. Topics Med Chem. doi:10.1007/7355_2016_29
134. Shi Z, Zhang J, Qian X, Han L, Zhang K, Chen L, Liu J, Ren Y, Yang M, Zhang A, Pu P, Kang C (2013) AC1MMYR2, an inhibitor of Dicer-mediated biogenesis of oncomir miR-21, reverses epithelial-mesenchymal transition and suppresses tumor growth and progression. Cancer Res 73:5519–5531
135. Ren Y, Zhou X, Liu X, Jia H-H, Zhao X-H, Wang Q-X, Han L, Song X, Zhu Z-Y, Sun T, Jiao H-X, Tian W-P, Yang Y-Q, Zhao X-L, Zhang L, Mei M, Kang C-S (2016) Reprogramming carcinoma associated fibroblasts by AC1MMYR2 impedes tumor metastasis and improves chemotherapy efficacy. Cancer Lett 374:96–106
136. Childs-Disney JL, Wu M, Pushechnikov A, Aminova O, Disney MD (2007) A small molecule microarray platform to select RNA internal loop-ligand interactions. ACS Chem Biol 2:745–754
137. Velagapudi SP, Seedhouse SJ, French J, Disney MD (2011) Defining the RNA internal loops preferred by benzimidazole derivatives via 2D combinatorial screening and computational analysis. J Am Chem Soc 133:10111–10118
138. Disney MD, Angelbello AJ (2016) Rational design of small molecules targeting oncogenic noncoding RNAs from sequence. Acc Chem Res 49:2698–2704
139. Costales MG, Childs-Disney JL, Disney MD (2017) Computational tools for design of selective small molecules targeting RNA: from small molecule microarray to chemical similarity searching. Topics Med Chem. doi:10.1007/7355_2016_21
140. Dambal S, Shah M, Mihelich B, Nonn L (2015) The microRNA-183 cluster: the family that plays together stay together. Nucleic Acids Res 43:7173–7188

141. Costales MG, Rzuczek SG, Disney MD (2016) Comparison of small molecules and oligonucleotides that target a toxic, non-coding RNA. Bioorg Med Chem Lett 26:2605–2609
142. Velagapudi SP, Cameron MD, Haga CL, Rosenberg LH, Lafitte M, Duckett DR, Phinney DG, Disney MD (2016) Design of a small molecule against an oncogenic noncoding RNA. Proc Natl Acad Sci U S A 113:5898–5903
143. Haga CL, Phinney DG (2012) MicroRNAs in the imprinted DLK1-DIO3 region repress the epithelial-to-mesenchymal transition by targeting the TWIST1 protein signaling network. J Biol Chem 287:42695–42707
144. Childs-Disney JL, Disney MD (2016) Small molecule targeting of a microRNA associated with hepatocellular carcinoma. ACS Chem Biol 11:375–380
145. Mei H-Y, Mack DP, Galan AA, Halim NS, Heldsinger A, Loo JA, Moreland DW, Sannes-Lowery KA, Sharmeen L, Truong HN, Czarnik AW (1997) Discovery of selective, small-molecule inhibitors of RNA complexes – I. The Tat protein/TAR RNA complexes required for HIV-1 transcription. Bioorg Med Chem 5:1173–1184
146. Mei H-Y, Cui M, Heldsinger A, Lemrow SM, Loo JA, Sannes-Lowery KA, Sharmeen L, Czarnik AW (1998) Inhibitors of protein-RNA complexation that target RNA: specific recognition of human immunodeficiency virus type 1 TAR RNA by small organic molecules. Biochemistry 37:14204–14212
147. Stelzer AC, Frank AT, Kratz JD, Swanson MD, Gonzalez-Hernandez MJ, Lee J, Andricioaei I, Markovitz DM, Al-Hashimi HM (2011) Discovery of selective bioactive small molecules by targeting an RNA dynamic ensemble. Nat Chem Biol 7:553–559
148. Naryshkin NA, Weetall M, Dakka A, Narasimhan J, Zhao X, Feng Z, Ling KKY, Karp GM, Qi H, Woll MG, Chen G, Zhang N, Gabbeta V, Vazirani P, Bhattacharyya A, Furia B, Risher N, Sheedy J, Kong R, Ma J, Turpoff A, Lee C-S, Zhang X, Moon Y-C, Trifillis P, Welch EM, Colacino JM, Babiak J, Almstead NG, Peltz SW, Eng LA, Chen KS, Mull JL, Lynes MS, Rubin LL, Fontoura P, Santarelli L, Haehnke D, McCarthy KD, Schmucki R, Ebeling M, Sivaramakrishnan M, Ko C-P, Paushkin SV, Ratni H, Gerlach I, Ghosh A, Metzger F (2014) Motor neuron disease. SMN2 splicing modifiers improve motor function and longevity in mice with spinal muscular atrophy. Science 345:688–693
149. Palacino J, Swalley SE, Song C, Cheung AK, Shu L, Zhang X, Van Hoosear M, Shin Y, Chin DN, Keller CG, Beibel M, Renaud NA, Smith TM, Salcius M, Shi X, Hild M, Servais R, Jain M, Deng L, Bullock C, McLellan M, Schuierer S, Murphy L, Blommers MJJ, Blaustein C, Berenshteyn F, Lacoste A, Thomas JR, Roma G, Michaud GA, Tseng BS, Porter JA, Myer VE, Tallarico JA, Hamann LG, Curtis D, Fishman MC, Dietrich WF, Dales NA, Sivasankaran R (2015) SMN2 splice modulators enhance U1-pre-mRNA association and rescue SMA mice. Nat Chem Biol 11:511–517
150. Tan GS, Chiu C-H, Garchow BG, Metzler D, Diamond SL, Kiriakidou M (2012) Small molecule inhibition of RISC loading. ACS Chem Biol 7:403–410
151. Schmidt MF, Korb O, Abell C (2013) MicroRNA-specific Argonaute 2 protein inhibitors. ACS Chem Biol 8:2122–2126
152. Masciarelli S, Quaranta R, Iosue I, Colotti G, Padula F, Varchi G, Fazi F, Del Rio A (2014) A small-molecule targeting the microRNA binding domain of Argonaute 2 improves the retinoic acid differentiation response of the acute promyelocytic leukemia cell line NB4. ACS Chem Biol 9:1674–1679
153. Hesse M, Arenz C (2016) A rapid and versatile assay for Ago2-mediated cleavage by using branched rolling circle amplification. Chembiochem 17:304–307
154. Lin S, Gregory RI (2015) Identification of small molecule inhibitors of Zcchc11 TUTase activity. RNA Biol 12:792–800
155. Roos M, Pradere U, Ngondo RP, Behera A, Allegrini S, Civenni G, Zagalak JA, Marchand J-R, Menzi M, Towbin H, Scheuermann J, Neri D, Caflisch A, Catapano CV, Claudo C, Hall J (2016) A small-molecule inhibitor of Lin28. ACS Chem Biol 11:2773–2781

156. Lightfoot HL, Miska EA, Balasubramanian S (2016) Identification of small molecule inhibitors of the Lin28-mediated blockage of pre-let-7g processing. Org Biomol Chem 14:10208–10216
157. Lim D, Byun WG, Koo JY, Park H, Park SB (2016) Discovery of a small-molecule inhibitor of protein-microRNA interaction using binding assay with a site-specifically labeled Lin28. J Am Chem Soc 138:13630–13638
158. Hermann T (2017) Viral RNA targets and their small molecule ligands. Topics Med Chem. doi:10.1007/7355_2016_20
159. Wirmer J, Westhof E (2006) Molecular contacts between antibiotics and the 30S ribosomal particle. Methods Enzymol 415:180–202

Top Med Chem (2018) 27: 111–134
DOI: 10.1007/7355_2016_20

Published online: 20 April 2017

Viral RNA Targets and Their Small Molecule Ligands

Thomas Hermann

Abstract RNA genomes and transcripts of viruses contain conserved structured motifs which are attractive targets for small molecule inhibitors of viral replication. Ligand binding affects conformational states, stability, and interactions of these viral RNA targets which play key roles in the infection process. Inhibition of viral RNA function by small molecule ligands has been extensively studied for human immunodeficiency virus (HIV) and hepatitis C virus (HCV) which provide valuable insight for the future exploration of RNA targets in other viral pathogens including severe respiratory syndrome coronavirus (SARS CoV), influenza A, and insect-borne flaviviruses (Dengue, Zika, and West Nile) as well as filoviruses (Ebola and Marburg). Here, I will review recent progress on the discovery and design of small molecule ligands targeting structured viral RNA motifs.

Keywords Antiviral drugs, Drug targets, Hepatitis C virus, Human immunodeficiency virus, Influenza A virus, Noncoding RNA, Viral inhibitors

Contents

1 Introduction 112
2 Viral RNA Targets 114
  2.1 Human Immunodeficiency Virus 114
  2.2 Hepatitis C Virus 117
  2.3 Influenza A Virus 118
  2.4 Severe Respiratory Syndrome Coronavirus 119
  2.5 Insect-Borne Flaviviruses 119

T. Hermann (✉)
Department of Chemistry and Biochemistry, University of California, San Diego, 9500 Gilman Drive, La Jolla, CA 92093, USA
Center for Drug Discovery Innovation, University of California, San Diego, 9500 Gilman Drive, La Jolla, CA 92093, USA
e-mail: tch@ucsd.edu

2.6 Filoviruses ... 120
2.7 Kaposi-Sarcoma Associated Herpesvirus ... 120
2.8 Hepatitis B Virus ... 121
3 Ligands Targeting Viral RNAs ... 121
3.1 Human Immunodeficiency Virus ... 121
3.2 Hepatitis C Virus ... 124
3.3 Influenza A Virus ... 126
3.4 Severe Respiratory Syndrome Coronavirus ... 127
4 Summary ... 128
References ... 129

Abbreviations

dsRNA Double-stranded RNA
ssRNA Single-stranded RNA

For abbreviations of virus names, see Table 1.

1 Introduction

The compact genomes of viruses offer a limited number of protein targets for the development of anti-infective therapy. Structured RNA elements in viral genomes and transcripts have the potential to expand the target space for antiviral drug discovery. Precedent for clinically approved RNA-binding drugs is found in natural product-derived antibiotics including macrolides, tetracyclins, oxazolidinones, and aminoglycosides which interact with ribosomal RNA (rRNA) of bacteria and block protein synthesis in the pathogens [1, 2]. The well-defined structure of rRNA provides selective binding sites for these antibiotics which serve as a paradigm for RNA recognition by small molecule ligands. RNA elements in viruses have been extensively explored as potential drug targets in the human immunodeficiency virus (HIV) and hepatitis C virus (HCV) [3, 4] whose genomes include conserved noncoding regions (ncRNA) that may present structured binding sites for small molecules [5, 6]. Challenges and successes in the discovery and design of compounds targeting RNA have been discussed in the previous comprehensive review articles which also provide a historic perspective on past efforts to explore viral RNA targets for small molecule inhibitors [7–12]. In the current chapter, I describe progress on discovery and investigation of small molecule ligands for viral RNA targets over the last 2–3 years and include perspectives on potential new viral RNA targets which have not yet been widely explored but may attract interest in pathogens of unmet or emerging medical needs (Table 1).

Table 1 Viral RNA targets

Family	Virus	Genome	RNA target	Small molecule ligands
Retrovirus	Human immunodeficiency virus (HIV)	(+) ssRNA	• Transactivation response (TAR) element • Rev response element (RRE) • Dimer initiation sequence (DIS) • Packaging signal (Ψ) stem-loop 3 (SL-3) • Frameshifting signal (FSS)	Reported for all HIV targets; previously reviewed [3, 13–15], and in this chapter
Flavivirus (genus hepacivirus)	Hepatitis C virus (HCV)	(+) ssRNA	• Internal ribosome entry site (IRES) • G-quadruplex in the C (nucleocapsid) gene (p22)	Previously reviewed [4], and here reviewed here
Insect-borne flavivirus (arbovirus; genus flavivirus)	Dengue (DENV) West Nile (WNV) Yellow fever (YFV) Zika (ZIKV) Tick-borne encephalitis (TBEV)	(+) ssRNA	• 5′ UTR (including RNA promoter in stem-loop A, SLA; RNA long-range interacting stem-loop B, SLB) • Structured elements in the coding region (including capsid coding region hairpin, cHP; pseudoknot C1) • 3′ UTR (including RNA long-range interacting structures) • 3′ UTR-derived ncRNA (including subgenomic flavivirus RNA, sfRNA, compromising host defense)	None published yet
Coronavirus	Severe acute respiratory syndrome coronavirus (SARS CoV)	(+) ssRNA	• Frameshifting pseudoknot (PK)	Reviewed here
Orthomyxovirus	Influenza A virus	(−) ssRNA	• RNA promoter for the viral RNA-dependent RNA polymerase (RdRp)	Previously reviewed [16], and reviewed here
Filovirus	Ebola (EBOV) Marburg (MARV)	(−) ssRNA	• RNA promoter for the viral RNA-dependent RNA	None published yet

(continued)

Table 1 (continued)

Family	Virus	Genome	RNA target	Small molecule ligands
			polymerase (RdRp) • Structured intergenic regions (IGR) of the viral genome • 5′ and 3′ UTR in viral transcripts	
Herpesvirus	Kaposi's sarcoma associated herpes-virus (KSHV)	dsDNA	• IRES in the transcript for the viral homolog of the FLICE inhibitory protein (vFLIP) • Polyadenylated nuclear (PAN) non-coding RNA	None published yet
Hepadnavirus	Hepatitis B (HBV)	ds/ssDNA	• Encapsidation signal epsilon of viral pregenomic RNA (pgRNA)	None published yet

2 Viral RNA Targets

While many RNA virus genomes and viral transcripts contain structured and conserved noncoding elements, not all RNA motifs may be accessible to selective targeting with drug-like small molecule ligands. In the following, I will discuss previously validated and new prospective RNA targets of viruses along with their structural properties.

2.1 Human Immunodeficiency Virus

The (+) ssRNA genome of HIV contains multiple regulatory elements that play key roles in transcriptional regulation, reverse transcription, viral protein translation, nucleocytoplasmic transport, genome dimerization, and virion packaging [3]. The HIV transactivation response (TAR) and Rev response (RRE) regulatory elements were among the first non-ribosomal RNA targets investigated for the discovery of small molecule inhibitors [17–23]. Other potential HIV RNA targets for small molecule ligands include the dimer initiation sequence (DIS), the packaging signal (Ψ), and the Gag/Pol frameshift site (FSS). Three-dimensional structures have been determined for all HIV RNA regulatory elements by NMR and crystallography studies (Fig. 1). Previous efforts targeting HIV RNA have been reviewed comprehensively

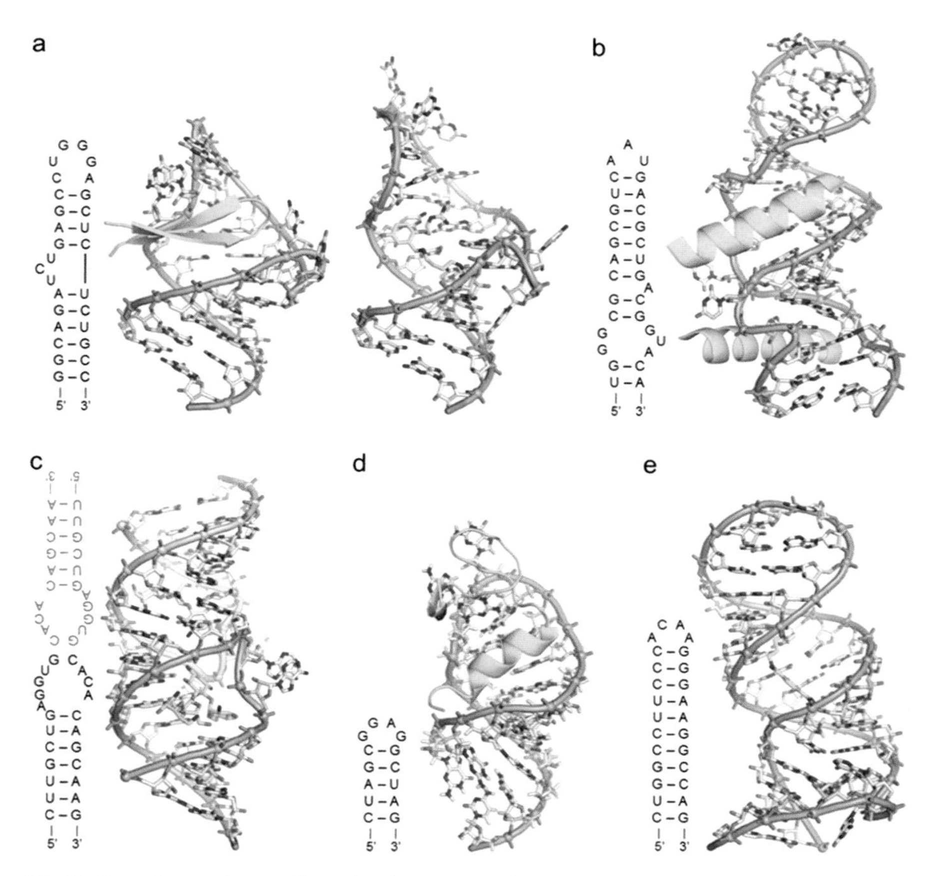

Fig. 1 Secondary and three-dimensional structures of HIV RNA elements which were previously explored as targets for small molecule ligands. Codes for atom coordinate files in the Protein Data Bank (PDB) are indicated. (**a**) The transactivation response (TAR) element in complex with a peptide mimetic of the Tat protein (PDB: 2KX5) [24] (*left*) and with a synthetic small molecule ligand (PDB: 1UUD) [25] (*right*). (**b**) Complex of the Rev response (RRE) RNA with Rev protein (PDB: 4PMI) [26]. (**c**) Kissing loop dimer of the dimer initiation sequence (DIS) in complex with the natural aminoglycoside neomycin (PDB: 2FCY) [27]. (**d**) Complex of the packaging signal (Ψ) stem-loop 3 (SL-3) with nucleocapsid protein (PDB: 1A1T) [28]. (**e**) The Gag/Pol frameshift site (FSS) in complex with a synthetic small molecule ligand (PDB: 2L94) [29]

(Table 1) [3, 10, 11, 13–15]. In Sect. 3, I will discuss more recent studies on discovery of inhibitors targeting the TAR and RRE RNA by screening and scaffold-based design.

Transcription of full-length HIV transcripts is stimulated by a complex mechanism that involves host cell factors and a complex of the viral Tat protein bound to the TAR element in the 5′ leader region of the virus genome [13]. Tat recognizes a conserved RNA stem-loop in TAR with a flexible pyrimidine-rich bulge which adopts a stable conformation in complex with the viral protein, peptide fragments, and small molecule ligands. Structures of TAR complexes determined by NMR revealed a relatively shallow and solvent-exposed ligand binding site in the widened RNA major groove, which in case of Tat-derived peptides extends to the terminal hairpin loop (Fig. 1a). Disruption of the Tat/TAR complex by competing RNA-binding ligands, including peptides, natural products such as aminoglycosides, and synthetic small molecules, blocks HIV replication [13, 18, 23].

Similarly, the viral Rev protein–RRE RNA complex has been extensively studied as a target for HIV inhibitors [3]. The RRE sequence of ~250 bases in the second intron of the viral RNA genome adopts a complex secondary structure which contains a stem-loop (SL-IIB) that serves as the binding site for Rev. Nucleocytoplasmic export of full-length and singly spliced viral transcripts depends on Rev binding to RRE. In contrast to Tat, which recognizes TAR RNA through a beta-sheet domain (Fig. 1a), the RNA binding of Rev is mediated by an alpha helix that inserts in a widened major groove at the purine-rich internal loop of RRE SL-IIB (Fig. 1b).

The packaging signal resides in the 5′ leader of the HIV genome, downstream of the TAR element, and directs selective packaging of unspliced viral RNA as a dimer into assembling virus particles. Genome dimerization initiates through kissing loop interaction between DIS hairpins and requires in addition the packaging signal (Ψ) stem-loop 3 (SL-3) which binds the viral nucleocapsid protein (NCp7). Both, the DIS and Ψ SL-3 have been explored as targets for ligands that affect viral genome packaging. Three-dimensional structures have been determined for the DIS kissing loop dimer in complexes with aminoglycoside ligands (Fig. 1c) and the Ψ SL-3 bound to NCp7 (Fig. 1d). The aminoglycoside binding site is located in the interface region between the kissing loops and resembles the structure of the internal loop of the bacterial ribosomal decoding site (A-site). The SL-3 interacts with NCp7 in the RNA major groove and residues of the loop. Recently, NMR analysis has revealed the three-dimensional structure of a 155-nucleotide region of the viral genome 5′ leader that contains the core encapsidation signal, including the Ψ SL-3 and DIS elements [30].

The Gag/Pol FSS regulates the transition of highly expressed HIV structural proteins to enzymes expressed at low levels by a programmed −1 frameshift during translation. Ribosomal frameshifting allows to maximize the coding content of viral genomes by giving access to overlapping reading frames [31]. Frameshifting depends on two distinct RNA motifs, including a slippery sequence for the reading frame change and a downstream motif whose relatively stable secondary structure stalls the ribosome. In HIV, an RNA hairpin with a long GC-rich stem serves as the frameshift motif. Ligands binding at the HIV FSS target may disrupt or stabilize the RNA hairpin and thereby affect the equilibrium between translation of structural and enzymatically functional viral proteins. The three-dimensional structure of the FFS RNA in complex with a synthetic compound has been determined by NMR, revealing ligand binding along the major groove of the hairpin stem (Fig. 1e).

2.2 *Hepatitis C Virus*

The HCV is a member of the genus hepacivirus in the flavivirus family. HCV proteins are translated by a cap-independent mechanism under the control of an internal ribosome entry site (IRES) in the 5′ untranslated region (UTR) of the viral (+) ssRNA genome. The HCV IRES adopts a structured fold comprised of four discrete domains which play key roles in the recruitment and assembly of host cell ribosomes. An RNA internal loop motif in subdomain IIa serves as a conformational switch during translation initiation and provides the binding site for selective inhibitors of viral translation. The small molecule ligands capture an extended conformation of the RNA switch and inhibit IRES-driven translation [32]. Discovery of the IRES binding HCV translation inhibitors and studies of their mechanism-of-action have been described in a comprehensive previous review [4]. Here, I will discuss the HCV IRES target (Fig. 2a) as well as a recently described G-quadruplex target in the C (nucleocapsid) gene. Progress in the discovery and characterization of HCV translation inhibitors will be outlined in Sect. 3.

The HCV IRES element recruits ribosomes to the translation start site of the viral genome, without the involvement of most canoncial eukaryotic initiation factors. Because of this crucial role for viral propagation and the high conservation of the IRES RNA sequence in clinical isolates, this ncRNA element has been recognized early as a potential drug target [35, 36]. Among the first inhibitors of IRES-driven translation described were phenazines [37] and biaryl guanidines [38] identified by high throughput screening against reporter translation in cells. IRES binding was not revealed in these studies but structural features of the two chemical series suggest that the compounds may interact with RNA. Screening for direct binding to the viral RNA was the basis of a high-throughput mass-spectrometry approach that identified 2-aminobenzimidazoles as ligands

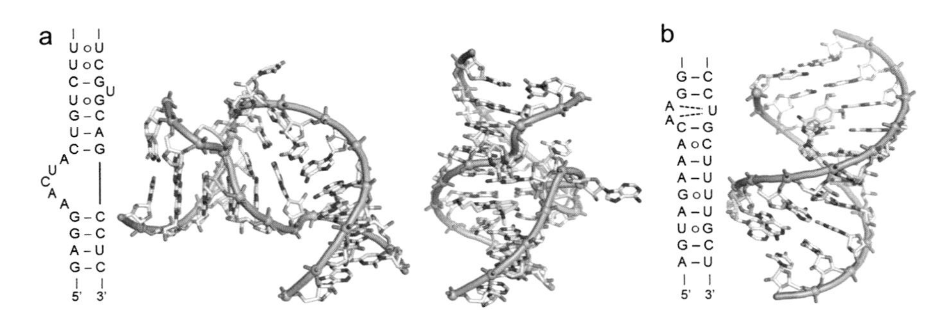

Fig. 2 Secondary and three-dimensional structures of RNA targets for small molecule inhibitors of HCV translation and influenza A virus replication. Codes for atom coordinate files in the PDB are indicated. (**a**) The HCV IRES subdomain IIa internal loop. Crystal structures have been determined for both, the free RNA and the target in complex with a benzimidazole translation inhibitor (*yellow sticks*) [33] (PDB: 1UUD) [25] (**b**) The influenza A virus RNA promoter. A three-dimensional model of a ligand–target complex was determined by NMR spectroscopy [34]. The ligand is shown in *yellow stick* representation. The added tetraloop is indicated in *grey* (PDB: 2LWK) [34]

of the subdomain IIa internal loop in the HCV IRES (Fig. 2a) [39]. Mechanism of action studies demonstrated that these compounds act as allosteric inhibitors of an RNA conformational switch [32]. Further investigations revealed that the HCV IRES subdomain IIa motif is the prototype of a new class of RNA conformational switches occurring in the IRES elements of flavi- and picornaviruses. Unlike traditional metabolite-sensing riboswitches, the viral RNA switches are structurally well-defined in both ligand-free and bound states and function as ligand-responsive, purely mechanical switches [40].

The structural signature of the IIa-like viral switches is an RNA internal loop flanked by two extended helices which adopt an overall bent conformation in the absence of bound ligand (Fig. 2a). The L-shaped fold provides a scaffold that directs the IRES subdomain IIa hairpin towards the ribosomal E site, at the interface of the small and large subunits. Crystal structure determination has provided insight into the conformational states of the HCV subdomain IIa switch in the absence [41] and presence [33] of ligands. It has been suggested that the 2-amino-benzimidazoles are fortuitous ligands of a guanosine binding site [42] which lock the subdomain IIa target in an extended conformation and thereby inhibit IRES function. In the RNA complex, the 2-aminobenzimidazole inhibitor binds in a deep solvent-excluded RNA pocket that resembles ligand interaction sites in aptamers and riboswitches (Fig. 2a) [33].

Recently, an RNA G-quadruplex (RG4) motif has been discovered in the HCV genome which may serve as a potential target for viral inhibitors [43]. A conserved guanine-rich sequence of the HCV core (C) nucleocapsid gene may transiently fold into an RG4 motif under physiological conditions. Porphyrin derivatives such as tetra-(N-methyl-4-pyridyl)porphyrin (TMPyP4) bind to the RG4 fold and stabilize the RNA motif sufficiently to inhibit viral replication and translation in HCV-infected cell culture [43]. While these findings support a potential RG4 motif in HCV as a new target for antivirals, a recent genome-wide study in yeast and human cells suggests that RG4 motifs are globally unfolded in eukaryotes, likely due to association with abundant single-stranded RNA-binding proteins [44]. However, it is conceivable that RG4-binding ligands may trap guanine-rich sequences in the quadruplex conformation and thereby affect biological processes.

2.3 *Influenza A Virus*

The (−) ssRNA genome of the influenza A virus contains eight protein-coding segments (vRNA) which are transcribed to mRNA and replicated to complementary sequences (cRNA). The viral replicase is an RNA-dependent RNA polymerase (RdRp) that recognizes a partial duplex motif [45] formed through hybridization of complementary sequences at the 5′ and 3′ end of each vRNA segment [46, 47]. Duplex formation between ends of segments leads to circularization of the vRNA and produces a promoter for transcription and replication [48]. NMR studies revealed the RNA promoter as an A-form duplex containing a noncanonical A•C base pair next to a uracil base that forms a bifurcated hydrogen bond interaction with two consecutive adenine residues in the opposite strand (Fig. 2b) [49]. These structural features induce widening of the RNA major groove in the promoter helix near the polymerase initiation site and may provide a selective recognition motif for small molecule inhibitors of influenza virus RNA replication. Ligands that interfere with replication by binding to the promoter structure would provide a novel route for the development of anti-influenza drugs. In Sect. 3, I will discuss recent studies of such RNA promoter-binding ligands [34].

2.4 *Severe Respiratory Syndrome Coronavirus*

In SARS CoV, the expression of viral replicase proteins such as the RdRp involves a −1 programmed frameshift during translation of the (+) ssRNA genome. Ribosomal frameshifting maximizes the coding content of the viral genome by regulating translation of overlapping reading frames [31]. In some RNA viruses such as SARS CoV and HIV, a −1 frameshift during translation enables a transition in the production of highly expressed structural proteins to viral enzymes expressed at low levels. Ribosomal frameshifting occurs at a slippery sequence and is triggered by a downstream structured RNA motif that stalls the ribosome. The frameshift in HIV translation is triggered by a stable RNA hairpin that has been explored as a target for ligands aimed at stabilizing or disrupting the RNA fold. These earlier efforts on targeting the HIV frameshift signal have been summarized in recent reviews [3, 15]. The SARS CoV frameshift motif is an RNA pseudoknot [50] which has recently been studied as a target for small molecule ligands that inhibit ribosomal frameshifting [51, 52]. Ligand discovery efforts will be discussed in Sect. 3.

2.5 *Insect-Borne Flaviviruses*

Insect-borne flaviviruses including West Nile, Dengue, and Zika viruses contain a (+) ssRNA genome. Unlike members of the hepacivirus family (e.g., HCV), these pathogens do not rely on an IRES element for translation but employ other structured RNA motifs for translational control, replication, and host defense suppression (Table 1) [53]. Conservation and structural features of flaviviral RNA elements suggest that they might be the viable targets for selective small molecule ligands interfering with the biological function of these RNAs. The best-studied motif among such elements is the replication promoter in the 5′ UTR of the Dengue virus (DENV) genome which recruits the viral RdRp [54]. During replication initiation, the viral RdRp binds at an RNA three-way junction, designated as stem-loop A (SLA), which comprises the first 70 nucleotides of the 5′ UTR [54, 55]. Replication of the DENV genome is preceded by circularization through complementary sequences in the 5′ and 3′ regions of the UTR and ORF, similar as in the influenza A virus. However, unlike in influenza A, circularization of the DENV genome does not involve the RNA promoter motif SLA [55]. The DENV SLA was discovered by secondary structure prediction and confirmed by enzymatic as well as chemical probing [56–58]. Key structural elements of the DENV SLA three-way junction are highly conserved in different serotypes and clinical isolates. Mutation studies demonstrated that structural and conformational integrity of the SLA is essential for the function of the RNA promoter [54, 56], which suggests that ligand binding at the RNA may achieve a similar inhibitory effect.

2.6 Filoviruses

The etiologic agents of Ebola and Marburg hemorrhagic fever are filoviruses which carry a (−) ssRNA genome. Similar as in the insect-borne flaviviruses and the influenza A virus, filovirus replication requires a structured RNA promoter but unlike in the former viruses the 5′ and 3′ termini of the filovirus genome do not interact during replication. Chemical probing has established an extended hairpin within the first 55 nucleotides of the filovirus genome which serves as the replication promoter in both EBOV and MARV [59, 60] and represents a potential target for small molecule inhibitors of viral replication. Other potential RNA targets in the filovirus genome occur in the long intergenic regions (IGR) which separate the reading frames for seven structural virus proteins. Specifically, the IGR between VP24 and VP30 has been proposed to adopt a two-armed stem-loop fold including an RNA four-way junction [61]. RNA hairpin motifs as potential targets in filoviruses are found at transcription start sites and the 5′ UTR of viral transcripts. A hairpin loop at the transcription start site binds the VP30 nucleocapsid protein which serves as an anti-termination factor [62, 63]. The 5′ UTR elements of viral transcripts derive from IGR and contain hairpin structures which have been proposed to regulate transcription and translation [64]. A recent bioinformatics analysis of the EBOV genome suggests that a conserved guanine-rich sequence within the L gene coding for the viral RdRp may fold into a G-quadruplex (RG4) that is stabilized by a porphyrin ligand (TMPyP4), similar to the RG4 motif discovered in the HCV core gene [65].

2.7 Kaposi-Sarcoma Associated Herpesvirus

Among DNA viruses that contain potential RNA targets for the development of antiviral drugs is the oncogenic Kaposi's sarcoma associated herpesvirus (KSHV) which causes malignancies in AIDS patients. KSHV-induced tumorigenesis involves the viral homolog of the FLICE inhibitory protein (vFLIP) [66]. Translation of this protein is driven by a structured RNA element that contains an IRES including a conserved segment of 252 nucleotides [67, 68]. The vFLIP IRES consists of an independently folding RNA core domain whose secondary structure has been determined by chemical probing and mutational analysis [69]. The IRES RNA, together with a series of hairpin motifs following immediately downstream, provides structurally well-defined sites for ligands that may suppress vFLIP expression. A second RNA target in KSHV for potential antiviral intervention with small molecule ligands is a conserved hairpin motif which acts as an enhancer of nuclear retention element (ENE) [70]. The ENE is a 79-nucleotide sequence in the 3′ terminus of the 1,077-nucleotide polyadenylated nuclear (PAN) ncRNA, which is the most abundant viral transcript during lytic KSHV replication [71]. PAN is an essential component required for viral propagation whose accumulation relies on posttranscriptional stabilization dependent on the cis-acting ENE RNA motif [70]. The ENE sequesters in cis the PAN poly(A) tail in an RNA triple helix that protects the ncRNA from decay and leads to PAN accumulation [72]. The ENE hairpin, which contains a U-rich internal loop, and the ENE–poly(A) triple helix complex are potential targets for small molecule ligands that may interfere with KSHV replication.

2.8 Hepatitis B Virus

The Hepatitis B virus (HBV) contains a DNA genome that is replicated through reverse transcription of an intermediate pregenomic RNA template (pgRNA) [73]. The HBV pgRNA is sequestered together with polymerase into subviral particles prior to reverse transcription. Initiation of reverse transcription requires a conserved sequence in the 5′ terminal region of the pgRNA which is also involved in virus encapsidation. The initiation and encapsidation motif adopts a stem loop structure with a uridine-rich internal loop referred to as the epsilon encapsidation signal [74–77]. Small molecule ligands of this RNA motif have not been reported yet, but RNA decoys of the epsilon sequence have been used to sequester reverse transcriptase, thereby providing proof-of-principle that disruption of the pgRNA–polymerase interaction suppresses HBV replication [78].

3 Ligands Targeting Viral RNAs

In the following, I will discuss recent progress on the discovery and design of small molecule ligands for RNA targets from viruses including HIV, HCV, influenza A, and SARS CoV, which were outlined in Sect. 2.

3.1 Human Immunodeficiency Virus

Among the RNA targets in HIV, the TAR element has been an early and primary focus for efforts to develop ligands that disrupt binding of the viral Tat protein [13, 18, 23]. Previously reported TAR-binding inhibitor ligands include synthetic molecules, natural products, and peptides whose discovery and design have been summarized in several previous reviews [3, 10, 11, 13, 14]. In a more recent study, small molecule microarray (SMM) screening of a TAR hairpin RNA conjugated with a fluorescent dye has been used to identify selective ligands from a library of ~20,000 drug-like immobilized synthetic molecules [79]. The thienopyridine derivative **1** (Fig. 3) was identified as a hit compound with a target affinity of 2.4 μM and anti-HIV activity in T-lymphoblasts (EC_{50} value of 12 μM). The ligand **1** represents a new and more drug-like chemotype compared to previously reported TAR binders, and lead candidates for the development of antiviral drugs may emerge from future improvement of similar thienopyridine derivatives.

In another recent effort to discover TAR-binding ligands, a fragment screen of 29 small molecules selected to represent molecular motifs beneficial for RNA recognition has been performed by applying a fluorimetric competition assay that measured ligand-induced displacement of a dye-labeled Tat peptide from a TAR complex [80]. The fragments were chosen to include hydrogen bond donors such as amines, guanidines, and amidines as well as aromatic rings to engage in stacking interactions. The most potent competitor ligands of the Tat–TAR interaction identified in the fragment screen were quinazoline derivatives (**2**; Fig. 3) which inhibited complex formation with IC_{50} values between 40 and 60 μM. Proton NMR spectroscopy confirmed the interaction of the quinazolines with the TAR RNA target as indicated by changes in imino-proton signals upon compound titration. While the ligands emerging from the fragment screening study were not tested for cellular activity, the quinazoline **2a**

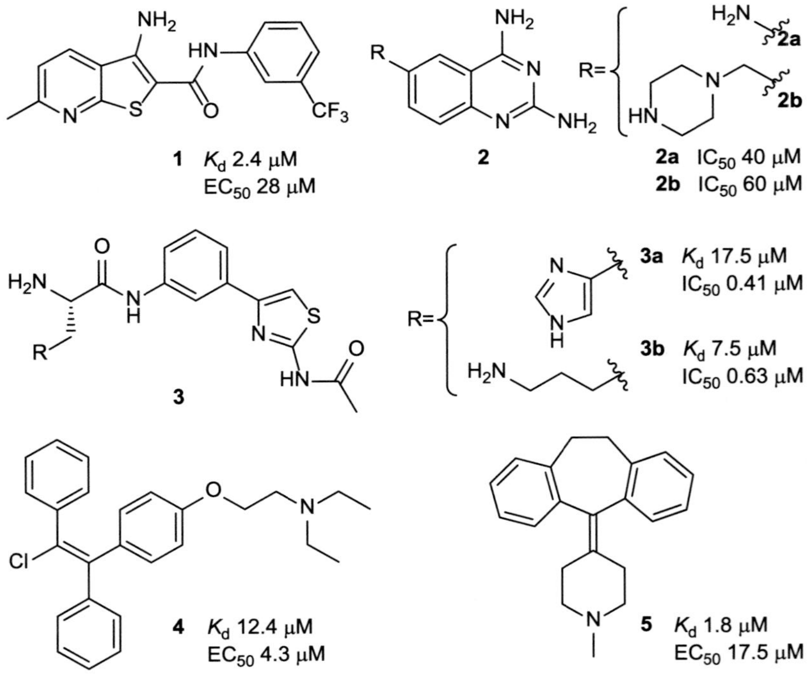

Fig. 3 Ligands for HIV TAR (**1–3**) and RRE (**4**, **5**) RNA targets

had previously been identified as an inhibitor of the Tat/TAR complex with biological activity to downregulate Tat transactivation in HIV-infected cells [23].

A previously reported approach of ligand discovery for the HIV TAR target focused on derivatives of amino-phenylthiazole (termed "S nucleobase") which had previously been developed as a scaffold designed to interact with A–U pairs through hydrogen bonding at the Hoogsteen edge of adenine [81, 82]. A set of 15 amino acid and dipeptide conjugates of the amino-phenylthiazole scaffold (**3**; Fig. 3) was tested for TAR target binding and antiviral activity in cell culture. While several derivatives showed binding to TAR in an assay that measured fluorescence changes upon compound titration to a terminally dye-conjugated RNA, only a histidine conjugate (**3a**; Fig. 3) was a selective ligand whose target interaction was not affected in the presence of competitor nucleic acids. A tighter binding lysine derivative (**3b**; Fig. 3) was compromised by promiscuous binding to other nucleic acids. Antiviral activity testing of the S nucleobase conjugates **3** in HIV-infected human cells resulted in IC_{50} values over tenfold lower than TAR binding affinity which suggests that these compounds may act also on targets other than TAR.

Structurally more complex ligands of TAR which have been reported recently include aminoglycoside-benzimidazole conjugates [83, 84] and nucleobase-linked aminoglycosides [85, 86] for which nanomolar affinity for the TAR RNA has been reported while antiviral activity in cells has not been tested yet.

For the HIV RRE–Rev complex target, inhibitors have mostly been explored by ligand-based design in the past, as summarized in previews reviews [10, 14], and two studies report small molecule high-throughput screens [87, 88]. However, these efforts have not resulted in confirmed inhibitors of the Rev–RRE complex that also showed antiviral activity in cells. Recent research suggests that post-transcriptional modification of HIV-1 RRE by N6-methylation of adenine bases in SL-IIB may play a key role in the activity of the RRE/Rev complex [89], indicating that authentic model systems are requisite for the study of RNA targets.

A binding competition screen for inhibitors of the RRE–Rev interaction has been used to identify inhibitors of HIV RNA biogenesis. Around 1,120 FDA-approved drugs were tested for the ability to block complex formation between the RRE SL-IIB RNA and a fluorescent dye-conjugated Rev peptide [90]. Two drugs, clomiphene and cyproheptadine (**4** and **5**; Fig. 3), were identified as inhibitors of HIV transcription that affected levels of spliced versus unspliced viral transcripts. It was shown that clomiphene (**4**), which is approved as a selective estrogen receptor modulator, bound to the RRE SL-IIB RNA with a K_d of 12.4 μM and had antiviral activity with an EC_{50} value of 4.3 μM in cells. Cyproheptadine (**5**), which is used as an antihistamine H1 receptor antagonist, bound the RRE RNA with a K_d of 1.8 μM and inhibited viral replication with an EC_{50} value of 17.5 μM. While the interaction of clomiphene (**4**) with the RRE RNA target was specific, target binding of cyproheptadine (**5**) was compromised in the presence of competitor nucleic acids. Interaction sites of the drugs **4** and **5** with the RRE target were investigated by NMR, revealing the G-rich internal loop in the lower stem of the SL-IIB RNA as the binding site. This region overlaps with the binding site of Rev, consistent with the proposed mechanism of inhibition by competition between the small molecule ligands and the viral protein. Interestingly, compounds **4** and **5** are quite hydrophobic and lack hydrogen bond donors which suggest that a large number of heteroatom hydrogen bond donors and acceptors are not required to confer RNA targeting properties to small molecule ligands.

3.2 *Hepatitis C Virus*

The HCV IRES subdomain IIa RNA was identified as a target for selective inhibitors of viral translation, as outlined above in Sect. 2.2, including 2-aminobenzimidazoles (**6**–**8**; Fig. 4) and diaminopiperidines (**10**; Fig. 4). The 2-aminobenzimidazole ligands, which were initially discovered in a high-throughput mass-spectrometry approach, were optimized for target binding by using structure–activity relationship data, resulting in inhibitors such as **6** (Fig. 4) which had an affinity of 0.9 μM (K_d) for the IRES target and showed anti-HCV activity in cell culture with an EC_{50} value of 3.9 μM [39]. Mechanism of action studies demonstrated that the 2-aminobenzimidazole compounds act as allosteric inhibitors of an RNA conformational switch in the subdomain IIa [32]. A FRET-based assay was developed to test compounds for the ability to bind and lock the conformation of subdomain IIa, leading to viral translation inhibition, and thereby identifying inhibitors that capture the IRES RNA switch in an extended state [32, 91]. Crystal structure analysis of the subdomain IIa target in complex with inhibitor **6a** revealed the ligand binding in a deep solvent-excluded pocket of the RNA [33]. Structural characteristics, depth, and complexity of the ligand binding pocket suggest that drug-like inhibitors may be developed that target this RNA as selective inhibitors of HCV translation.

A different fluorescence assay, which did not rely on FRET, was used to identify diaminopiperidines (**10**; Fig. 4) as ligands of the HCV IRES which lock the RNA conformational switch in a bent state and thereby inhibit viral translation initiation [92]. An abundance of polar groups renders the diaminopiperidines hydrophilic compounds whose binding affinity for the subdomain IIa RNA decreases in the presence of salt, including physiological concentrations of sodium or magnesium [92]. The discovery, optimization, structure, and mechanism of action studies of 2-aminobenzimidazole and diaminopiperidine HCV translation inhibitors have been comprehensively reviewed recently [4].

In attempts to optimize the synthesis of 2-aminobenzimidazoles such as inhibitor **6**, which required a lengthy route to construct the pyran ring, we designed second-generation ligands for the IRES subdomain IIa target. We synthesized N1-coupled aryl derivatives (**7**) in which sterical hindrance of the aryl substituent induces a non-planar conformation of the resulting compounds [93]. To address the basicity of the 2-aminobenzimidazole ligands, which increases the overall charge of the inhibitors under physiological conditions, we replaced the imidazole ring with the less basic oxazole ring to obtain compounds such as **8a** and **8b** [94]. Neither the N1-coupled aryl benzimidazoles (**7**) [93] nor the oxazoles (**8**) [94] had an affinity for the IIa RNA target better than the original 2-aminobenzimidazoles (Fig. 4).

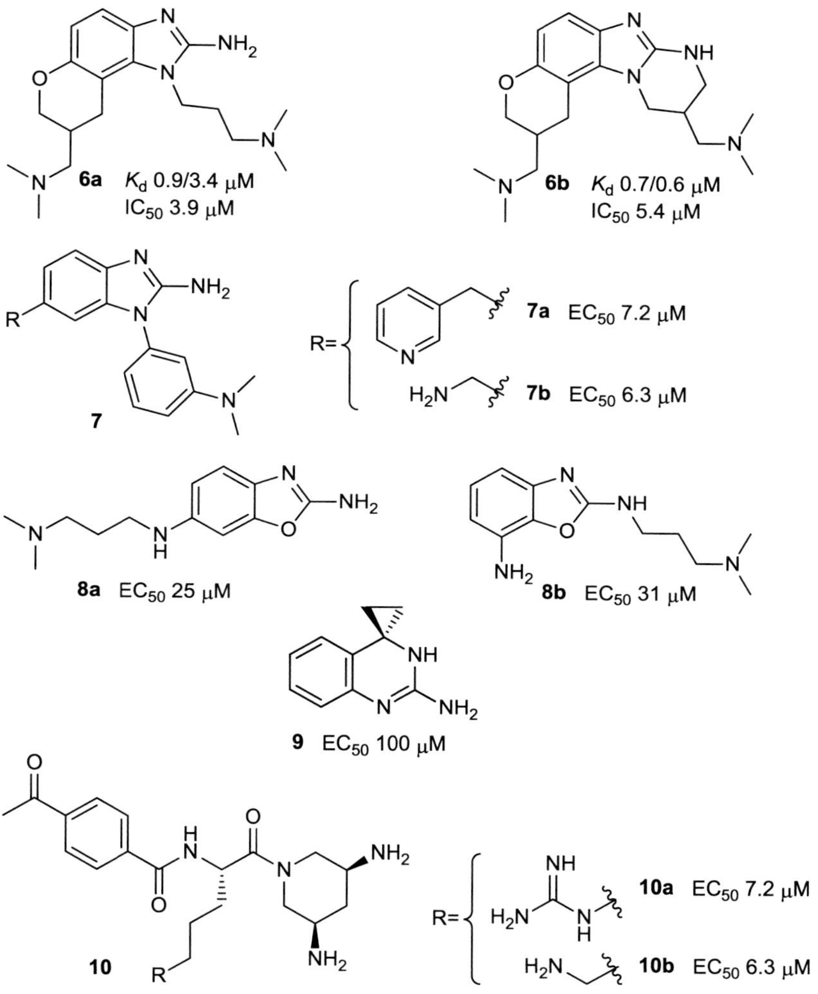

6a K_d 0.9/3.4 μM
IC_{50} 3.9 μM

6b K_d 0.7/0.6 μM
IC_{50} 5.4 μM

7 R=
7a EC_{50} 7.2 μM
7b EC_{50} 6.3 μM

8a EC_{50} 25 μM

8b EC_{50} 31 μM

9 EC_{50} 100 μM

10 R=
10a EC_{50} 7.2 μM
10b EC_{50} 6.3 μM

Fig. 4 Ligands for the HCV IRES subdomain IIa RNA target

Based on the finding that 2-aminobenzimidazoles are fortuitous ligands of a guanosine binding site in the subdomain IIa RNA switch, we explored amino-quinazoline derivatives as more drug-like scaffolds to develop ligands for the HCV IRES target. Closer analysis of the ligand binding site in the subdomain IIa led us to the design of the amino-quinazoline fragment **9** (Fig. 4) whose spiro-cyclopropyl modification targets a small pocket at the backside of the inhibitor interaction site [95]. While the fragment **9** showed only moderate binding affinity to the HCV IRES target, the positive impact of the hydrophobic spiro-cyclopropyl substituent on ligand binding suggests that inclusion of carefully placed nonpolar groups that improve shape complementarity is a promising strategy for optimization of compounds binding to RNA. Compared to the 2-aminobenzimidazole and oxazole compounds **6–8**, the fragment **9** stands out for the simplicity of synthesis which is achieved in only two steps from commercial starting material, thereby allowing straightforward preparation of more potent derivatives in the future.

Ligands for the recently described putative G-quadruplex (R4G) motif in a guanine-rich sequence of the HCV core gene include the porphyrin derivative **11** (TMPyP4) [96] and the pyridostatin derivative **12** (PDP) [97] (Fig. 5) which stabilize the R4G RNA fold sufficiently to inhibit viral replication and translation [43]. Porphyrins such as **11** have been used before to target DNA G-quadruplexes, for example, in telomeric regions and the c-MYC promoter [98–100] and were recently found to stabilize an RG4 motif in the EBOV genome [65]. Efficacy studies of TMPyP4 (**11**) in rodent xenograft tumor models revealed that despite the cationic nature of the porphyrin derivative, intraperitoneal administration of the compound resulted in systemic distribution and decreased tumor growth, presumably by action on the c-MYC G-quadruplex DNA target [96, 100].

In addition to small molecule ligands of the HCV IRES, copper-binding metallopeptides have been reported recently which bind at IRES domains and are proposed to inhibit viral translation by damaging the RNA through metal-catalyzed cleavage [101–103]. The IRES-targeting metallopeptides were 7–27 amino acids in length, including a Cu-binding Gly-Gly-His motif followed by an RNA-binding sequence, and inhibited HCV in cell culture with sub-micromolar activity [101–103].

3.3 *Influenza A Virus*

The RNA promoter motif, which provides the initiation site for the influenza A virus replicase, has been proposed as a target for ligands that inhibit viral replication. While aminoglycoside antibiotics were shown to bind the promoter RNA with micromolar affinity, the impact of these promiscuously RNA-binding natural products on replication

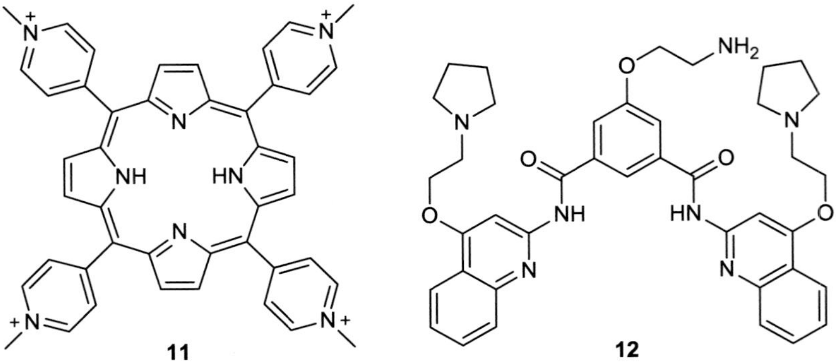

Fig. 5 Ligands for RNA G-quadruplex (RG4) targets, the porphyrin derivative TMPyP4 (**11**) and PDP (**12**)

was not reported [104]. In a recent study, NMR fragment screening of an oligonucleotide representing the RNA promoter against over 4,000 compounds identified the amino-quinazoline derivative **13a** (Fig. 6) as a selective ligand with a target affinity of 50 μM [34, 105]. Modeling of the RNA–ligand complex based on NMR NOE distance constraints suggested binding of the quinazoline ligand **13a** in the major groove at a motif including a bifurcated U < A/A motif (Fig. 2b). While the NMR model of the promoter complex shows the quinazoline **13a** interacting with the RNA target by close contacts of ligand methoxy substituents, it is not clear how much contribution to binding may be attributed to hydrogen bonds involving C–H donor groups which are weak and quite rare but not without precedent [106].

Antiviral activity of compound **13a** against different strains of influenza A was demonstrated by measuring inhibition of virus cytopathic effect, with the highest activity achieved on H1N1 with an EC_{50} value of 72 μM [34]. However, a cell-based viral replication assay returned the antiviral potency of **13a** corresponding to an EC_{50} value in the range of 430–550 μM [34, 105]. Synthesis of analogs derived from **13a** furnished compounds **13b** and **13c** which had slightly better binding affinity for the RNA promoter and improved activity as inhibitors of viral replication (Fig. 6) [105]. However, the investigators noted that direct inhibition of the viral RdRp may contribute to the antiviral activity of the quinazoline derivatives **13** [105].

3.4 Severe Respiratory Syndrome Coronavirus

An RNA pseudoknot in the genome of SARS CoV which triggers a −1 frameshift during translation and thereby enables the transition from production of structural proteins to viral enzymes has been proposed as a target for small molecule ligands that inhibit ribosomal frameshifting. A three-dimensional structure model of the RNA

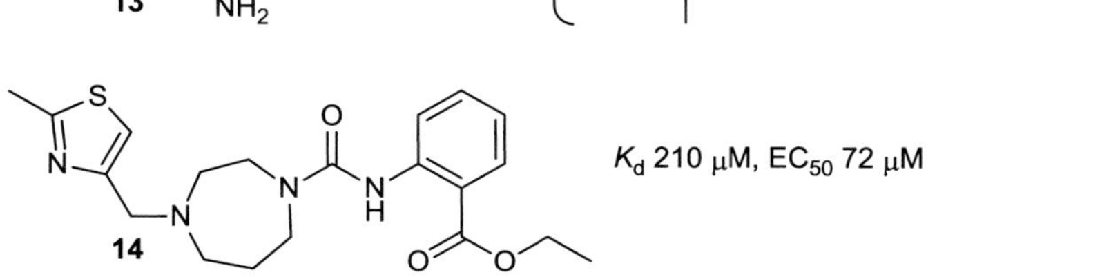

Fig. 6 Ligands of the SARS CoV frameshifting pseudoknot RNA (**13**) and the influenza A virus RNA promoter (**14**)

pseudoknot was used for in silico docking which identified the 1,4-diazepane **14** (Fig. 6) as a potential ligand [50]. Subsequent testing revealed **14** as an inhibitor of SARS CoV translational frameshifting in vitro and in virus-infected cells [50]. More recently, binding of **14** at the viral pseudoknot was confirmed by surface plasmon resonance (SPR), however with a relatively weak K_d of 210 μM [107]. Comparison of single-molecule unfolding of the SARS CoV pseudoknot RNA in the absence and presence of **14** suggested that ligand binding reduces the conformational plasticity of the RNA fold which, in turn, affects ribosomal frameshifting [108]. The ability of the RNA fold to adopt alternate conformations and structures are determinants of frameshifting efficiency rather than thermodynamic stability of the RNA fold or its impact on ribosomal pausing. Therefore, ligand-induced frameshifting modulation may only partially rely on stabilization of an RNA fold. Previously described inhibitors of HIV translational frameshifting may affect ribosome function through promiscuous RNA binding rather than by binding to the viral genomic frameshifting signal [15]. Similarly, the SARS CoV pseudoknot-binding ligand **14** may interact with other RNA targets as well, which may explain the over 450-fold higher potency of this compound as a frameshifting inhibitor in a cell-based assay [50] compared to its binding affinity for the pseudoknot RNA [107].

4 Summary

While viruses show high genetic variability, regulatory motifs in viral transcripts and RNA genomes are often conserved in clinical isolates and, therefore, may provide potential drug targets with a high barrier to resistance development. Development of small molecule inhibitors is challenging for structured RNA, however, as target drugability and ligand selectivity have to be carefully evaluated. RNA folds rarely contain deep and structurally rigid binding pockets which are the most promising targets for drug-like ligands. Among viral RNAs, such characteristics are most prominently found in the HCV IRES subdomain IIa which offers additional advantages in targeting as a switch motif whose conformational states may be affected by ligand binding in a deep RNA pocket. Similar well-defined ligand binding sites are present in bacterial riboswitches which have been explored as antibiotic targets [109, 110]. Just recently, a novel class of synthetic antibacterial compounds has been discovered, which exert their activity through an unprecedented mechanism of action that involves targeting a bacterial riboswitch involved in cofactor metabolism [111]. This success story of antibiotic discovery for a bacterial RNA target sets a promising precedent for ligand discovery directed at viral RNAs which provide future therapeutic opportunities defined by the targets' structural complexity, participation in key processes of infection as well as high conservation in the pathogens.

References

1. Hermann T (2005) Drugs targeting the ribosome. Curr Opin Struct Biol 15:355–366
2. McCoy LS, Xie Y, Tor Y (2011) Antibiotics that target protein synthesis. Wiley Interdiscip Rev RNA 2:209–232
3. Le Grice SF (2015) Targeting the HIV RNA genome: high-hanging fruit only needs a longer ladder. Curr Top Microbiol Immunol 389:147–169
4. Dibrov SM, Parsons J, Carnevali M, Zhou S, Rynearson KD, Ding K, Garcia Sega E, Brunn ND, Boerneke MA, Castaldi MP et al (2014) Hepatitis C virus translation inhibitors targeting the internal ribosomal entry site. J Med Chem 57:1694–1707
5. Gallego J, Varani G (2002) The hepatitis C virus internal ribosome-entry site: a new target for antiviral research. Biochem Soc Trans 30:140–145
6. Jubin R (2003) Targeting hepatitis C virus translation: stopping HCV where it starts. Curr Opin Investig Drugs 4:162–167
7. Hermann T, Westhof E (1998) RNA as a drug target: chemical, modelling, and evolutionary tools. Curr Opin Biotechnol 9:66–73
8. Hermann T (2000) Strategies for the design of drugs targeting RNA and RNA-protein complexes. Angew Chem Int Ed Engl 39:1890–1904
9. Gallego J, Varani G (2001) Targeting rna with small-molecule drugs: therapeutic promise and chemical challenges. Acc Chem Res 34:836–843
10. Thomas JR, Hergenrother PJ (2008) Targeting RNA with small molecules. Chem Rev 108:1171–1224
11. Guan L, Disney MD (2012) Recent advances in developing small molecules targeting RNA. ACS Chem Biol 7:73–86
12. Disney MD, Yildirim I, Childs-Disney JL (2014) Methods to enable the design of bioactive small molecules targeting RNA. Org Biomol Chem 12:1029–1039
13. Mousseau G, Mediouni S, Valente ST (2015) Targeting HIV transcription: the quest for a functional cure. Curr Top Microbiol Immunol 389:121–145
14. Blond A, Ennifar E, Tisne C, Micouin L (2014) The design of RNA binders: targeting the HIV replication cycle as a case study. ChemMedChem 9:1982–1996
15. Brakier-Gingras L, Charbonneau J, Butcher SE (2012) Targeting frameshifting in the human immunodeficiency virus. Expert Opin Ther Targets 16:249–258
16. Shortridge MD, Varani G (2015) Structure based approaches for targeting non-coding RNAs with small molecules. Curr Opin Struct Biol 30:79–88
17. Zapp ML, Stern S, Green MR (1993) Small molecules that selectively block RNA binding of HIV-1 Rev protein inhibit Rev function and viral production. Cell 74:969–978
18. Mei H-Y, Galan AA, Halim NS, Mack DP, Moreland DW, Sanders KB, Truong HN, Czarnik AW (1995) Inhibition of an HIV-1 Tat-derived peptide binding to TAR RNA by aminoglycoside antibiotics. Bioorg Med Chem Lett 5:2755–2760
19. Ratmeyer L, Zapp ML, Green MR, Vinayak R, Kumar A, Boykin DW, Wilson WD (1996) Inhibition of HIV-1 Rev-RRE interaction by diphenylfuran derivatives. Biochemistry 35:13689–13696
20. Park WKC, Auer M, Jaksche H, Wong C-H (1996) Rapid combinatorial synthesis of aminoglycoside antibiotic mimetics: use of a polyethylene glycol-linked amine and a neamine-derived aldehyde in multiple component condensation as a strategy for the discovery of new inhibitors of the HIV RNA Rev responsive element. J Am Chem Soc 118:10150–10155
21. Wang S, Huber PW, Cui M, Czarnik AW, Mei HY (1998) Binding of neomycin to the TAR element of HIV-1 RNA induces dissociation of Tat protein by an allosteric mechanism. Biochemistry 37:5549–5557
22. Mei HY, Cui M, Heldsinger A, Lemrow SM, Loo JA, Sannes-Lowery KA, Sharmeen L, Czarnik AW (1998) Inhibitors of protein-RNA complexation that target the RNA: specific recognition of human immunodeficiency virus type 1 TAR RNA by small organic molecules. Biochemistry 37:14204–14212

23. Mei HY, Mack DP, Galan AA, Halim NS, Heldsinger A, Loo JA, Moreland DW, Sannes-Lowery KA, Sharmeen L, Truong HN et al (1997) Discovery of selective, small-molecule inhibitors of RNA complexes – I. The Tat protein/TAR RNA complexes required for HIV-1 transcription. Bioorg Med Chem 5:1173–1184
24. Davidson A, Patora-Komisarska K, Robinson JA, Varani G (2011) Essential structural requirements for specific recognition of HIV TAR RNA by peptide mimetics of Tat protein. Nucleic Acids Res 39:248–256
25. Davis B, Afshar M, Varani G, Murchie AI, Karn J, Lentzen G, Drysdale M, Bower J, Potter AJ, Starkey ID et al (2004) Rational design of inhibitors of HIV-1 TAR RNA through the stabilisation of electrostatic "hot spots". J Mol Biol 336:343–356
26. Jayaraman B, Crosby DC, Homer C, Ribeiro I, Mavor D, Frankel AD (2014) RNA-directed remodeling of the HIV-1 protein Rev orchestrates assembly of the Rev-Rev response element complex. Elife 3:e04120
27. Ennifar E, Paillart JC, Bodlenner A, Walter P, Weibel JM, Aubertin AM, Pale P, Dumas P, Marquet R (2006) Targeting the dimerization initiation site of HIV-1 RNA with aminoglycosides: from crystal to cell. Nucleic Acids Res 34:2328–2339
28. De Guzman RN, Wu ZR, Stalling CC, Pappalardo L, Borer PN, Summers MF (1998) Structure of the HIV-1 nucleocapsid protein bound to the SL3 psi-RNA recognition element. Science 279:384–388
29. Marcheschi RJ, Tonelli M, Kumar A, Butcher SE (2011) Structure of the HIV-1 frameshift site RNA bound to a small molecule inhibitor of viral replication. ACS Chem Biol 6:857–864
30. Keane SC, Heng X, Lu K, Kharytonchyk S, Ramakrishnan V, Carter G, Barton S, Hosic A, Florwick A, Santos J et al (2015) RNA structure. Structure of the HIV-1 RNA packaging signal. Science 348:917–921
31. Dinman JD (2012) Mechanisms and implications of programmed translational frameshifting. Wiley Interdiscip Rev RNA 3:661–673
32. Parsons J, Castaldi MP, Dutta S, Dibrov SM, Wyles DL, Hermann T (2009) Conformational inhibition of the hepatitis C virus internal ribosome entry site RNA. Nat Chem Biol 5:823–825
33. Dibrov SM, Ding K, Brunn ND, Parker MA, Bergdahl BM, Wyles DL, Hermann T (2012) Structure of a hepatitis C virus RNA domain in complex with a translation inhibitor reveals a binding mode reminiscent of riboswitches. Proc Natl Acad Sci U S A 109:5223–5228
34. Lee MK, Bottini A, Kim M, Bardaro MF Jr, Zhang Z, Pellecchia M, Choi BS, Varani G (2014) A novel small-molecule binds to the influenza A virus RNA promoter and inhibits viral replication. Chem Commun (Camb) 50:368–370
35. Jubin R (2001) Hepatitis C IRES: translating translation into a therapeutic target. Curr Opin Mol Ther 3:278–287
36. Tan SL, Pause A, Shi Y, Sonenberg N (2002) Hepatitis C therapeutics: current status and emerging strategies. Nat Rev Drug Discov 1:867–881
37. Wang W, Preville P, Morin N, Mounir S, Cai W, Siddiqui MA (2000) Hepatitis C viral IRES inhibition by phenazine and phenazine-like molecules. Bioorg Med Chem Lett 10:1151–1154
38. Jefferson EA, Seth PP, Robinson DE, Winter DK, Miyaji A, Osgood SA, Swayze EE, Risen LM (2004) Biaryl guanidine inhibitors of in vitro HCV-IRES activity. Bioorg Med Chem Lett 14:5139–5143
39. Seth PP, Miyaji A, Jefferson EA, Sannes-Lowery KA, Osgood SA, Propp SS, Ranken R, Massire C, Sampath R, Ecker DJ et al (2005) SAR by MS: discovery of a new class of RNA-binding small molecules for the hepatitis C virus: internal ribosome entry site IIA subdomain. J Med Chem 48:7099–7102
40. Boerneke MA, Hermann T (2015) Ligand-responsive RNA mechanical switches. RNA Biol 12:780–786
41. Dibrov SM, Johnston-Cox H, Weng YH, Hermann T (2007) Functional architecture of HCV IRES domain II stabilized by divalent metal ions in the crystal and in solution. Angew Chem Int Ed Engl 46:226–229

42. Boerneke MA, Dibrov SM, Gu J, Wyles DL, Hermann T (2014) Functional conservation despite structural divergence in ligand-responsive RNA switches. Proc Natl Acad Sci U S A 111:15952–15957
43. Wang SR, Min YQ, Wang JQ, Liu CX, Fu BS, Wu F, Wu LY, Qiao ZX, Song YY, Xu GH et al (2016) A highly conserved G-rich consensus sequence in hepatitis C virus core gene represents a new anti-hepatitis C target. Sci Adv 2:e1501535
44. Guo JU, Bartel DP (2016) RNA G-quadruplexes are globally unfolded in eukaryotic cells and depleted in bacteria. Science 353
45. Pflug A, Guilligay D, Reich S, Cusack S (2014) Structure of influenza A polymerase bound to the viral RNA promoter. Nature 516:355–360
46. Fodor E, Pritlove DC, Brownlee GG (1994) The influenza virus panhandle is involved in the initiation of transcription. J Virol 68:4092–4096
47. Flick R, Neumann G, Hoffmann E, Neumeier E, Hobom G (1996) Promoter elements in the influenza vRNA terminal structure. RNA 2:1046–1057
48. Noble E, Mathews DH, Chen JL, Turner DH, Takimoto T, Kim B (2011) Biophysical analysis of influenza A virus RNA promoter at physiological temperatures. J Biol Chem 286:22965–22970
49. Bae SH, Cheong HK, Lee JH, Cheong C, Kainosho M, Choi BS (2001) Structural features of an influenza virus promoter and their implications for viral RNA synthesis. Proc Natl Acad Sci U S A 98:10602–10607
50. Park SJ, Kim YG, Park HJ (2011) Identification of RNA pseudoknot-binding ligand that inhibits the -1 ribosomal frameshifting of SARS-coronavirus by structure-based virtual screening. J Am Chem Soc 133:10094–10100
51. Plant EP, Perez-Alvarado GC, Jacobs JL, Mukhopadhyay B, Hennig M, Dinman JD (2005) A three-stemmed mRNA pseudoknot in the SARS coronavirus frameshift signal. PLoS Biol 3: e172
52. Su MC, Chang CT, Chu CH, Tsai CH, Chang KY (2005) An atypical RNA pseudoknot stimulator and an upstream attenuation signal for -1 ribosomal frameshifting of SARS coronavirus. Nucleic Acids Res 33:4265–4275
53. Villordo SM, Carballeda JM, Filomatori CV, Gamarnik AV (2016) RNA structure duplications and flavivirus host adaptation. Trends Microbiol 24(4):270–283
54. Filomatori CV, Iglesias NG, Villordo SM, Alvarez DE, Gamarnik AV (2011) RNA sequences and structures required for the recruitment and activity of the dengue virus polymerase. J Biol Chem 286:6929–6939
55. Gebhard LG, Filomatori CV, Gamarnik AV (2011) Functional RNA elements in the dengue virus genome. Viruses 3:1739–1756
56. Lodeiro MF, Filomatori CV, Gamarnik AV (2009) Structural and functional studies of the promoter element for dengue virus RNA replication. J Virol 83:993–1008
57. Sztuba-Solinska J, Le Grice SF (2014) Insights into secondary and tertiary interactions of dengue virus RNA by SHAPE. Methods Mol Biol 1138:225–239
58. Sztuba-Solinska J, Teramoto T, Rausch JW, Shapiro BA, Padmanabhan R, Le Grice SF (2013) Structural complexity of dengue virus untranslated regions: cis-acting RNA motifs and pseudoknot interactions modulating functionality of the viral genome. Nucleic Acids Res 41:5075–5089
59. Crary SM, Towner JS, Honig JE, Shoemaker TR, Nichol ST (2003) Analysis of the role of predicted RNA secondary structures in Ebola virus replication. Virology 306:210–218
60. Weik M, Enterlein S, Schlenz K, Muhlberger E (2005) The Ebola virus genomic replication promoter is bipartite and follows the rule of six. J Virol 79:10660–10671
61. Neumann G, Watanabe S, Kawaoka Y (2009) Characterization of Ebolavirus regulatory genomic regions. Virus Res 144:1–7
62. Weik M, Modrof J, Klenk HD, Becker S, Muhlberger E (2002) Ebola virus VP30-mediated transcription is regulated by RNA secondary structure formation. J Virol 76:8532–8539

63. Enterlein S, Schmidt KM, Schumann M, Conrad D, Krahling V, Olejnik J, Muhlberger E (2009) The Marburg virus 3′ noncoding region structurally and functionally differs from that of ebola virus. J Virol 83:4508–4519
64. Brauburger K, Boehmann Y, Krahling V, Muhlberger E (2015) Transcriptional regulation in Ebola virus: effects of gene border structure and regulatory elements on gene expression and polymerase scanning behavior. J Virol 90:1898–1909
65. Wang SR, Zhang QY, Wang JQ, Ge XY, Song YY, Wang YF, Li XD, Fu BS, Xu GH, Shu B et al (2016) Chemical targeting of a G-quadruplex RNA in the Ebola virus L gene. Cell Chem Biol 23:1113–1122
66. Guasparri I, Keller SA, Cesarman E (2004) KSHV vFLIP is essential for the survival of infected lymphoma cells. J Exp Med 199:993–1003
67. Bieleski L, Talbot SJ (2001) Kaposi's sarcoma-associated herpesvirus vCyclin open reading frame contains an internal ribosome entry site. J Virol 75:1864–1869
68. Bieleski L, Hindley C, Talbot SJ (2004) A polypyrimidine tract facilitates the expression of Kaposi's sarcoma-associated herpesvirus vFLIP through an internal ribosome entry site. J Gen Virol 85:615–620
69. Othman Z, Sulaiman MK, Willcocks MM, Ulryck N, Blackbourn DJ, Sargueil B, Roberts LO, Locker N (2014) Functional analysis of Kaposi's sarcoma-associated herpesvirus vFLIP expression reveals a new mode of IRES-mediated translation. RNA 20:1803–1814
70. Tycowski KT, Shu MD, Borah S, Shi M, Steitz JA (2012) Conservation of a triple-helix-forming RNA stability element in noncoding and genomic RNAs of diverse viruses. Cell Rep 2:26–32
71. Conrad NK (2016) New insights into the expression and functions of the Kaposi's sarcoma-associated herpesvirus long noncoding PAN RNA. Virus Res 212:53–63
72. Mitton-Fry RM, DeGregorio SJ, Wang J, Steitz TA, Steitz JA (2010) Poly(A) tail recognition by a viral RNA element through assembly of a triple helix. Science 330:1244–1247
73. Beck J, Nassal M (2007) Hepatitis B virus replication. World J Gastroenterol 13:48–64
74. Kramvis A, Kew MC (1998) Structure and function of the encapsidation signal of hepadnaviridae. J Viral Hepat 5:357–367
75. Jones SA, Boregowda R, Spratt TE, Hu J (2012) In vitro epsilon RNA-dependent protein priming activity of human hepatitis B virus polymerase. J Virol 86:5134–5150
76. Feng H, Chen P, Zhao F, Nassal M, Hu K (2013) Evidence for multiple distinct interactions between hepatitis B virus P protein and its cognate RNA encapsidation signal during initiation of reverse transcription. PLoS One 8:e72798
77. Cao F, Jones S, Li W, Cheng X, Hu Y, Hu J, Tavis JE (2014) Sequences in the terminal protein and reverse transcriptase domains of the hepatitis B virus polymerase contribute to RNA binding and encapsidation. J Viral Hepat 21:882–893
78. Feng H, Beck J, Nassal M, Hu KH (2011) A SELEX-screened aptamer of human hepatitis B virus RNA encapsidation signal suppresses viral replication. PLoS One 6:e27862
79. Sztuba-Solinska J, Shenoy SR, Gareiss P, Krumpe LR, Le Grice SF, O'Keefe BR, Schneekloth JS Jr (2014) Identification of biologically active, HIV TAR RNA-binding small molecules using small molecule microarrays. J Am Chem Soc 136:8402–8410
80. Zeiger M, Stark S, Kalden E, Ackermann B, Ferner J, Scheffer U, Shoja-Bazargani F, Erdel V, Schwalbe H, Gobel MW (2014) Fragment based search for small molecule inhibitors of HIV-1 Tat-TAR. Bioorg Med Chem Lett 24:5576–5580
81. Joly JP, Mata G, Eldin P, Briant L, Fontaine-Vive F, Duca M, Benhida R (2014) Artificial nucleobase-amino acid conjugates: a new class of TAR RNA binding agents. Chemistry 20:2071–2079
82. Duca M, Malnuit V, Barbault F, Benhida R (2010) Design of novel RNA ligands that bind stem-bulge HIV-1 TAR RNA. Chem Commun (Camb) 46:6162–6164
83. Ranjan N, Kumar S, Watkins D, Wang D, Appella DH, Arya DP (2013) Recognition of HIV-TAR RNA using neomycin-benzimidazole conjugates. Bioorg Med Chem Lett 23:5689–5693

84. Kumar S, Ranjan N, Kellish P, Gong C, Watkins D, Arya DP (2016) Multivalency in the recognition and antagonism of a HIV TAR RNA-TAT assembly using an aminoglycoside benzimidazole scaffold. Org Biomol Chem 14:2052–2056
85. Watanabe K, Katou T, Ikezawa Y, Yajima S, Shionoya H, Akagi T, Hamasaki K (2007) Nucleobase modified neamines, their synthesis and binding specificity for HIV TAR RNA. Nucleic Acids Symp Ser (Oxf):209–210
86. Inoue R, Watanabe K, Katou T, Ikezawa Y, Hamasaki K (2015) Nucleobase modified neamines with a lysine as a linker, their inhibition specificity for TAR-Tat derived from HIV-1. Bioorg Med Chem 23:2139–2147
87. Chapman RL, Stanley TB, Hazen R, Garvey EP (2002) Small molecule modulators of HIV Rev/Rev response element interaction identified by random screening. Antiviral Res 54:149–162
88. Shuck-Lee D, Chen FF, Willard R, Raman S, Ptak R, Hammarskjold ML, Rekosh D (2008) Heterocyclic compounds that inhibit Rev-RRE function and human immunodeficiency virus type 1 replication. Antimicrob Agents Chemother 52:3169–3179
89. Lichinchi G, Gao S, Saletore Y, Gonzalez GM, Bansal V, Wang Y, Mason CE, Rana TM (2016) Dynamics of the human and viral m6A RNA methylomes during HIV-1 infection of T cells. Nat Microbiol 1:16011
90. Prado S, Beltran M, Coiras M, Bedoya LM, Alcami J, Gallego J (2016) Bioavailable inhibitors of HIV-1 RNA biogenesis identified through a Rev-based screen. Biochem Pharmacol 107:14–28
91. Zhou S, Rynearson KD, Ding K, Brunn ND, Hermann T (2013) Screening for inhibitors of the hepatitis C virus internal ribosome entry site RNA. Bioorg Med Chem 21:6139–6144
92. Carnevali M, Parsons J, Wyles DL, Hermann T (2010) A modular approach to synthetic RNA binders of the hepatitis C virus internal ribosome entry site. Chembiochem 11:1364–1367
93. Ding K, Wang A, Boerneke MA, Dibrov SM, Hermann T (2014) Aryl-substituted aminobenzimidazoles targeting the hepatitis C virus internal ribosome entry site. Bioorg Med Chem Lett 24(14):3113–3117
94. Rynearson KD, Charrette B, Gabriel C, Moreno J, Boerneke MA, Dibrov SM, Hermann T (2014) 2-Aminobenzoxazole ligands of the hepatitis C virus internal ribosome entry site. Bioorg Med Chem Lett 24:3521–3525
95. Charrette BP, Boerneke MA, Hermann T (2016) Ligand optimization by improving shape complementarity at a hepatitis C virus RNA target. ACS Chem Biol 11(12):3263–3267
96. Grand CL, Han H, Munoz RM, Weitman S, Von Hoff DD, Hurley LH, Bearss DJ (2002) The cationic porphyrin TMPyP4 down-regulates c-MYC and human telomerase reverse transcriptase expression and inhibits tumor growth in vivo. Mol Cancer Ther 1:565–573
97. Muller S, Kumari S, Rodriguez R, Balasubramanian S (2010) Small-molecule-mediated G-quadruplex isolation from human cells. Nat Chem 2:1095–1098
98. Huppert JL (2008) Four-stranded nucleic acids: structure, function and targeting of G-quadruplexes. Chem Soc Rev 37:1375–1384
99. Parkinson GN, Ghosh R, Neidle S (2007) Structural basis for binding of porphyrin to human telomeres. Biochemistry 46:2390–2397
100. Hurley LH, Von Hoff DD, Siddiqui-Jain A, Yang D (2006) Drug targeting of the c-MYC promoter to repress gene expression via a G-quadruplex silencer element. Semin Oncol 33:498–512
101. Bradford S, Cowan JA (2012) Catalytic metallodrugs targeting HCV IRES RNA. Chem Commun (Camb) 48:3118–3120
102. Bradford SS, Ross MJ, Fidai I, Cowan JA (2014) Insight into the recognition, binding, and reactivity of catalytic metallodrugs targeting stem loop IIb of hepatitis C IRES RNA. ChemMedChem 9:1275–1285
103. Ross MJ, Bradford SS, Cowan JA (2015) Catalytic metallodrugs based on the LaR2C peptide target HCV SLIV IRES RNA. Dalton Trans 44:20972–20982
104. Kim H, Lee MK, Ko J, Park CJ, Kim M, Jeong Y, Hong S, Varani G, Choi BS (2012) Aminoglycoside antibiotics bind to the influenza A virus RNA promoter. Mol Biosyst 8:2857–2859

105. Bottini A, De SK, Wu B, Tang C, Varani G, Pellecchia M (2015) Targeting influenza A virus RNA promoter. Chem Biol Drug Des 86:663–673
106. Hermann T (2016) Small molecules targeting viral RNA. Wiley Interdiscip Rev RNA 7:726–743
107. Ritchie DB, Soong J, Sikkema WK, Woodside MT (2014) Anti-frameshifting ligand reduces the conformational plasticity of the SARS virus pseudoknot. J Am Chem Soc 136:2196–2199
108. Ritchie DB, Foster DA, Woodside MT (2012) Programmed -1 frameshifting efficiency correlates with RNA pseudoknot conformational plasticity, not resistance to mechanical unfolding. Proc Natl Acad Sci U S A 109:16167–16172
109. Blount KF, Breaker RR (2006) Riboswitches as antibacterial drug targets. Nat Biotechnol 24:1558–1564
110. Matzner D, Mayer G (2015) (Dis)similar analogues of riboswitch metabolites as antibacterial lead compounds. J Med Chem 58:3275–3286
111. Howe JA, Wang H, Fischmann TO, Balibar CJ, Xiao L, Galgoci AM, Malinverni JC, Mayhood T, Villafania A, Nahvi A et al (2015) Selective small-molecule inhibition of an RNA structural element. Nature 526:672–677

Top Med Chem (2018) 27: 135–176
DOI: 10.1007/7355_2017_12

Published online: 21 April 2017

Drugging Pre-mRNA Splicing

Matthew G. Woll, Nikolai A. Naryshkin, and Gary M. Karp

Abstract The splicing of precursor messenger RNA (pre-mRNA) requires the precise cleavage and formation of multiple phosphodiester bonds in nascent pre-mRNA polymers in order to produce a protein coding message that can be properly translated by the ribosome. Despite the precision of this process, the spliceosome maintains considerable flexibility to include, or not include, defined segments in the final message, thus allowing for the production of diverse transcripts with distinct functions from a single gene sequence. The combination of control and flexibility displayed by the spliceosome, in conjunction with input from *cis*-acting sequences and *trans* factors, presents a unique opportunity for molecular intervention during gene expression. Various chemical agents have the capacity to alter the natural process of pre-mRNA splicing, thereby producing levels of splicing products different than those found under natural conditions. This approach has powerful therapeutic utility where mutation has caused certain splice variants to be under- or over-represented. The following chapter highlights the exceptional advances that have been achieved recently in splicing modulation with splice switching oligonucleotides (SSOs) and small molecules, the two leading therapeutic modalities in this field.

Keywords Alternative splicing, Duchenne muscular dystrophy, Exon skipping, Familial dysautonomia, Pre-mRNA splicing, Spinal muscular atrophy, Splice switching oligonucleotides, Spliceosome

---

M.G. Woll (✉), N.A. Naryshkin, and G.M. Karp
PTC Therapeutics, 100 Corporate Ct, South Plainfield, NJ 07080, USA
e-mail: mwoll@ptcbio.com

Contents

1 Introduction 136
2 Exon Inclusion 141
2.1 SSO Approach to Treat Spinal Muscular Atrophy 142
2.2 Small Molecule Approach to Treat SMA 147
2.3 Small Molecule Approach to Treat FD 152
3 Exon Skipping 155
3.1 Treatment of Duchenne Muscular Dystrophy 156
3.2 Clinical Development of Exon Skipping SSOs 163
3.3 Small Molecules that Enhance Exon Skipping 165
4 Splicing Inhibition 166
4.1 Targeting SF3b to Inhibit Splicing 166
4.2 Spliceosome Stalling with Isoginkgetin 168
4.3 Spliceosome Stalling with Madrasin 169
5 Conclusion 170
References 171

1 Introduction

Eukaryotic genes are discontinuous, with protein coding sequences (expressed regions or exons) being disrupted by non-coding sequences (intragenic regions or introns) [1]. Before the protein coding message can be properly translated by the ribosome, the introns must be removed from the precursor messenger RNA (pre-mRNA) and the remaining exons must be ligated together into a contiguous coding sequence in a process known as pre-mRNA splicing. When reduced to its simplest form, the splicing process is merely two sequential transphosphoesterification reactions (Fig. 1). And although the transformation seems unimposing from a purely chemical standpoint, the idea of facilitating a ring closing event (step 1) between two atoms that may be thousands of bonds apart in sequence space, amongst numerous functional groups with similar reactivity seems unfathomable. It is no wonder that nature employs a very complex cellular machine, known as the spliceosome, to attain high selectivity in this process. The spliceosome achieves great precision by recognizing conserved pre-mRNA sequence elements. Several small nuclear ribonucleoproteins (snRNPs) play an important role in defining the locations for exon junctions, organizing the pre-mRNA and forming the catalytically active spliceosome (Fig. 2). Additionally, dozens of proteins participate in the various stages and contribute to the high selectivity of this remarkable catalytic process [2].

Despite the high precision of the spliceosome, very often there are clearly marked exon junctions that are passed over and ignored. This apparent sloppiness is actually an essential element of a highly regulated system. The presence of nuclear *trans*-acting factors and pre-mRNA sequence elements (*cis*-acting elements) can either encourage or discourage the splicing machinery to choose a particular exon junction. These sequence elements are termed exonic and intronic splicing enhancers (ESEs and ISEs) and silencers (ESSs and ISSs). Nearly all

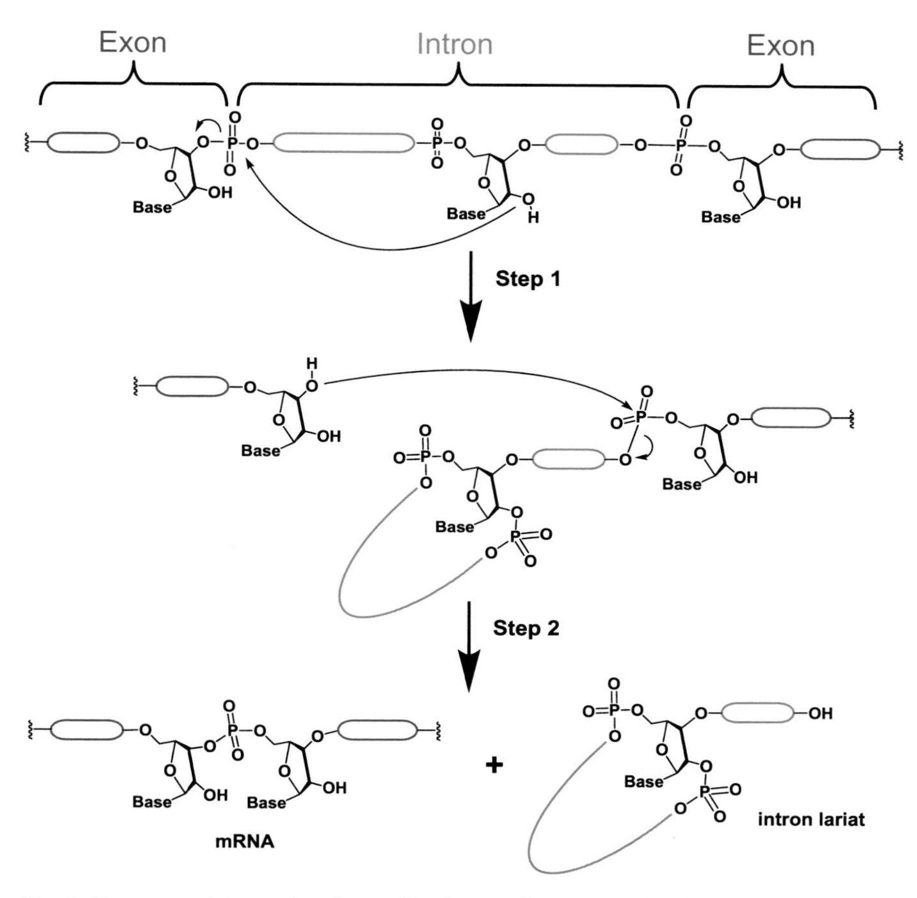

Fig. 1 Two sequential transphosphoesterification reactions are required for exon splicing

human genes that contain multiple exons can produce multiple distinct mRNA transcripts by alternative splicing [3]. Overall, the conceptual framework for mRNA production is relatively simple – the order of the exons is defined in the gene sequence. *Cis* and *trans* regulatory inputs determine the identity of exons (e.g., constitutive, cassette, or pseudo/cryptic) included in or excluded from the mature RNA transcript. Transcription start site selection and polyadenylation site selection can also be influenced by splicing control.

The flexibility of alternative splicing and the diversity of its outcomes are remarkable. It is thought to increase the diversity of the human proteome [4, 5] although typically a single major mRNA isoform is generated and alternative protein isoforms, expected to be expressed by minor alternative mRNA variants, are often undetectable. This coupled with the fact that the majority of alternative splicing lacks inter-species conservation have led to an on-going debate about the extent of splicing contribution to proteome complexity [6–8] and its role in gene evolution [9]. Nevertheless, many fundamental biological processes, such as

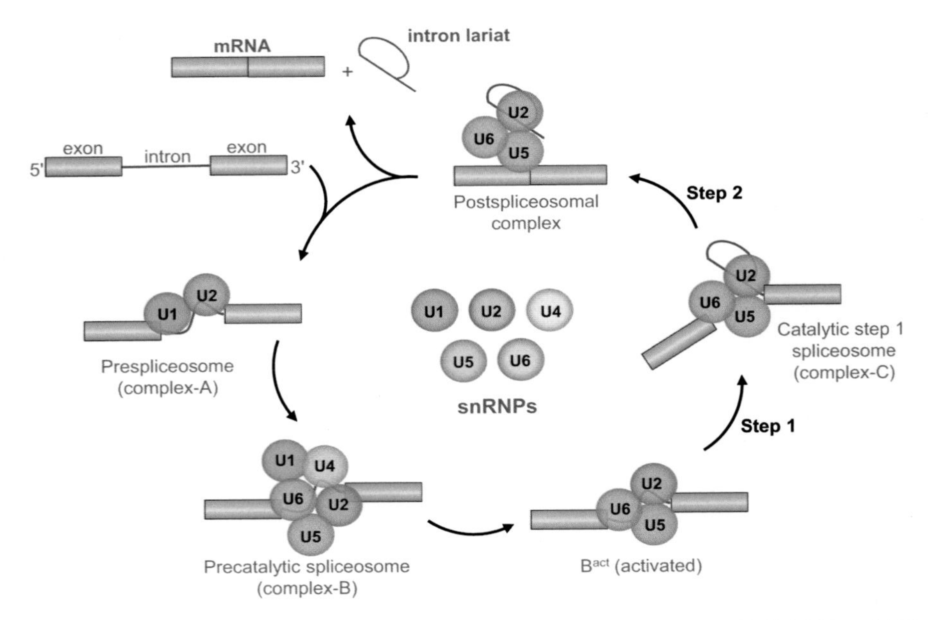

Fig. 2 The catalytic cycle for pre-mRNA splicing by the spliceosome. U1, U2, U4, U5, and U6 are snRNPs involved in the catalytic cycle and formation of the spliceosome. Many additional protein factors that are required for each of the complexes involved in the cycle are not shown for simplicity

organogenesis and development, apoptosis, signal transduction, cellular metabolism, and gene expression rely on alternative splicing to achieve a fast and sensitive switch in biological function (e.g., from a pro-apoptotic to an anti-apoptotic protein isoform) in response to developmental and environmental cues. For example, signal transduction via a single pass cell surface receptor can be attenuated by skipping the exon encoding the transmembrane domain (e.g., [10]); metabolic pathways can be modified by enzyme-inactivating exon skipping [11]; gene expression homeostasis can be maintained using a negative feedback loop that relies on an inclusion of a nonsense mediated decay-inducing "poison" exon [12].

The diversity of alternative outcomes also requires a precise control of pre-mRNA splicing to avoid potential deleterious effects. The dysregulation of splicing factors or gene mutations can lead to over- or under-representation of certain splice variants, which ultimately affects the level of proteins encoded by the corresponding sequences. Tissue-specific degeneration leading to disease can occur from the lack or overabundance of certain proteins required for normal tissue maintenance. The flexibility of pre-mRNA splicing also makes it a promising target for therapeutic intervention that may include switching between naturally occurring mRNA isoforms and the generation of novel isoforms with desired properties.

There are several potential scenarios where therapeutic modulation of splicing could be beneficial in disease therapy (also reviewed in [13, 14]). Abnormal exclusion or inclusion of a cassette exon could lead to overexpression of an undesired mRNA isoform. This increased skipping of an essential exon can be the result of an intronic or translationally synonymous exonic mutation. An external stimulus that shifts splicing toward the inclusion of the cassette exon could be therapeutically beneficial (Fig. 3a). In another case, the overexpression of an isoform produced from a mutually exclusive splicing event may have detrimental effects on the cell. Treatment with a compound that shifts the splicing, thereby decreasing the expression of the overexpressed isoform would be desirable (Fig. 3b). In cases where an open reading frame is disrupted by an internal deletion or (in some cases) insertion, a stimulus that induces skipping of the exon flanking the mutation locus may restore the protein-encoding open reading frame and preserve protein function (Fig. 3c). An exon skipping approach could also be used in cases where a point mutation that disrupts protein function is present in an internal exon, provided that the resulting internally truncated mRNA maintains the protein-encoding open reading frame and that the expressed truncated protein maintains proper function (Fig. 3d). Finally, and in contrast to the previous examples, a splice altering stimulation may be used to create a faulty mRNA to reduce the expression level of a toxic or undesired mRNA or protein by generating a frameshifted message that undergoes subsequent degradation via nonsense-mediated decay or a related mRNA surveillance process and/or degradation of protein due to the presence of an unnatural C-terminus (Fig. 3e).

Several important questions arise when thinking about targeting pre-mRNA splicing as a means to treat human disease. What are the possible biomolecular targets for splicing modulation? How can selectivity be achieved? What are the practical limits for shifting an alternative splicing event? Therapeutics that target pre-mRNA splicing can be broadly classified into two groups: those that encourage the formation of a splice junction and those that discourage the formation of a splice junction. The practical output of these methods is to favor the expression of the splice variant that would be beneficial to address a particular disease.

Conceptually there are several conceivable modes of action for a molecule to modify pre-mRNA splicing. A drug may bind to a spliceosomal component or *trans* regulatory element to inhibit its interaction with pre-mRNA (*trans* blocker, Fig. 4a). The same biological outcome may result from a drug that directly engages the pre-mRNA, thereby blocking the binding of a spliceosomal component or *trans* regulatory element (*cis* blocker, Fig. 4b). Type B would also cover drug–RNA interactions that disrupt (or stabilize) a pre-mRNA structure that serves as a *cis* regulatory element. A type B scenario has the potential for greater specificity than type A if the drug has the ability to distinguish different pre-mRNAs by sequence. In another scenario, a drug may indirectly influence a spliceosomal or *trans* regulatory element by targeting upstream transcription factors, kinases, phosphatases, etc. (upstream regulator, Fig. 4c). In this case, no direct contacts are made with pre-mRNA, thus limiting the potential for high specificity. Lastly, there is the potential for a molecule to stabilize or activate a complex composed of a pre-mRNA transcript bound to a spliceosomal component or *trans* regulatory element (complex stabilizer, Fig. 4d). In this last scenario, since pre-mRNA is directly engaged there is potential for pre-mRNA sequence specificity. While molecules

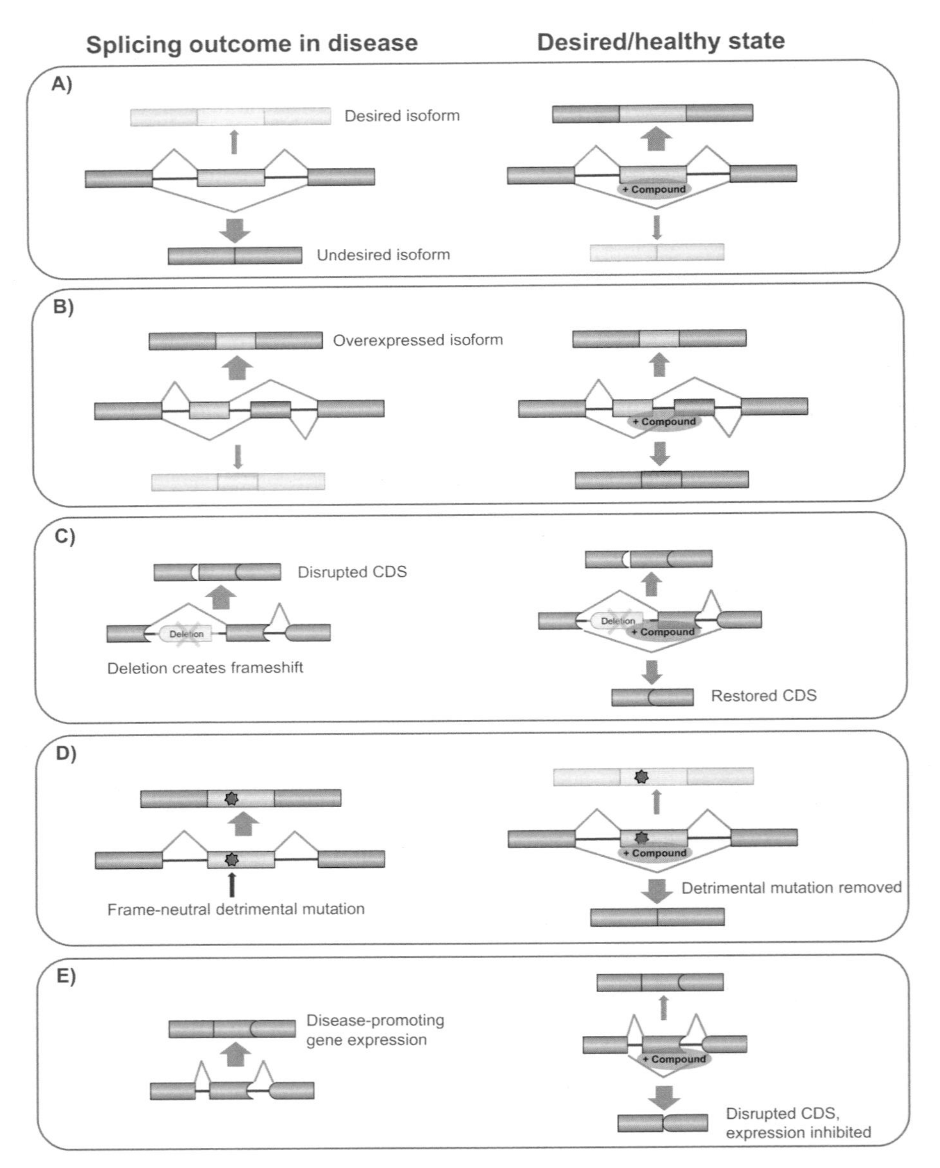

Fig. 3 Potentially druggable splicing events. (**a**) Inclusion (or exclusion) of a cassette exon. (**b**) Switch between mutually exclusive exons. (**c**) Exon skipping to restore reading frame. (**d**) Exon skipping to remove mutation. (**e**) Exon skipping to inhibit expression

that act by each of the described modes may elicit profound biological effects, the advancement of a molecule into the clinic will require a suitable therapeutic index, which is often directly linked to specificity. As is the case with any therapy that

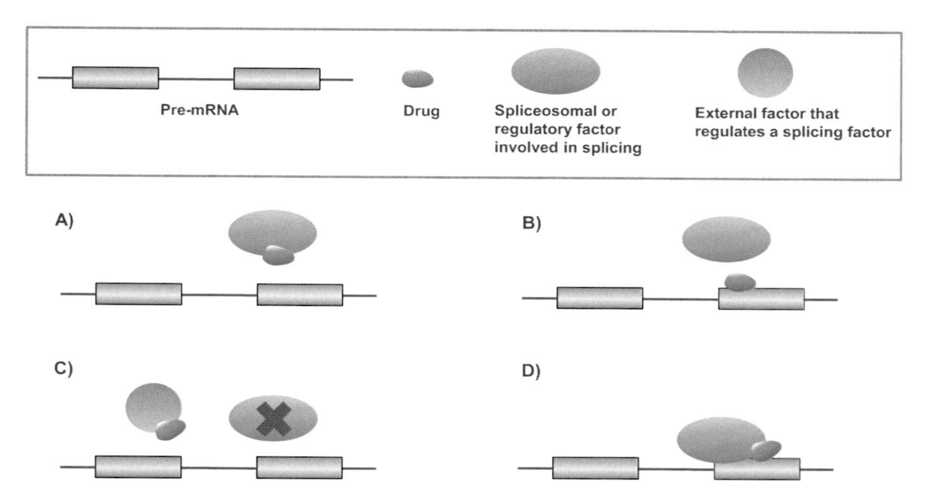

Fig. 4 Conceptual modes of action for drugs targeting pre-mRNA splicing. (**a**) *Trans* blocker. (**b**) *Cis* blocker. (**c**) Upstream regulator. (**d**) Stabilizer

modifies gene expression levels, the greater the number of off-target alterations, the more likely the chance of promoting undesired side effects.

In the last few years, novel therapeutic approaches utilizing different treatment modalities, including splice switching oligonucleotides (SSOs) and small molecule therapeutics, have targeted diseases caused by known splicing defects. New chemical entities (including SSOs and small molecules) targeting several distinct disease areas have been tested in human clinical studies. In 2016, two splice-switching oligonucleotides received regulatory approval. EXONDYS 51™ (eteplirsen), an SSO that induces the skipping of exon 51 of the *DMD* gene, was approved (under the accelerated approval mechanism) by the FDA for treatment of Duchenne muscular dystrophy in patients who have a confirmed mutation of the *DMD* gene that is amenable to exon 51 skipping (See [15]). Spinraza™ (nusinersen), an SSO that promotes the inclusion of exon 7 of the *SMN2* gene, was approved for the treatment of spinal muscular atrophy (SMA) in pediatric and adult patients (See [16]). Recent examples of methods used to promote exon inclusion induce exon skipping or inhibit some general aspect of splicing are described in the following sections.

2 Exon Inclusion

In some human diseases the underlying cause of disease can be directly linked to the reduced expression of a critical protein due to an impaired splicing event. Indeed, a single translationally synonymous exonic point mutation or intronic point mutation could lead to excessive skipping of an exon that is essential for the translation of fully functional protein product. Since some, albeit insufficient, desired isoform is produced, a reasonable and rational approach to treat the disease would require the discovery of molecules that could take advantage of the mechanisms that impart spliceosomal flexibility to induce exon inclusion and hopefully ameliorate disease phenotype. Some exciting advances using this type of methodology are found in this section.

2.1 SSO Approach to Treat Spinal Muscular Atrophy

Patients with spinal muscular atrophy (SMA) have a deficit of the survival motor neuron (SMN) protein, which causes degeneration of α-motor neurons [17, 18]. This deficiency is almost always the result of a homozygous deletion of or mutation in the *SMN1* gene. Humans usually have two or more copies of the paralogous *SMN2* gene, but this gene produces much less functional SMN protein due to alternative splicing at exon 7. This alternative pathway is activated by a translationally synonymous C to T mutation at position 6 in exon 7. The seemingly innocuous mutation abrogates an exonic splicing enhancer sequence and creates an inhibitory motif. As a result, the majority of transcripts produced from *SMN2* lack exon 7 (*SMN2*-Δ7), and encode a truncated *SMN*-Δ7 protein that is rapidly degraded [19, 20]. Because *SMN2* contains all of the necessary coding information for wild type SMN protein, but just lacks the correct splicing impetus, any treatment that has the potential to shift the splicing of *SMN2* pre-mRNA to favor the desired full length transcript (*SMN2*-FL) should hold considerable promise for treating SMA.

One approach to modify a cellular splicing event is to treat cells with a splice switching oligonucleotide (SSO) that is complementary to a regulatory sequence within the pre-mRNA. The selective binding event that occurs between a specific pre-mRNA sequence and the SSO could have various consequences, depending on the location of the sequence within the pre-mRNA. To enhance exon inclusion, one might target an ESS or ISS proximal to the exon of interest. Another approach would be to slow a competing splicing event by binding to the competing exon junction. Recruiting splicing enhancers to a splicing region using a bifunctional SSO that contains a pre-mRNA binding motif and a splicing enhancer motif would be another means of influencing splicing events.

All of the above-mentioned techniques have been successfully employed to shift *SMN2* splicing in favor of the *SMN2*-FL transcript (Fig. 5). SSOs directed to repressor regions within intron 6 (element 1) [21], exon 7 (two ESSs) [22], and intron 7 (ISS-N1) [23, 24] have demonstrated *SMN2* splicing correction in patient-derived cells. Additionally, SSOs that bind to the intron 7–exon 8 junction have been shown to discourage splicing at the 3′ splice site of exon 8 and improve the ratio of *SMN2*-FL to *SMN2*-Δ7 [25]. In some cases a bifunctional oligomer, incorporating both a pre-mRNA binding motif and a splicing enhancer/repressor motif, has shown improved activity over the SSO alone [26, 27]. An even more elaborate example of a bifunctional activator of exon 7 inclusion is a U7 snRNA derivative delivered using a viral vector [28]. To date, SSOs targeting ISS-N1 have

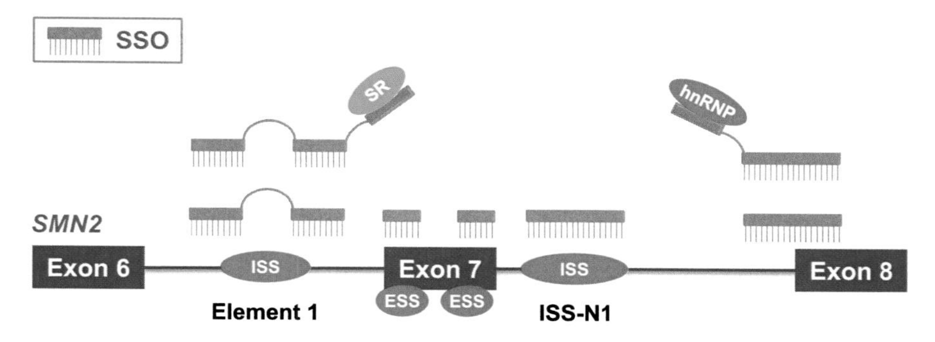

Fig. 5 Strategies for using SSOs to enhance exon 7 inclusion in *SMN2*

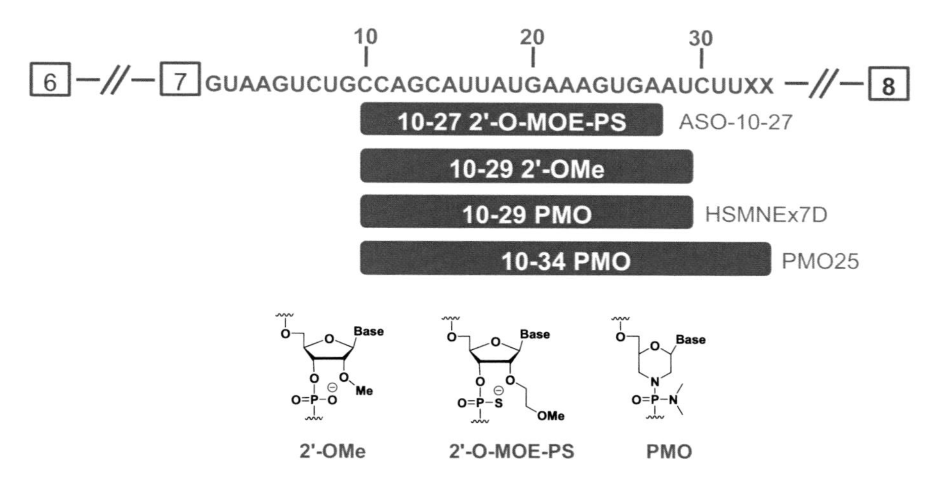

Fig. 6 SSOs targeting the ISS-N1 region of intron 7 of *SMN2*

had the most profound effect on splicing in vitro, and have shown the greatest therapeutic potential in animal models and SMA patients.

SSOs with varying length and backbone chemistry have demonstrated pharmacodynamic responses in SMA mouse models (Fig. 6) [23, 29–32]. The long-term benefit to motor function and survival depend on the delivery method and distribution of the SSO. An SSO complementary to positions 10–27 (targeting the ISS-N1 region) of intron 7 with 2′-*O*-methoxyethyl ribose phosphorothioate (2′-*O*-MOE-PS) modified backbone (ASO-10-27) was administered to *hSMN2* mice (mild SMA phenotype) intravenously twice a week at a dose of 25 mg/kg [23]. Tissues were harvested after 1, 2, 3, and 4 weeks. An increase in exon 7 inclusion was observed in the liver, kidney, and thigh muscle, but not in the spinal cord, suggesting that the SSO did not cross the blood–brain barrier (BBB).

Methodology for injecting SSOs intracerebroventricularly (ICV) to neonatal Δ7 SMA mice (severe SMA phenotype) was developed using a 2′-*O*-methyl ribose (2′-OMe) phosphodiester SSO spanning positions 10–29 of *SMN2* intron 7 (Fig. 6) [24, 29]. After injecting 1 μg SSO at post-natal days 1 (P1), 3, 5, 7, and 10, CNS tissue was harvested on P12. Western blot analysis indicated that SMN protein was increased greater than twofold in brain and spinal cord, with SMN levels in cervical spinal cord reaching as high as 3.6-fold. Analysis of the brain tissue using semi-quantitative reverse-transcription real-time polymerase chain reaction (RT-qPCR) indicated that *SMN2*-FL had significantly increased by ~1.4-fold in the brain. Modest improvement in body weight and righting reflex were observed, however the study was terminated at P12, thus no long-term benefits including survival were reported.

A subsequent study in a mouse model of severe SMA (median survival time 10 days) [33] demonstrated that ICV injection of 20 μg of ASO-10-27 on P1 produced significant increase of *SMN2*-FL and SMN protein in the brain (Fig. 7b, c), but a very modest increase in median survival time (MST) (16 vs. 10 d for untreated) (Fig. 7a) [30]. Strikingly, when the mice were administered 50 μg/g of ASO-10-27 by subcutaneous (SC) injections at P0 and P3 they had a median survival time of 108 days. Combining the ICV and SC injections further increased the median survival time to 173 days (Fig. 7d). Additional studies showed that increasing the SC dose at P0 and P3 to 160 μg/g increased the median survival time to 248 days, with some mice living longer than 500 days (Fig. 7e).

The SC injections increased the SMN levels not only in peripheral tissues (kidney, liver, muscle, and heart), but in the CNS (spinal cord and brain) as well. When the *SMN2* FL/Δ7 mRNA ratio was measured over time in the liver, a significant decrease in exon 7 inclusion was seen at P30 as compared to P10 after the 160 μg SC dose. Interestingly, exon 7 inclusion at P180 was still significantly higher than that observed for vehicle treated mice at P10. Since all of the vehicle treated mice die very early (<20 days), there is no direct vehicle comparison for P180. It is not entirely clear how much CNS penetration of the SSO is required for survival benefit. The SSO clearly achieves CNS penetration in mice after SC dosing at P0 and P3 when the blood–brain barrier is not completely developed. SC dosing at P5 and P7, when the BBB is more mature, produces a very modest survival benefit (Fig. 7d). The diminished benefit could also be rationalized by the need to treat these severe mice very early in their developmental cycle. Nonetheless, it seems clear that in this model system an increase of SMN in the CNS and periphery is preferable to SMN increase in the CNS alone.

Consistent with the dramatic increase in survival, histological examination of the mice treated SC with 160 μg/g of ASO-10-27 revealed α-motor neuron counts and neuromuscular junction integrity comparable to heterozygous littermates (Fig. 8a, b). Additionally, muscle fiber mean area cross sections and heart weight were nearly normalized for treated mice (Fig. 8c, d). Muscle strength, balance, and coordination were evaluated using the rotarod test in 3-month-old mice treated with 40, 80, or 160 μg/g ASO-10-27. The mice in the 80 and 160 μg/g groups performed better than the 40 μg/g group, but not as well as heterozygotes (Fig. 8e, f).

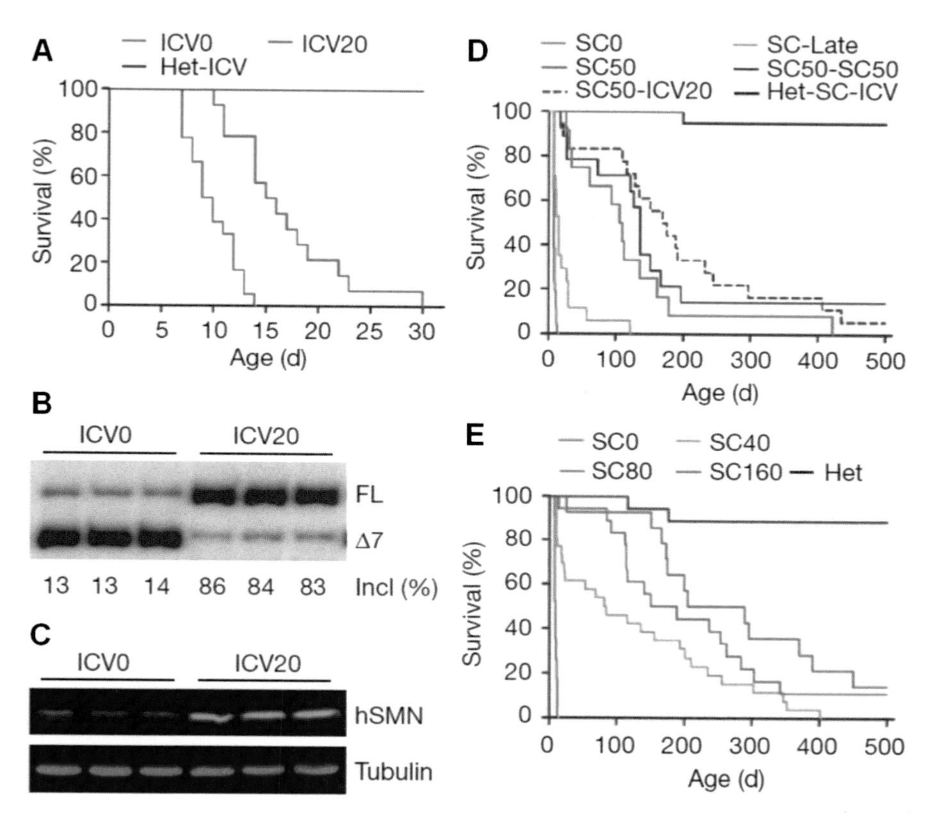

Fig. 7 (**a**) Survival curves for mice after ICV administration of 20 μg ASO-10-27 or vehicle on P1. (**b**) RT-PCR detecting *SMN*-FL and *SMN*-Δ7 transcripts in spinal cord tissue analyzed on P7 after ICV administration of 20 μg ASO-10-27 at P1. (**c**) Western blot detecting human SMN (hSMN) in spinal cord tissue analyzed on P7 after ICV administration of 20 μg ASO-10-27 at P1. (**d**) Survival curves after SC administration of saline (SC0) or 50 μg/g ASO-10-27 (SC50) twice between P0 and P3. SC50-SC50 mice received two additional SC injections on P5 and P7. Het-SC-ICV and SC50-ICV20 were heterozygous and SMA mice, respectively, that received combined P1 ICV and P0–P3 SC injections. SC-Late were SMA mice that received only two SC injections, on P5 and P7. (**e**) Dose-escalating SC injections at P0 and P3 with 40 (SC40), 80 (SC80), or 160 (SC160) μg/g of ASO-10-27. Saline-treated SMA (SC0) or heterozygous mice (Het) served as controls. Adapted from [30]. Copyright 2011 Macmillan Publishers Ltd

Superior in vivo efficacy can be achieved by switching from 2′-*O*-MOE-PS oligomers to phosphorodiamidate morpholino oligonucleotides (PMOs). A PMO targeting positions 10–29 of intron 7 (Fig. 6, HSMNEx7D), masking ISS-N1, showed long-term survival benefit in Δ7 mice after a single ICV injection (MST >100 days) [31, 32]. This is in contrast to the minimal survival benefit achieved with an ICV injection of 2′-*O*-MOE-PS oligomers of similar length [34]. When the

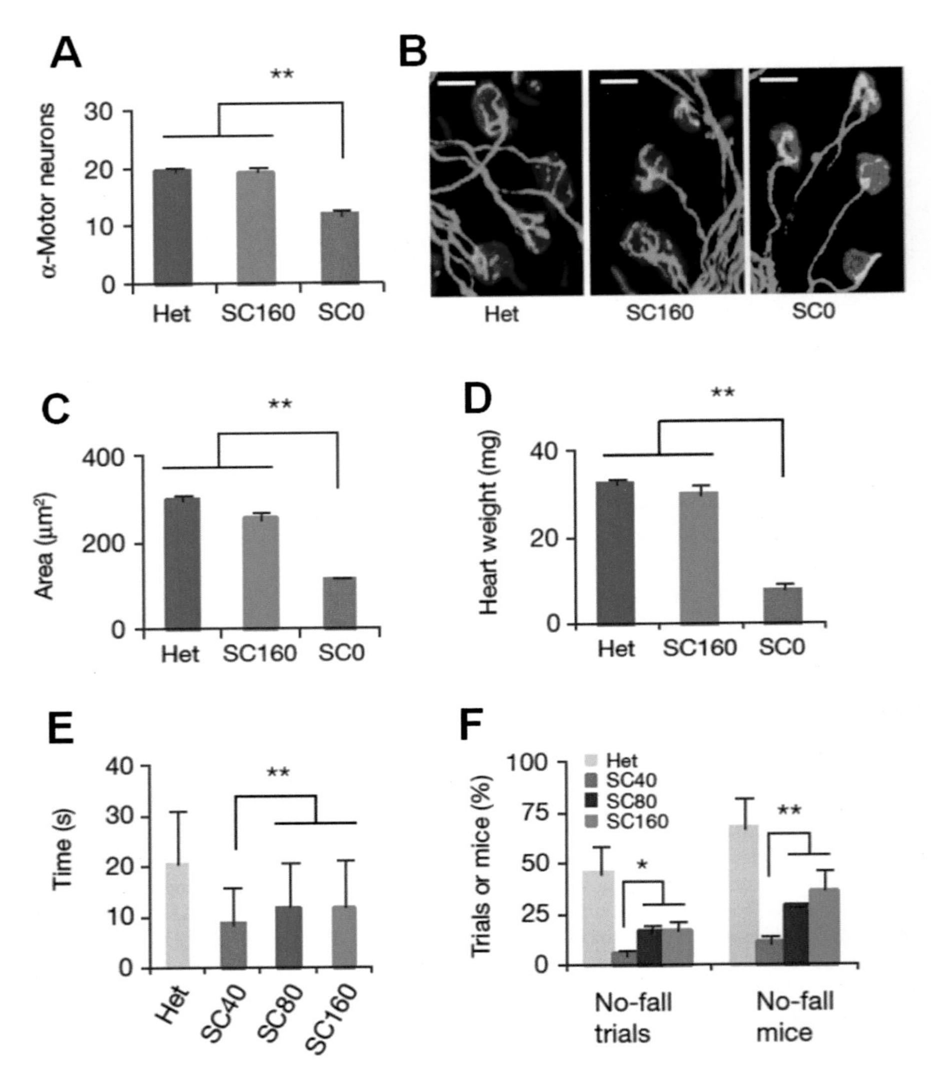

Fig. 8 SMA mice were treated with ASO-10-27 (SC160, two SC injections at 160 μg/g at P0 and P3), saline controls (SC0), and untreated heterozygotes (Het). (**a**) α-Motor neuron counts in each cross-section of the L1–L2 spinal cord. (**b**) Arborization complexity of neuromuscular junctions (*red* endplates; *green* neurofilament medium). (**c**) Mean fiber cross-sectional area of the rectus femoris muscle; (**d**) heart weight. (**e, f**) At P90, SC40, SC80, SC160 and untreated heterozygous (Het) mice were tested three to five times per day for 3 days on a rotarod, using an acceleration profile. The mean times for staying on the spinning rod (**e**) and the percentage of no-fall trials and of mice with ≥1 no-fall trial (**f**) are shown. Adapted from [30]. Copyright 2011 Macmillan Publishers Ltd

PMO was extended by five units (Fig. 6, PMO25), complementary to positions 10–34 of intron 7, additional activity improvement was observed both in vitro and in vivo [35]. No data were reported for SC dosing of PMOs targeting ISS-N1 as a comparison to results reported for ASO-10-27.

The very promising results achieved in pre-clinical animal studies with SSOs targeting *SMN2* splicing has led to multiple clinical trials in SMA patients with ASO-10-27 (nusinersen, Spinraza™, IONIS-SMN$_{Rx}$, ISIS 396443) being the most advanced. An open label phase 2 clinical trial in infants with Type I SMA receiving nusinersen via intrathecal injections showed increases in median event-free survival and muscle function scores (Available from: http://clinicaltrials.gov/show/NCT01494701). A 13-month phase 3 randomized, double-blind, sham-controlled study in 121 patients with infantile-onset SMA evaluated the efficacy and safety of a 12 mg dose of nusinersen (delivered via an intrathecal injection). A planned interim analysis of the trial revealed that a greater percentage of infants treated with nusinersen achieved a motor milestone response compared to those who did not receive treatment (40 versus 0%; $p < 0.0001$) as measured by the Hammersmith Infant Neurological Examination (HINE). At the end of the study, Spinraza also demonstrated a statistically significant 47% reduction in the risk of death or permanent ventilation ($p < 0.01$). A separate 15-month phase 3 study in 126 non-ambulatory patients with later-onset SMA also met the primary endpoint at the interim analysis, achieving a mean improvement of 4.0 points in the Hammersmith Functional Motor Scale Expanded (HFMSE). The results from these two phase 3 studies along with data from multiple other clinical trials were sufficient for the FDA to grant approval of Spinraza™ for the treatment of SMA in pediatric and adult patients (Available from: http://clinicaltrials.gov/show/NCT02193074; http://www.accessdata.fda.gov/drugsatfda_docs/nda/2016/209531Orig1s000MedR.pdf).

2.2 *Small Molecule Approach to Treat SMA*

Orally bioavailable small molecules are easier to administer and generally have much broader tissue distribution than SSOs. Despite these advantages, the advancement of small molecule drugs that target pre-mRNA splicing has been hampered by the difficulty of targeting specific sequences. Synthetic molecules have been elegantly designed to recognize specific DNA sequences through minor grove interactions [36], however, the same level of specificity has been more challenging to achieve when targeting RNA. There is no doubt that small molecules have the capacity to bind specific RNA structures with high affinity based on numerous examples of high affinity ligands designed to bind the bacterial ribosome and riboswitches [37]. Other synthetic molecules have shown promise for targeting specific mRNA sequences, such as trinucleotide repeats [38, 39]. However, as is typically the case with small molecule RNA interactions, high affinity comes at the cost of increased molecular weight, making the molecules less than ideal for oral delivery and broad distribution. The challenge therefore remains: Can a small molecule with optimal pharmaceutical properties be designed to target specific mRNA sequences? One area that may be underexplored, perhaps due to the complexity of the target, is the design of small molecules that interact with RNA–protein complexes, where RNA sequence defined structural elements may be targeted (*vide infra*).

To identify small molecules that induce the inclusion of exon 7 in *SMN2* pre-mRNA, a HEK293 cell line was modified to contain an *SMN2* minigene (from exon 6 to the 5′ region of exon 8), followed by the firefly luciferase coding sequence (Fig. 9) [40]. The transcript would express the luciferase enzyme and increased chemoluminescence only when exon 7 was included. The exclusion of exon 7 results in a frameshift such that the upstream initiation codon and luciferase coding sequence reside in different reading frames. This minigene construct was used to conduct a high throughput screen of ~200,000 discrete compounds which led to the discovery of a weakly active coumarin compound, **2-1**, that increased the luminescence signal by >50% at 10 μM. The activity was confirmed using RT-qPCR, quantifying the change in *SMN2*-FL mRNA with increasing compound concentration. The concentration required to increase *SMN2*-FL mRNA by 50% ($EC_{1.5X\ RNA}$) was 9 μM. However, no measurable increase in SMN protein was observed upon compound treatment of SMA patient-derived fibroblasts.

Through an activity-based optimization process guided by structure–activity relationships (SAR), potency was steadily improved by more than 1,000-fold in the reporter assay. Although the initial lead (**2-1**) had no activity in patient-derived fibroblasts, the optimized compounds (e.g., **2-2**) demonstrated low nanomolar $EC_{1.5X}$ values for *SMN2*-FL mRNA ($EC_{1.5X\ RNA}$) in the minigene construct in HEK293 cells and SMN protein ($EC_{1.5X\ PRO}$) in patient fibroblasts. The dramatic increase in potency was linked to a few key structural modifications. An important pharmacophore was discovered by modifying the right-side heterocycle of **2-3**

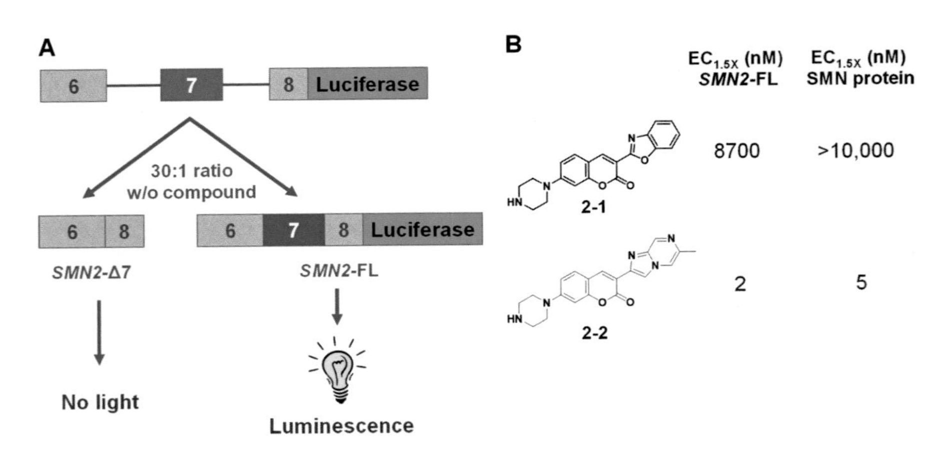

Fig. 9 (**a**) SMN2 minigene reporter construct in HEK293 cells was used to screen small molecules for exon 7 inclusion. (**b**) $EC_{1.5X}$ values for *SMN2*-FL increase in HEK293 cells and SMN protein in patient fibroblasts for a small molecule lead **2-1** and activity optimized compound **2-2**

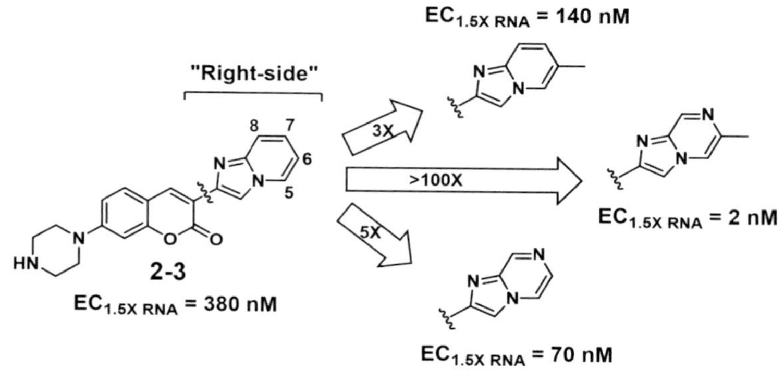

Fig. 10 Modification of the "right-side" heterocycle leads to >100-fold improvement in potency

(Fig. 10). The potency of **2-3** ($EC_{1.5X\ RNA}$ = 380 nM) was increased threefold by introducing a methyl group at the 6-position of the imidazopyridine. A fivefold potency improvement was achieved by replacing the ring carbon at the 7-position with nitrogen (7-aza). Interestingly, when both modifications were introduced (i.e., 6-methyl and 7-aza) a synergistic effect leading to >100-fold improvement in potency was observed. This "nitrogen-methyl pair" motif was incorporated into additional analogs during the optimization process.

Improved pharmaceutical properties could be achieved by replacing the 3,7-substitued coumarin core with a 4,8-substituted pyridopyrimidinone core (e.g., molecules **SMN-C3** and **RG7800**, Fig. 11) [41, 42]. The optimized molecules demonstrated a dose-dependent increase of SMN protein in the brain and spinal cord of Δ7 mice when administered by intraperitoneal (IP) injection. Sustained daily dosing of the compounds (IP from P3 to P23, then orally from P24 to P60) led to increased body weight, improved motor function, and longer survival (Fig. 12 for **SMN-C3**). In mice with a mild SMA phenotype (C/C-allele), both a shift in *SMN2* splicing and an increase in SMN protein were observed in the brain and whole blood. **RG7800**, a structural analog, was selected as the first small molecule *SMN2* splicing modifier to enter human clinical trials. A pharmacodynamic proof-of-concept was demonstrated by changes in the splicing ratio of *SMN2* mRNA in whole blood of healthy volunteers dosed with **RG7800**. In a subsequent study in SMA patients (Available from: http://clinicaltrials.gov/show/NCT02240355), RG7800, dosed once per day for 12 weeks, demonstrated an exposure-dependent increase in *SMN2*-FL mRNA with a corresponding decrease in *SMN*-Δ7 mRNA, as expected for a splicing modifier. SMN protein levels in whole blood increased up to 100% [43].

The specificity for **SMN-C3** was addressed in an RNA sequencing study, since one might surmise that a molecule exhibiting widespread changes in pre-mRNA splicing would be limited in its application due to potential toxicity related to the downstream effects. Comparing the differences in total transcript expression for 11,714 human genes in type 1 SMA patient-derived fibroblasts, treated with DMSO or **SMN-C3** (500 nM), demonstrated relatively few changes greater than twofold upon compound treatment (*SMN2* showed no significant change in total mRNA

SMN-C3 RG7800 LMI070

Fig. 11 Small molecule *SMN2* splicing modifiers

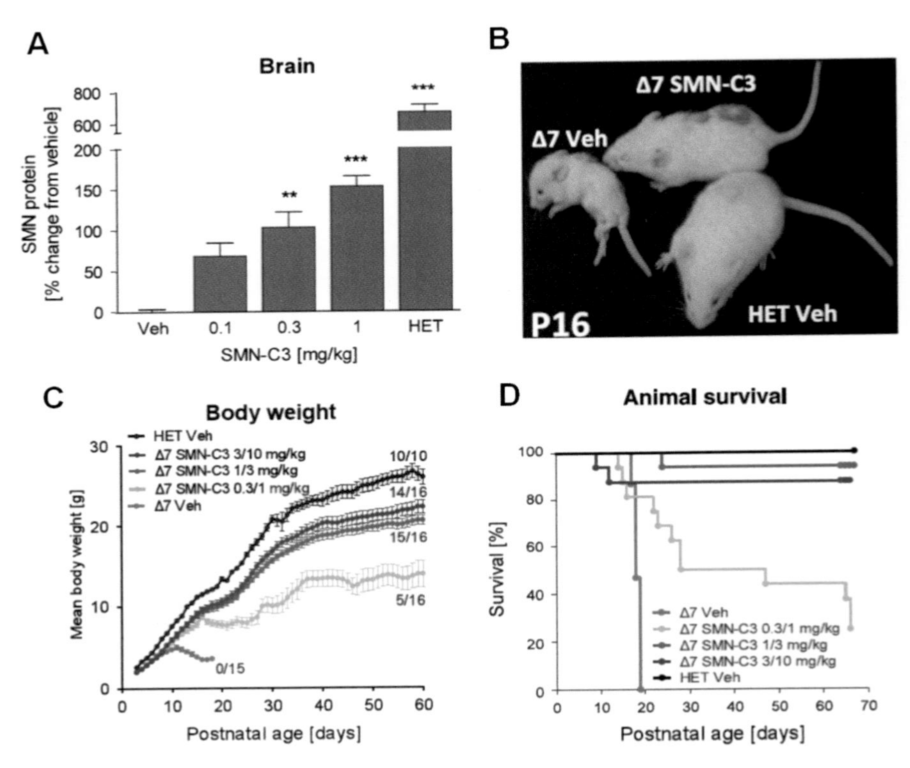

Fig. 12 (**a**) SMN protein levels in the brain of Δ7 mice after seven daily intraperitoneal doses (P3 through P9) of vehicle or SMN-C3 (0.1, 0.3, or 1 mg/kg). (**b**) Appearance of a vehicle-treated Δ7 mouse (Δ7 Veh), a SMN-C3–treated Δ7 mouse (Δ7 SMN-C3), and a vehicle-treated heterozygous mouse (HET Veh). (**c**) Body weight from P3 through P60. Numbers at right indicate survivors at P60 among 10 (HET) or 16 (Δ7) mice per group. (**d**) Kaplan–Meier survival curves from P3 to P65. Adapted from [41]

abundance), suggesting that compound treatment did not cause widespread changes in the transcription of genes or gene families (Fig. 13).

An additional analysis of changes to annotated splice junctions within the observed transcripts revealed that only a small group of splice junctions were

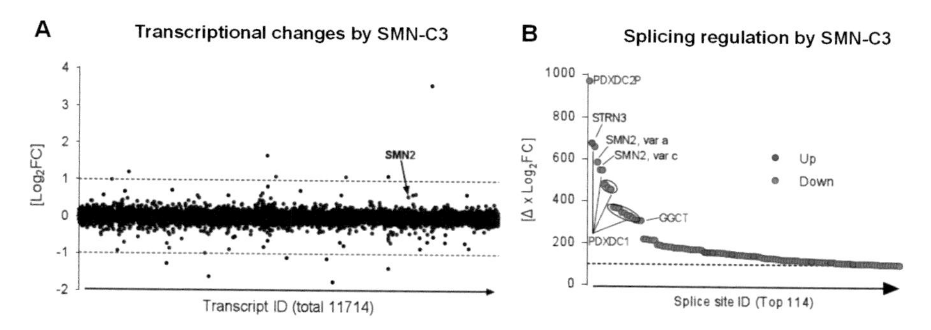

Fig. 13 (**a**) Difference in total transcript expression of **SMN-C3** (500 nM) versus DMSO-treated SMA type I patient fibroblasts for 11,714 human genes. (**b**) Differential effects of treatment on individual splice junctions in human transcripts. Affected splice junctions are characterized by either absolute difference in counts (Δ) or relative changes (Log2FC). The product $p = \Delta \times$ Log2FC was used to rank splice junctions (up-regulated in blue, downregulated in *red*). The top 114 splice junctions with $p > 100$ are shown (~300,000 splice junctions analyzed in total). Adapted from [41]

highly affected by compound treatment. Among these splice junctions were two mRNA variants of *SMN2* lacking exon 7 and several belonging to *PDXC1*. In effort to determine the molecular target, mutational analysis of sequences known to be involved with *SMN2* exon 7 splicing regulation was performed. **SMN-C3** appeared to work independently of the known splicing regulators.

Another class of small molecule *SMN2* splicing modifiers based on a pyridazine scaffold was independently discovered (e.g., **LMI070**, Fig. 11) [44]. SAR-based optimization aimed at maximizing cellular potency was driven by in vitro SMN ELISA activity in SMA patient-derived fibroblasts and Δ7 mouse-derived myoblasts. Oral dosing of **LMI070** resulted in changes in the splicing ratio of *SMN2* transcripts to favor *SMN2*-FL and increased SMN protein in C/+ mouse model of SMA as well as increased levels of SMN protein in the brain of Δ7 mice. A dose of 3 mg/kg/day of **LMI070** in Δ7 mouse model of SMA resulted in a MST >35 days compared to 14 days for mice treated with vehicle only. **LMI070** was the first small molecule splicing modifier to enter the clinic for studies in infants with type 1 SMA.

RNA sequencing analysis of **LMI070** at 100 nM identified 39 splice junction change events in 35 genes. Close inspection of the sequences of the affected splice junctions indicated a preferred 5′ splice site sequence, containing G and A at the −2 and −1 positions, respectively. This exon terminal "GA" sequence differs from the canonical "AG" sequence and forms a less stable pairing to the U1-snRNP. The hypothesis was proposed that the compounds bind to and stabilize U1-snRNP/mRNA complexes that contain the less common "GA" sequence. Additional binding experiments using size exclusion chromatography, surface plasmon resonance (SPR), and nuclear magnetic resonance (NMR) spectroscopy provided additional evidence of an interaction between **LMI070** and U1 snRNP in the presence of mRNA. No high resolution spectroscopic data have been published.

This hypothetical mode of action provides a rational explanation for compound-mediated splicing in the case of *SMN2* exon 7 (Fig. 14). As described previously,

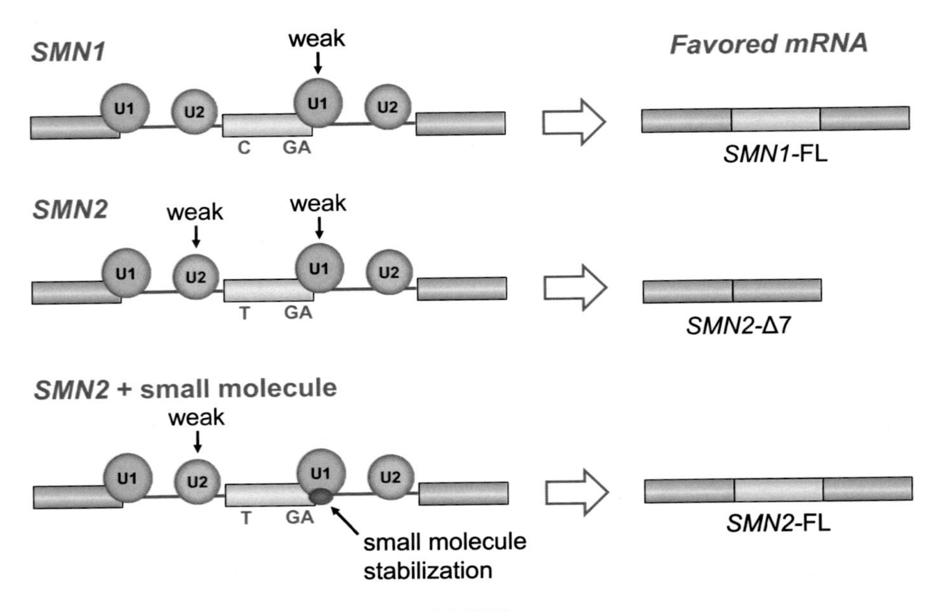

Fig. 14 Proposed mechanism of action for **LMI070**

the C to T transition in exon 7 of *SMN2* causes less favorable U2 recognition and, in conjunction with the less favorable "GA" sequence at the end of exon 7, impairs exon definition. This change is sufficient to cause exon 7 skipping 50–90% of the time. Compound treatment causes the stabilization of the U1/pre-mRNA complex, thus overcoming the weakness of the U2 recognition near the 3′ splice site. It is interesting to note that compound treatment also stabilizes the U1/exon 7–intron 7 complex in *SMN1* pre-mRNA and even though exon 7 is included in the final transcript in 95% of the *SMN1* mRNA, these small molecule splicing modifiers increase the fraction of full length *SMN1* mRNA to 100% (A. Dakka, N. A. Naryshkin, PTC Therapeutics, South Plainfield, NJ, unpublished work, 2012). This would suggest that on a global level additional sequence context around the 5′ terminal "GA" will largely determine the responsiveness to the treatment and could explain why the compounds show such remarkable selectivity.

2.3 Small Molecule Approach to Treat FD

Familial dysautonomia [FD, also known as Riley–Day syndrome or hereditary sensory and autonomic neuropathy (HSAN) III] is a recessive neurodegenerative disorder that is almost always caused by a noncoding point mutation at nucleotide 6 of the intron 20 donor splice site (IVS20+6T-C) of the *IKBKAP* gene (Fig. 15a) [46, 47]. In >99% of FD patients both alleles contain this same mutation, which

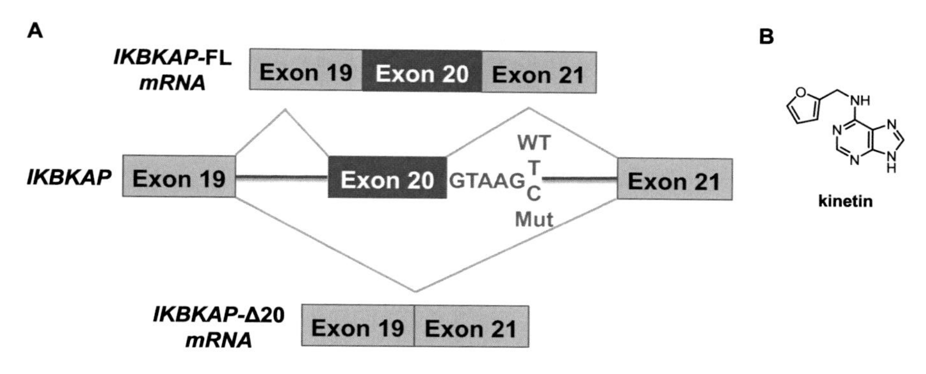

Fig. 15 (**a**) Alternative splicing of *IKBKAP* pre-mRNA. Mutation at nucleotide +6 of intron 20 causes a shift toward *IKBKAP*-Δ20. (**b**) Structure of the plant growth hormone kinetin

weakens the 5′ splice site. When mutant *IKBKAP* pre-mRNA is processed by the spliceosome, exon 20 is skipped much more frequently than would normally occur for the wild type *IKBKAP*, thus producing less full-length *IKBKAP* (*IKBKAP-FL*) and more exon-20-skipped IKBKAP (*IKBKAP-Δ20*) [48]. The frameshifted *IKBKAP-Δ20* transcript contains a premature termination codon in exon 21, which causes rapid degradation through the nonsense mediated decay (NMD) pathway. This is confirmed by the observation that *IKBKAP-Δ20* levels are increased in patient cells treated with cycloheximide, an NMD inhibitor [49]. Reduced levels of *IKBKAP*-FL lead to lower expression of functional IKAP, a protein that has roles in transcription elongation, histone acetylation, DNA methylation, and tRNA modification. The absolute amount of IKAP protein needed for normal cell function may vary by cell type. It is also known that the ratio of *IKBKAP-FL* to *IKBKAP-Δ20* varies between tissues, but tends to be lowest in the cells of the central and peripheral nervous system [50]. The level of IKAP protein produced in neuronal tissues of FD patients is not sufficient for normal function.

Because this devastating disease hinges on the balance of a single splicing event, several attempts have been made to identify a small molecule that could shift the splicing balance in favor of *IKBKAP*-FL [51–53]. The Slaugenhaupt lab developed an assay in FD patient lymphoblasts using RT-PCR to quantify the ratio of *IKBKAP*-FL/*IKBKAP*-Δ20 [49]. The assay was used to screen a small library of 1,040 bioactive compounds (mostly FDA approved drugs). A plant growth hormone, kinetin (Fig. 15b), demonstrated a shift in splicing at 10 μM. A full dose-response curve demonstrated that increasing the concentration of kinetin increases the ratio of *IKBKAP*-FL to *IKBKAP*-Δ20 from ~1:1 in the absence of compound to 12:1 in the presence of 400 μM kinetin. A corresponding dose dependent increase in IKAP protein was observed by Western blot.

To demonstrate the selectivity of kinetin on *IKBKAP* splicing, additional genes that either demonstrate alternative splicing in vivo or have been shown to have altered expression in the presence of kinetin were chosen for analysis [54]. Of the seven genes analyzed, the splicing ratios were unchanged in five of them upon kinetin treatment. The two splicing responsive genes had a common feature with exon 20 of *IKBKAP*. Mutational analysis in and around exon 20 revealed the importance of the sequence CAA at the −3 through −1 positions of the 5′-splice site of exon 20. It is estimated that ~9% of internal exons contain this sequence at the 5′-splice site [54]. Three additional genes that contain the CAA sequence were shown to be kinetin sensitive; *ABI2* exon 2, *BMP2K* exon 14, and *NF1* exon 36 (*NF1* exon 36 ends in CAAG).

To test the efficacy of kinetin in vivo, a humanized transgenic mouse line carrying the complete human *IKBKAP* locus with the FD mutation was generated [55]. The tgFD mice appropriately express the human *IKBKAP* from the transgene with tissue-specific splicing patterns that match those in FD patients. In pharmacokinetic studies in mice, kinetin was observed in all tissues including those within the central nervous system [56]. Mice dosed orally with kinetin at 400 mg/kg/day achieved serum concentrations >30 μM, well above the 10 μM concentration required for measurable activity in cultured cells. After 30 days of dosing, harvested tissues were analyzed. RT-PCR analysis showed a shift in splicing toward *IKBKAP-FL* mRNA in all tissues analyzed. This shift translated to higher levels of *IKBKAP-FL* mRNA transcript and increased IKAP protein.

To determine whether orally administered kinetin could alter mRNA splicing in FD patients, eight patients that were homozygous for the splice mutation were administered 23.5 mg/kg/day of kinetin for 28 days [57]. Plasma concentrations consistent with the levels required for activity in vitro were achieved in most of the patients at this dose. The levels of *IKBKAP-FL* in patient peripheral leukocytes after 8 and 28 days of dosing were compared to baseline levels measured prior to dosing. At baseline the mean percent inclusion of exon 20 was 54 ± 10%. After 8 days of dosing, six of eight individuals showed increased levels of *IKBKAP-FL*, with a mean percent inclusion of 57 ± 10%. After receiving kinetin for 28 days, patients achieved an even greater percent inclusion of 71 ± 9% ($p = 0.002$, Fig. 16). No serious adverse effects related to treatment were noted for the study. Based on proof of concept results both in preclinical and clinical studies, additional kinetin analogs have been designed by the Slaugenhaupt lab with improved potency and pharmaceutical properties [58].

Roughly a decade after the initial discovery of kinetin as an *IKBKAP* pre-mRNA modifier, a screening effort in the Hagiwara lab led to the discovery of a more potent molecule mediating *IKBKAP* exon 20 inclusion [59]. A dual color reporter assay was designed and used to quantify the ratio of *IKBKAP-FL* to *IKBKAP-Δ20*. Exon 21 of the *IKBKAP* minigene construct contained sequences for enhanced green fluorescent protein (EGFP) and monomeric red fluorescent protein (mRFP). EGFP is placed in the proper translational reading frame only when exon 20 is included in the mRNA transcript. Conversely, when exon 20 is excluded, the resulting transcript places mRFP in the proper translational reading frame (Fig. 17a). The relative amounts of the two reporters can be quantified using fluorescence microscopy. The construct was used to screen 638 molecules from a chemical library and some additional approved pharmaceuticals.

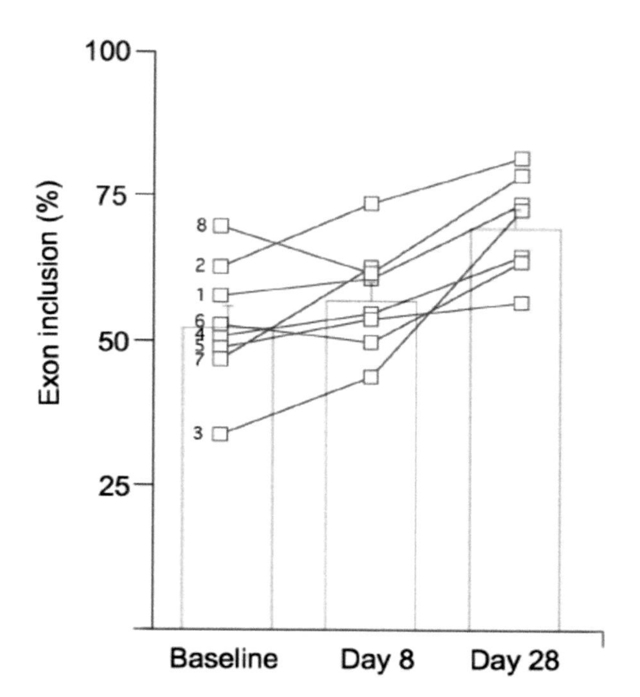

Fig. 16 The percent of the IKBKAP transcript that includes exon 20 in FD patients treated with 23.5 mg/kg/day kinetin at baseline, after 8 days of dosing, and after 28 days of dosing. Individual data points shown with open squares. Adapted from [57]. Copyright 2011 Macmillan Publishers Ltd

Interestingly, the most active molecule identified from the screen was a close structural analog of kinetin, differing only by the addition of chlorine at the 2-position of the purine ring (Fig. 17b, RECTAS). RECTAS (*rect*ifier of *a*berrant *s*plicing) demonstrated roughly a 25-fold improvement in activity over kinetin in the fluorescence reporter assay (Fig. 17c). In addition, RECTAS increased the inclusion of exon 20 in patient fibroblasts over kinetin when cells were treated at similar concentrations. The increased levels of *IKBKAP-FL* led to a corresponding increase in IKAP protein as demonstrated by western blot.

3 Exon Skipping

Skipping of an exon during pre-mRNA splicing can be exploited in several different genetic circumstances. The most common applications include (1) generation of a shorter isoform with altered function, (2) skipping of a downstream exon in order to restore an open reading frame with at least partial functional recovery, (3) skipping of a deleterious mutation-containing exon (provided that the shorter isoform retains at least some biological function), and (4) skipping to disrupt an open reading frame in order to suppress the expression of the target gene. The utility of exon skipping has now been documented for many therapeutically relevant genes, some of which are reviewed in detail below.

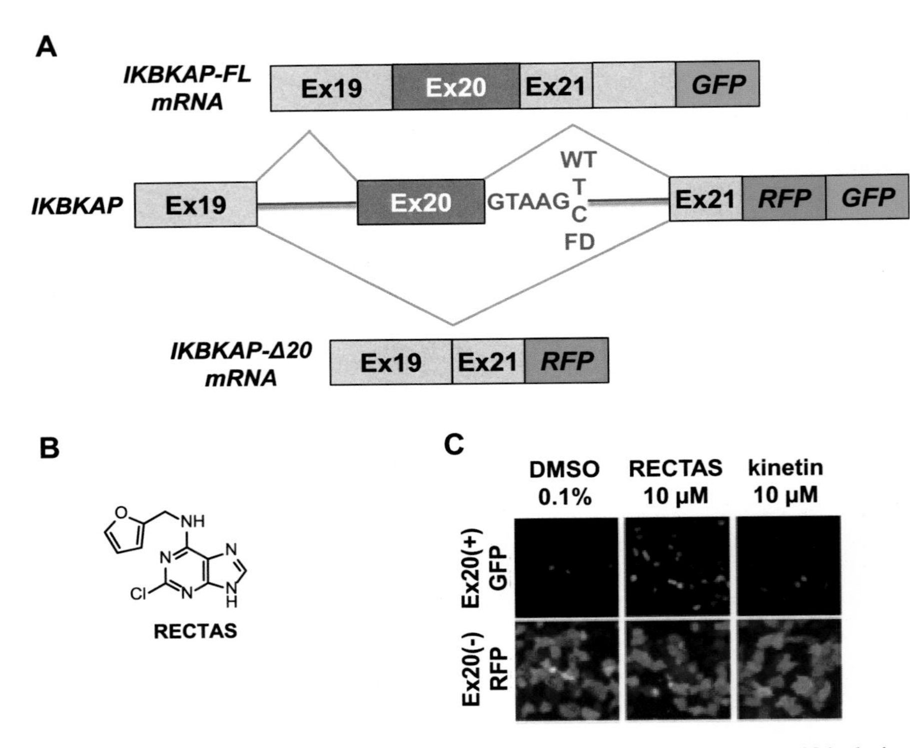

Fig. 17 (**a**) *IKBKAP-FD* reporter construct used to screen small molecules for exon 20 inclusion. (**b**) Small molecule identified in a screen that shifts *IKBKAP* pre-mRNA splicing toward *IKBKAP-FL*. (**c**) Microscopic analysis of HeLa cells expressing the *IKBKAP-FD* reporter treated with the indicated small molecules. Adapted with permission from [59]

3.1 Treatment of Duchenne Muscular Dystrophy

The application and clinical advancement of exon skipping for disease therapy has received the most attention in the field of Duchenne muscular dystrophy (DMD; MIM #310200). DMD is an X chromosome-linked progressive degenerative myopathy caused by mutations in the dystrophin gene, whose main function in muscle tissue is to connect cytoskeleton with the sarcolemma to maintain the structural integrity of the muscle fiber. The N-terminal domain of dystrophin binds actin filaments and, via 24 spectrin-like (SR) domains and four hinge regions, is connected to the C-terminal domain which interacts with the inner side of the sarcolemma where it joins a large multiprotein assembly called the dystrophin-associated protein complex (DAP). The complete list of biological functions for dystrophin is still being elucidated. In the muscle, the main role of dystrophin is to act as both a linker and a mechanical stress absorber during the cycles of contraction and relaxation. This function requires the presence of both attachment domains and at least some of the internal domains (much like a car suspension's strut system). Therefore, mutations that maintain this general three-part structure would be predicted to retain some function, whereas mutations that result in the absence of either terminal domains or the internal "stress absorber" would be expected to lose all muscle-supporting properties. Indeed, following the discovery of the *DMD* gene in 1986 [60] and subsequent genotyping of a sufficient number of DMD patients, it was realized that a sizable fraction of *DMD* gene mutations do result in a considerably milder Becker muscular dystrophy (BMD; MIM #300376) in which affected individuals have reduced dystrophin function due to an internal truncation or an amino acid substitution. Thus, "the ORF rule" was put forward stating that genetic alterations retaining the open reading frame (ORF) of dystrophin are much more likely to result in the milder Becker dystrophy (Fig. 18) [61].

The ORF rule suggests a practical approach to the treatment of Duchenne muscular dystrophy in which the detrimental effect of a frame-disrupting deletion could be partly reversed during pre-mRNA splicing by skipping an exon flanking the deletion if such skipping restores the ORF (see Fig. 19 for a graphical overview highlighting exons that would maintain the ORF when skipped). A deleterious point

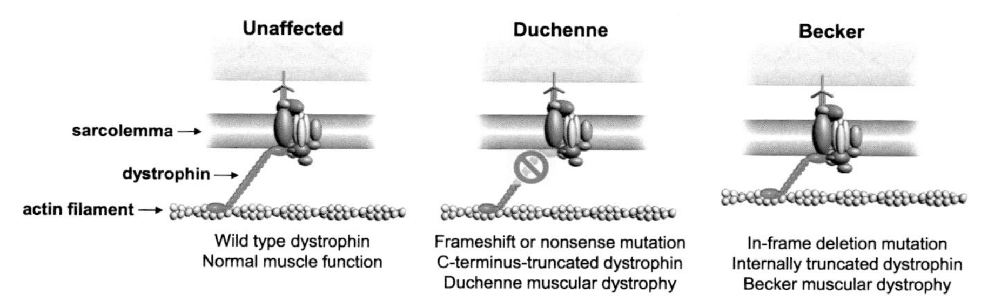

Fig. 18 Cartoon depiction of the role of dystrophin in wild type muscle fiber (*left*), muscle fiber with nonfunctional terminally truncated dystrophin leading to Duchenne muscular dystrophy (*middle*) and muscle fiber with an internally truncated dystrophin protein leading to Becker muscular dystrophy (*right*)

Fig. 19 Overview of the exons coding for the *DMD* gene. In-frame exons are shown in light blue and out-of-frame exons are shown in dark blue. Exon deletions lead to in-frame messages when the exons flanking the deletion have matching shapes. When the flanking exons do not match, the deletions disrupt the reading frame. Adapted from [62]

mutation in *DMD* could also be partly ameliorated by skipping the mutation-carrying exon provided the ORF is maintained in the shortened transcript (Fig. 3c, d). This phenomenon may be responsible for the production of dystrophin-positive ("revertant") fibers in human DMD and BMD patients and in animal models of Duchenne, although the origin of revertant fibers is still being elucidated [63–66]. The unexpectedly mild phenotype of the commonly used mdx mouse model of DMD [67] carrying an inactivating nonsense mutation in exon 23 is driven by massive sporadic exon skipping that generates functional internally deleted dystrophin isoforms [68, 69].

Additional appeal for utilizing an exon skipping strategy to treat DMD arises from the fact that a high prevalence of *DMD* mutations are amenable to this approach. Dystrophin is the third largest human gene comprising at least 7 characterized promoters, 79 exons, and introns whose size reaches over 248,000 nucleotides. Dystrophin's pre-mRNA takes about 20 h to be synthesized and undergoes complex splicing. According to the recent analysis of the TREAT-NMD DMD Global database, 80% of deletions and 55% of all *DMD* mutations can be potentially corrected using exon skipping therapy [70]. The top ten target exons for skipping, nine of which are located in the mutational "hot spot" area of exons 45–55, are shown in Table 1.

A naturally occurring mutation provided some of the initial insight leading to the development of methods for pharmacologically induced exon skipping. The so-called dystrophin Kobe, in which 52 nucleotides spanning an exonic splicing enhancer (ESE) region in the middle of *DMD* exon 19 are deleted, leads to native skipping of exon 19 [71]. Using a splice switching oligonucleotide (2′-OMe backbone) complementary to the first 31 nucleotides of the sequence deleted in the Kobe mutation, the Matsuo group first demonstrated, in 1995, the inhibition of *DMD* intron 18 removal from a *DMD* minigene pre-mRNA in HeLa cell nuclear extract [72]. Then, in 1996, the Matsuo group demonstrated that a DNA-PS oligomer (**AO19**, Fig. 20) complementary to the same region caused the skipping of DMD exon 19 during the splicing of wild type *DMD* pre-mRNA in cultured human lymphoblastoid cells [73]. Cellular delivery of **AO19** required the use of a

Table 1 Prevalence of mutations in *DMD* exons in Duchenne muscular dystrophy patients (top ten shown)

DMD exon	Mutation (%)	Deletion (%)
51	14.0	20.5
53	9.0	13.1
45	8.1	11.8
44	7.6	11.1
43	3.8	5.6
46	3.1	4.5
50	2.0	2.9
52	1.7	2.5
55	0.9	1.3
8	0.9	1.3

Column 2 contains the percentage of all mutations, while column 3 contains only the percentage of deletions [70]

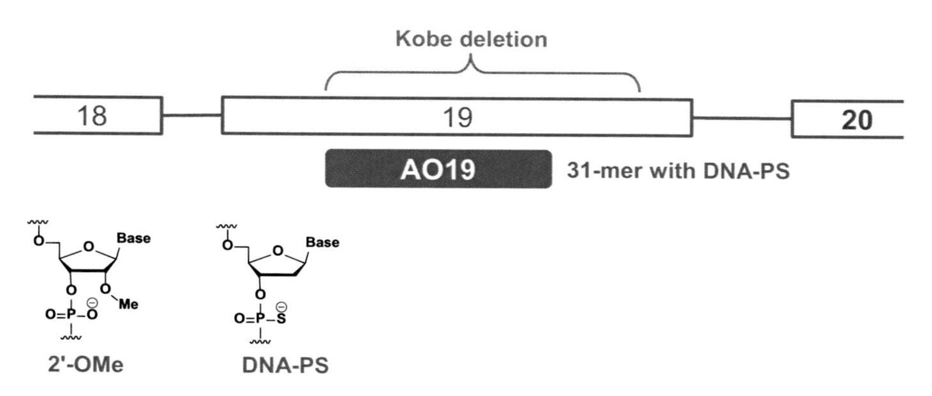

Fig. 20 SSO developed to mimic the Kobe deletion in *DMD*

cationic lipid vehicle. When cells were treated with 200 nM of **AO19**, only *DMD*-Δex19 mRNA was observed, demonstrating for the first time the feasibility of completely skipping a constitutive human exon through SSO treatment. Importantly, other *DMD* exons were not affected and the negative control, a DNA-PS sense-strand, did not induce exon skipping. Utilizing **AO19**, *DMD* exon 19 was skipped in myotubes generated from myogenic cells isolated from a DMD patient harboring a deletion of the entire exon 20 region of the dystrophin gene [74]. On this genetic background, skipping of exon 19 restores the open reading frame and is predicted to restore, at least partially, the expression of dystrophin. The authors observed up to 20% of dystrophin-positive fibers in cultures that were treated with the SSO.

When a single dose of **AO19** was delivered intraperitoneally as an aqueous solution to mdx mice, measurable skipping of exon 19 was detected, with an effect lasting up to 14 days post injection [75]. When a fluorescent label was linked to **AO19**, the authors were able to detect the compound in interstitial tissue, cytoplasm, and myocyte nuclei, including centered nuclei representing the regenerating myocytes. Thus, in vivo, **AO19** delivered as a naked molecule in solution was reaching the intended sites of action in skeletal muscle cells. These results gave the Matsuo group confidence to test **AO19** in a DMD patient with an exon 20 deletion [74]. Four doses of the **AO19**, given once a week, delivered as an intravenous infusion resulted in skipping of exon 19 in both lymphocytes and biceps muscle, achieving ~6% skipping in the muscle. Immunostaining of the biopsied muscles with antibodies detecting the N- and C-termini and the rod region of dystrophin showed a modest increase in dystrophin production in post-treatment samples [76].

In the late 1990s–early 2000s, several other groups initiated research efforts in exon skipping for DMD. The Wilton and Dunckley groups were inspired by another natural phenomenon observed in mdx mice. This mouse model carries a premature translation termination codon in exon 23 of the dystrophin gene, which would be expected to result in a severe dystrophic phenotype. However, mdx mice have a nearly normal lifespan and only mild dystrophy, primarily attributable to the presence of a small but measurable number (typically less than 1%) of dystrophin-positive muscle fibers, thought to be caused by intrinsic exon skipping during pre-mRNA splicing or due to somatic mutation [69, 77, 78]. SSOs were designed to bind to the 3′ and 5′ splice sites at the ends of exon 23 and the branch point in intron 22. Effective exon skipping was achieved upon transfection of both mouse myoblast cell line C2C12 and primary mdx mouse myoblasts [79, 80]. Interestingly, while Wilton et al. found blocking of the 5′ splice site to be most effective in eliciting exon 23 skipping, Dunckley et al. achieved maximal skipping with an SSO blocking the 3′ splice site (the nucleotide sequences of the SSOs were different in the two studies). Dunckley et al. also observed increased levels of dystrophin protein at the sarcolemma in about 1% of cultured myotubes. Around the same time, van Deutekom et al. demonstrated the utility of 2′-OMe-PS oligomers in skipping murine and human *DMD* exon 46. A series of oligonucleotides that bind to predicted exonic splicing enhancers in exon 46 were used to transfect murine C2C12 cells and human primary muscle cells isolated from a healthy individual and two unrelated DMD patients. Efficient skipping of exon 46 was achieved and, in DMD patients-derived myotubes, skipping resulted in the appearance of dystrophin staining in the sarcolemma region [81]. Human *DMD*-specific skipping of exons 44, 46, and 49 was demonstrated in the hDMD mouse model carrying the full 2.4 Mb human dystrophin gene [82].

This seminal work clearly established the feasibility of exon skipping during *DMD* pre-mRNA splicing and confirmed the hypothesis that a translation-compatible open reading frame can be generated by skipping an exon carrying a point mutation or flanking a frame-disrupting deletion. To determine the general applicability of exon skipping as a potential therapy for DMD, several groups performed comprehensive systematic surveys of "skippability" of all internal exons in dystrophin using splice switching oligonucleotides [83, 84]. Every exon was found to be skippable; however, the ease of eliciting the skipping varied considerably between exons. For some exons treatment with two or more oligonucleotides was required. Simultaneous skipping of multiple exons, necessary for certain classes of *DMD* mutations, was also shown to be possible [85, 86].

Initial exon skipping studies in DMD relied mostly on oligoribonucleotides with a 2′-methoxy substituted phosphorothioate backbone (2′-OMe-PS). These molecules satisfy two key requirements for an SSO – they are nuclease-resistant and do not support RNase H-mediated cleavage. However, due to their negative charge, crossing biological membranes (which are also negatively charged) presents a challenge which is particularly severe in cardiac muscle and the CNS. Indeed, the extent of exon skipping observed in the in vivo studies presented to this point was typically a few percent at best.

Several additional oligonucleotide chemistries (Fig. 21), including those based on locked nucleic acids (LNA), peptide nucleic acids (PNA), phosphorodiamidate morpholino oligonucleotides (PMO) and, recently, tricycle-DNA (tcDNA) have been tested for DMD skipping (reviewed in [87, 88]). The presence of LNA units results in much stronger hybridization energy, with each monomer adding at least 5 °C to the duplex melting temperature [89]. The stronger binding has effectively

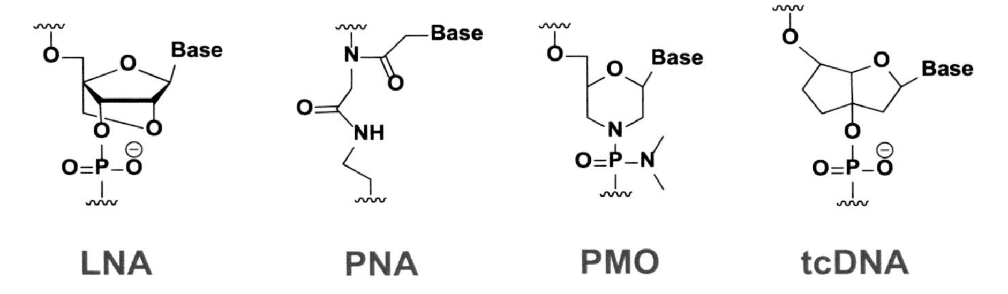

Fig. 21 Alternative SSO backbone chemistries

precluded the use of pure LNA oligomers as splice-switching reagents, because oligomers with mismatches maintain high binding affinity, raising concerns with regard to specificity. Shorter LNA oligomers (14-mers are typical) suffer from the same consequence of reduced specificity. Copolymers comprised of LNA and 2′-OMe-PS sequences may hold promise for maintaining the high nuclease stability afforded by the LNA segment while achieving higher specificity from the increased length permitted by the addition of the lower affinity 2′-OMe-PS segment [90]. However, the application of such chimeras has not produced the expected improvement in exon skipping activity in vivo [91]. The application of PNA oligomers showed early promise in modulating the splicing of pre-mRNAs for several membrane receptors, but the activity in *DMD* exon skipping proved to be much less consistent [89, 92, 93]. This inconsistency combined with higher reagent costs has caused PNAs to receive less attention than other classes.

The tcDNA backbone may be emerging as the next promising scaffold for SSO design due to its ability to induce exon skipping not only in skeletal muscle but also in the heart and even, to a small but measurable extent, in the brain of the mdx mouse [94] without signs of overt toxicity. In this study the tcDNA was designed to induce *Dmd* exon 23 skipping, which is one of the most easily skipped exons. Future work will need to address both the general utility of tcDNA for exon skipping and the tolerability of this backbone in vivo.

PMOs have emerged as the leading type of SSO owing to a unique combination of efficient hybridization, stability to nucleases, lack of immune response, and aqueous solubility (reviewed in [95]). This class of SSO has shown robust *DMD* exon skipping in cultured cells [96, 97]; in the mdx and hDMD mouse models [98–100], and in the CXMD dog model [101]. Aoki and coworkers innovatively proposed using a cocktail of PMOs to cause the skipping of the entire region covering DMD exons 45–55 [102]. This approach would be theoretically applicable to ~60% of DMD patients carrying a frame-changing deletion. The corresponding internally truncated protein is known to be associated with mild myodystrophy. The approach showed early promise in the mdx52 mouse model [102, 103]. Several review articles provide excellent surveys of the efforts directed toward multiple exon skipping in DMD [85, 86, 104].

There are divergent estimates, likely driven by the differences in sequence and biological models used, of unassisted cellular uptake of backbone-neutral PMOs vs. negatively charged 2-OMe-PS oligonucleotides, with some groups demonstrating better uptake of PMOs and others giving the edge to 2-OMe-PS oligonucleotides [91, 98]. A considerable advantage of the PMO backbone is its compatibility with covalent attachment of cellular uptake-promoting moieties.

Cell-penetrating peptide-conjugated oligonucleotides have emerged as another very exciting class of splicing modifiers due to their enhanced systemic bioavailability (reviewed in [91, 105, 106]). This approach turned a challenge – the presence of negative charges on the cell membrane and their interference with the uptake of negatively charged oligonucleotides – into an opportunity. By the mid-1990s several short peptides containing a high proportion of positively charged amino acids (arginines and lysines) were shown to facilitate the delivery of various kinds of molecular payload to the cytoplasm and nucleus of the cell. Research from several laboratories has now expanded the arsenal to several dozens of cell-penetrating peptides (CPPs) and work in this area continues. The mechanism of CPP-mediated internalization is not completely understood and, depending on the structure and cell type, can involve either the direct membrane translocation or an endocytic pathway. The positive charge of CPPs constrains the type of molecular cargo that can be covalently conjugated; only charge-neutral backbones such as PNA and PMO are compatible. Oligomers carrying negative charge (e.g., siRNA duplexes) can only be delivered using non-cationic peptides or as noncovalent complexes with amphipathic peptides. Non-charged homing peptides are being developed for oligonucleotides with charged backbones [107]. CPP-modified SSOs offer enhanced potency and tissue penetration upon systemic delivery. A very recent example is the work by Hammond and colleagues in which considerable efficacy was achieved upon treating severe SMA mice with systemically delivered PMO conjugated with Pip6a delivery peptide [108]. Of relevance to DMD, the intravenously delivered Pip6a-PMO conjugate induced efficient *SMN2* splicing correction in several tissues including brain and spinal cord. For treatment of DMD, the challenges of tissue distribution for SSO therapies are particularly acute with respect to penetration into the cardiac muscle. Arginine-rich CPPs have been the first SSO class to show DMD exon skipping in the heart. Additionally, they confer benefit at doses lower than those required for unconjugated PMOs, and prevent exercise-induced myopathy in mdx mice [106, 109–112]. Pharmacokinetics, biodistribution, stability, and toxicity of an early version of a cell-penetrating peptide attached to a PMO were studied in rats by Amantana and coworkers [113]. Intravenous administration of the conjugate in rats produced an LD_{50} of 210–250 mg/kg. Doses of 15 and 30 mg/kg were well tolerated, with no evidence of clinical abnormalities. The target organ for toxicity appeared to be the kidney, adding a cautionary note to this class of SSO. However, it should also be noted that CPP-PMO conjugates are biologically active at much lower doses (typically in the 1–10 mg/kg range) than PMOs by themselves. Active research focused on CPP optimization, including the reduction of kidney damage, is ongoing in several laboratories opening the way to commercial and therapeutic applications.

Two additional splice-switching modalities that rely on nucleic acid hybridization have been developed to achieve exon skipping. They are based on U1 [114–117] and U7 [118–120] small nuclear RNAs engineered to interact with various *cis* elements essential for splicing and interfere with the formation of a splicing-competent spliceosome. These RNAs are delivered using lentivirus- and AAV-based vectors and therefore achieve more consistent expression, proper cellular localization, and readily associate with spliceosome components. The downsides of this approach as a therapy are inherent to virus-mediated therapies and include the need for production of large number of viral particles and existing and induced immune response.

A comprehensive treatment for DMD would require not only a dystrophin-restoring compound but also pharmacological agents addressing other aspects such as chronic inflammation, fibrosis, the depletion of satellite cells and reduced muscle mass. Myostatin (*MSTN*) is a well-characterized muscle growth suppressor [121]. Animals with two null *MSTN* alleles develop overtly hypertrophic muscles. The *MSTN* gene consists of three exons. Skipping of the second exon disrupts the open reading frame and inhibits myostatin protein synthesis. Several groups achieved successful skipping of *MSTN* exon 2 using both 2-OMe-PS oligonucleotides and PMOs, typically in conjunction with frame-restoring exon skipping in DMD pre-mRNA [122–124]. It is important to note that in many myopathies the inhibition of myostatin alone, while increasing muscle mass, may not contribute substantially to muscle strength. To be effective, MSTN inhibition needs to be coupled with agents that restore muscle structure [125].

3.2 Clinical Development of Exon Skipping SSOs

Three classes of splice switching oligonucleotides, DNA-PS oligomers, 2′-OMe-PS oligomers and PMOs, have been undergoing extensive clinical development, although the bulk of the work has involved the 2-OMe-PS and PMO classes (reviewed in [126–128]). As described earlier, the first human trial of a *DMD* exon skipping oligonucleotide was performed by the Matsuo group using AO19 in a patient with a *DMD* exon 20 deletion [74]. This study served as a proof of concept, but left several open questions. What role does RNase H-driven degradation of *DMD* mRNA play? Will exon skipping studies in the mdx mouse translate to humans? And what cells and tissues are relevant for the analysis of drug efficacy (summarized in [87]). Since the skipping of *DMD* exon 51 can restore dystrophin's open reading frame in the largest cohort (13–15%) of DMD patients [62, 70], the largest clinical effort has been devoted to SSOs targeting this exon.

Drisapersen (PRO051, Kyndrisa™), a 20-mer 2′-OMe-PS oligonucleotide from Prosensa/GSK/BioMarin, was identified by van Deutekom and co-workers as a candidate for testing in DMD patients with mutations amenable to ORF restoration by *DMD* exon 51 skipping. The first clinical trial, at a single-center, was an open label study of a single intramuscular dose of drisapersen in four DMD patients. No adverse events were observed. A variable degree of exon 51 skipping was detected in extracted muscle samples as measured by RT-PCR. Samples displayed an increase in dystrophin-positive fibers with a sarcolemma-localized signal and an upregulation of dystrophin production shown by western blotting; however, the increase in protein did not appear to correlate with the relative fraction of Δ51 mRNA [129]. Encouraged by the initial results, several phase 2 and 3 studies on drisapersen were initiated to investigate dose response, dosing regimens, and patient cohorts. In total more than 300 DMD patients were dosed over the course of the trials. The primary outcome measure was the distance traveled during a 6 min walk (6 min walk test or 6MWT). Drisapersen, while showing positive trends in subgroups of patients, did not achieve a statistically significant difference in the primary endpoint, which in combination with drug-related side effects made regulatory approval challenging. Further development of drisapersen and other DMD exon skipping candidates in BioMarin's pipeline was halted.

With similar timing to the development of drisapersen, eteplirsen (AVI-4658, EXONDYS 51™), a 30-mer PMO, was identified by researchers at Sarepta Therapeutics (AVI BioPharma at the time). Eteplirsen also targets patients with mutations amenable to ORF restoration by *DMD* exon 51 skipping. The first human clinical trial for eteplirsen was a single-blind placebo-controlled dose-escalation study of a single intramuscular dose in seven DMD patients. No-treatment-related adverse events were reported. Extensor digitorum brevis muscle biopsies were analyzed for *DMD* gene expression. The high dose (0.9 mg) resulted in measurable increases in Δ51 *DMD* mRNA, dystrophin-positive fibers and dystrophin protein. There appeared to be a correlation between the level of Δ51 *DMD* mRNA and dystrophin protein [130]. Several phase 2 studies of eteplirsen were conducted to select a dosing regimen and a therapeutic dose. Based on the results from one of these trials, a phase 2 study (including the study extension) in 12 patients, the company sought a regulatory approval in the USA. The primary outcome measure was an increase in the number of dystrophin positive fibers, while the 6 min walk test was a secondary outcome. Following an extended regulatory review that included additional analyses of dystrophin protein levels in an ongoing phase 3 study using western blotting, in September 2016 EXONDYS 51™ (eteplirsen) received accelerated approval for the treatment of Duchenne muscular dystrophy in patients who have a confirmed mutation of the *DMD* gene that is amenable to exon 51 skipping (Available from: https://www.accessdata.fda.gov/drugsatfda_docs/nda/2016/206488Orig1s000Approv.pdf). Using a similar technology, SSOs that induce the skipping of exons 45, 51, and 53 are currently in clinical development for the corresponding DMD patients (Available from: http://www.sarepta.com/our-pipeline). Additional SSOs targeting DMD exons 8, 44, 50, 52, and 55, as well as a CPP-modified PMO to skip exon 51, are in preclinical development.

3.3 *Small Molecules that Enhance Exon Skipping*

Antisense oligonucleotide and virally delivered small nucleic acids dominate the *DMD* exon skipping field; nonetheless, small molecule splicing modifiers have been pursued as a potential therapeutic modality for this cause. Nishida and co-workers screened a library of over 100,000 compounds for inhibitors of kinases responsible for phosphorylation of SR proteins which regulate splicing. One of the hits, TG003 (Fig. 22), was shown to be an inhibitor of Cdc2-like kinases and to inhibit SF2/ASF-dependent splicing [131]. Upon treatment with TG003, a *DMD* exon 31 minigene was responsive to increased exon skipping in HeLa nuclear extracts. Additionally, when myotubes generated from DMD patient cells carrying a nonsense mutation in *DMD* exon 31 were treated with TG003, the inherent sporadic exon 31 skipping was increased [132].

Dantrolene, a postsynaptic muscle relaxant and ryanodine receptor antagonist, was identified in a chemical screen of 300 compounds as a potentiator of SSO-driven exon skipping [133]. In addition to the screening assay, SSO-enhancing activity was demonstrated in vitro in mdx mouse myoblasts (exon 23 skipping) and human DMD patient myoblasts (exon 51 skipping) and in vivo in mdx mice. Both 2′-OMe-PS oligonucleotides and PMOs were responsive to dantrolene enhancement.

Hu and coworkers performed a screen of 2,000 compounds using a construct that contained an interrupted green fluorescent protein coding sequence that carried either murine *Dmd* exon 23 or human *DMD* exon 50, both flanked by their corresponding intronic sequences [134]. The small molecule 6-thioguanine (6TG) was found to be a weak inducer of exon skipping when administered alone. When used in combination with a PMO, 6TG enhances the exon skipping activity of the PMO in C2C12 mouse myoblasts. The combination therapy was also reported to enhance exon skipping in mdx mice, after an intramuscular injection. Verhaart and Aartsma-Rus further examined the effect of 6TG on *DMD* exon skipping in cultured human muscle cells derived from a healthy donor, in C2C12 murine myoblasts, and in mdx mice [135]. They concluded that indeed 6TG enhances the effect of both 2′-OMe-PS oligomers and PMOs. Additionally, it was found that the activity of 6TG in the absence of SSO was non-selective, causing exon skipping of various exons throughout the transcript, suggesting a general splicing disruption and not an activity directly coupled to the binding of the SSO.

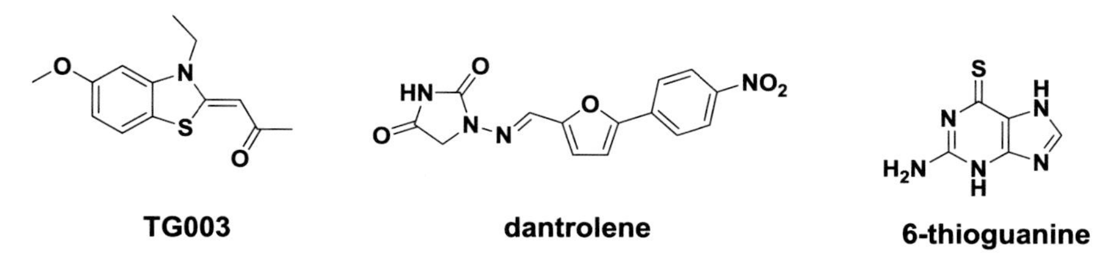

Fig. 22 Small molecules found to increase exon skipping

4 Splicing Inhibition

Various small molecules that inhibit some general aspect of the splicing process have been discovered in screening efforts with assays that were either designed specifically to identify splicing inhibitors or designed to identify cytotoxic molecules [136]. In the latter case, detailed mechanistic work was required to link the cytotoxic behavior of the molecule to the inhibition of a splicing process. Because these molecules often stall the assembly of a catalytically competent spliceosome or the catalytic splicing cycle at specific stages, they have served as useful tools in understanding certain aspects of pre-mRNA splicing. The therapeutic application of such molecules has largely been linked to cancer, since most display strong antiproliferative activity [137]. However, other applications may be possible depending on the splicing target, since some splicing events may be more strongly inhibited than others.

4.1 *Targeting SF3b to Inhibit Splicing*

The natural product FR901464 (Fig. 23) was isolated from the fermentation broth of *Pseudomonas* sp. due to its potent cytotoxicity. It was determined that FR901464 caused cell cycle arrest at the G1 and G2/M phases [138]. A close analysis of the

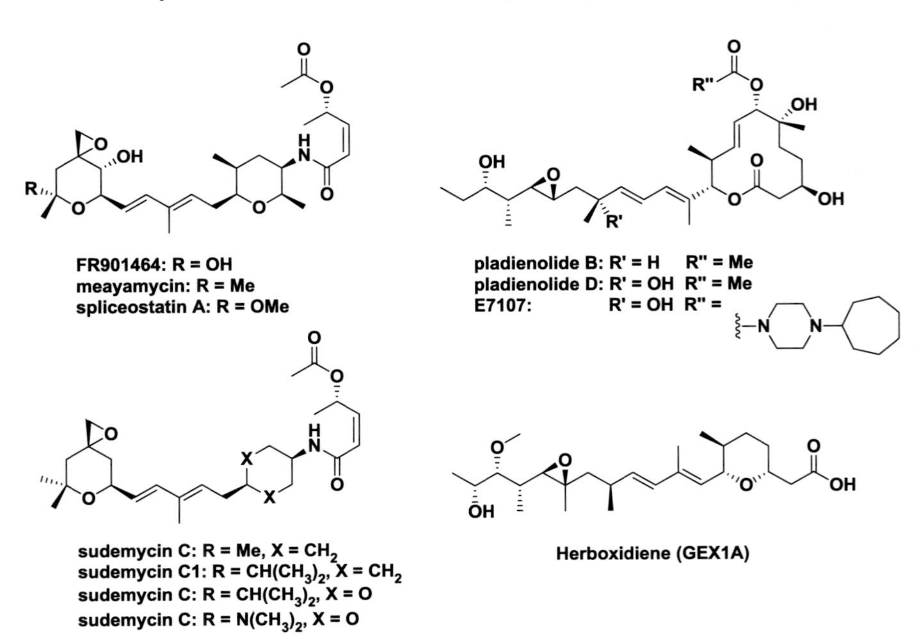

Fig. 23 Small molecules that inhibit pre-mRNA splicing by binding the SF3b complex

proteins involved in cell cycle regulation, after treatment with FR901464 or a more stable methyl ketal analog spliceostatin A, indicated that the protein p27 was upregulated [139]. Further analysis indicated that the upregulated protein was a C-terminally truncated variant of p27, and that the truncated variant did not arise from proteolytic processing. A biotinylated derivative of SSA was used in conjunction with streptavidin beads to pull out possible binding partners. Three proteins were enriched in the 130–160 kDa range. Using LC-MS these proteins were shown to be SAP155, SAP145, and SAP130, all of which are components of the SF3b sub-complex of the U2 snRNP. Although FR901464 and SSA both have reactive epoxides, they do not appear to form covalent attachments to the binding partners as evidenced by wash out experiments.

A class of macrolides, with similar structural features as spliceostatin A, revealed a similar mode of action. Pladienolides B and D (Fig. 23) were isolated from culture of *Streptomyces platensis* [140]. Treatment of cells with ^{3}H and fluorescently labeled analogs of pladienolide B showed that the compound was localized to the nucleus and present in nuclear speckles [141]. The nuclear fraction of cells treated with the ^{3}H labeled pladienolide B was subjected to immunoprecipitation using antibodies that bind nuclear speckle proteins. The results suggested compound binding to components of the U2 snRNP. Of the various antibodies raised against the components of the U2 snRNP, anti-SAP155 most efficiently coprecipitated the ^{3}H labeled pladienolide B. Using a similar coprecipitation strategy with anti-SAP155 antibody and the nuclear fraction, a pladienolide B analog containing a diazirine photo-reactive group was crosslinked to its binding partner after UV irradiation. The pladienolide B analog was also linked to biotin, allowing detection using streptavidin horseradish peroxidase (streptavidin-HRP). A single prominent band of approximately 140 kDa was seen only after treatment with both probe and UV irradiation. SAP130 was determined to be the target of the crosslinking agent.

Spliceostatin A, pladienolide B, and other molecules with a similar chemical architecture (Fig. 23) [142–144] influence splicing by binding to proteins in the SF3b complex, particularly SF3b1 (SAP155). It has been suggested that the binding event alone is not responsible for the activity, since inactive analogs effectively outcompete active compounds at the binding site to restore splicing activity in treated cells [145], leaving room for speculation that active molecules induce a conformation change upon binding the complex. Mechanistic studies have shown that SF3b interacting splicing inhibitors may impede the transition between complex A and B [146], compromise the fidelity of branch point recognition by U2 snRNP [147], and/or impede exon ligation [145]. Regardless of the exact mechanism, it is clear that although a core component of the spliceosome is being compromised, not all splicing events are equally affected [147]. The underlying principle may be rooted in the fact that not all pre-mRNA substrates have the same affinity for spliceosomal components (i.e., there is competition between introns for U2 snRNP association). RNA sequence and abundance, and associated splicing factors all play a role in determining the favorability of the pre-mRNA/U2 snRNP interaction. Ultimately, in an inhibitory environment pre-mRNA sequences that lead to more weakly associated complexes may experience greater changes in splicing than their more stable counterparts [148].

So far, the therapeutic application of SF3b binders has been focused on cancer, since many analogs arrest cancer cell growth at nanomolar concentration. An analog of pladienolide D with improved pharmaceutical properties, E7107 (Fig. 23), showed promising results in oncology-related preclinical studies with several xenograft models [149], leading to entry into a phase 1 clinical trial [150]. The trial was halted due to vision loss experienced by two patients. The exact underlying mechanism for the visual toxicity remains unclear. Additional therapeutic applications of these molecules remain to be seen.

4.2 Spliceosome Stalling with Isoginkgetin

Whereas the SF3b binding compounds were identified from screening methods designed to identify cytotoxic or cytostatic compounds, followed by detailed mechanistic work to identify the target, a more direct approach of screening molecules in assays designed to detect splicing inhibition has led to the discovery of various other splicing inhibitors. In one such assay developed in HEK293 cells, the open reading frame of firefly luciferase was placed downstream of a minigene cassette containing exon 6-intron 6-exon 7 of human *TPI* (Fig. 24a) [151]. All the in-frame stop codons were removed from intron 6 plus an additional base was added to exon 7 such that luciferase only remains in the proper translational reading frame

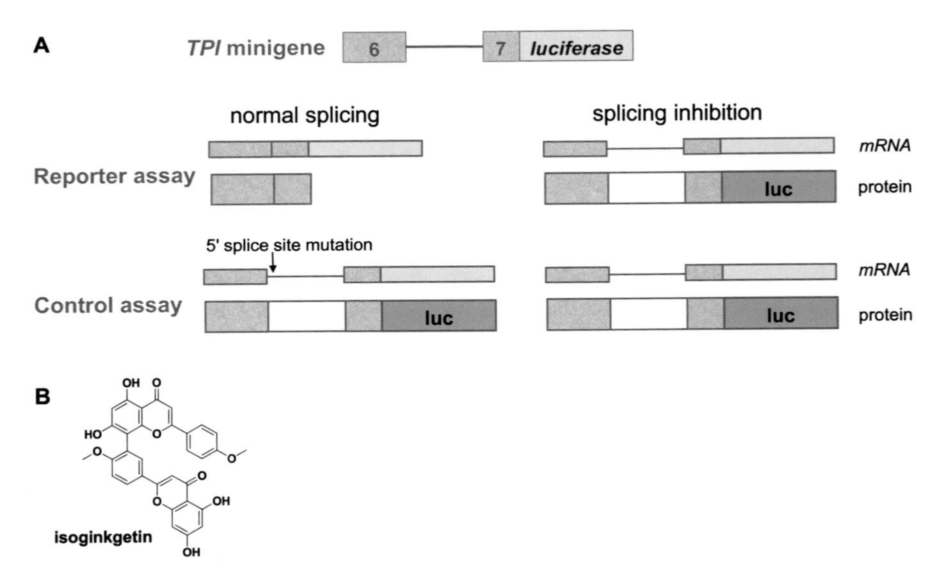

Fig. 24 (**a**) Construct designed to identify pre-mRNA splicing inhibitors. (**b**) Small molecule isoginkgetin identified to inhibit pre-mRNA splicing

in the intron-containing mRNA. Although transcripts that retain introns are not typically exported from the nucleus for translation, a small amount of exported transcript is sufficient to generate a luminescence signal that distinguishes changes in splicing. An additional control construct was created with a mutation at the 5′ splice site of exon 6 that effectively abolishes splicing. Comparing the effect of compound treatment in the reporter assay vs. the control assay distinguishes splicing changes from other effects (e.g., transport). Approximately 8,000 compounds from natural product and synthetic libraries were screened in the assay. A bisflavanoid natural product, isoginkgetin, stood far above the rest of the compounds in its ability to inhibit splicing (Fig. 24b). Isoginkgetin had been previously described as a potential anti-cancer agent due to its unique activity [152]. To test whether the activity was unique to the reporter assay, several endogenous transcripts were evaluated, including those encoding β-tubulin, actin, DNAJB1, TPI, glyceraldehyde-3-phosphate dehydrogenase, and RIOK3. All of these transcripts exhibited pre-mRNA accumulation (2–20-fold) upon treatment with isoginkgetin. It appears that isoginkgetin blocks spliceosome transition from complex A to complex B. The binding target for isoginkgetin has not been determined. One possible candidate is the SRPK2 kinase that phosphorylates the DEAD-box ATPase PRP28 responsible for complex B formation.

4.3 Spliceosome Stalling with Madrasin

A splicing inhibition assay was developed in HeLa cell nuclear extract, utilizing a FLAG-tagged spliceosomal protein and anti-FLAG peroxidase conjugated antibody [153]. The pre-mRNA substrate in the assay lacks the 3′ exon thus allowing spliceosome assembly and step 1, but not step 2 in the catalytic cycle (Fig. 2). The quantification of complex C formation is possible due to the stalled splicing process (Fig. 25). Any inhibition of early spliceosome formation or of step 1 in the catalytic cycle would be detected by a lower abundance of complex C. Screening a library of 30,000 compounds at 50 μM concentration revealed two cyclic esters, psoromic acid and norstictic acid, which inhibited splicing by more than 50%. In a separate study, a very similar assay was used to screen a library of ~72,000 compounds [154]. Madrasin (i.e., 2-((7-methoxy-4-methylquinazolin-2-yl)*a*mino)-5,6-*d*imethylpyrimidin-4(3H)-one *R*NA *s*plicing *in*hibitor) was identified as the most active compound. Madrasin completely inhibits splicing of the Ad1 pre-mRNA at 62.5 μM, whereas splicing of the MINX pre-mRNA was completely inhibited at 150 μM. Analysis using native agarose gel electrophoresis indicates that madrasin interferes with one or more of the early steps in spliceosome assembly, allowing the formation of the complex A but blocking the formation of subsequent larger spliceosome complexes. When HeLa and HEK293 cells were treated with up to 30 μM madrasin, pre-mRNA splicing inhibition was observed for several transcripts, ultimately causing cell cycle arrest. As is the case with the other general

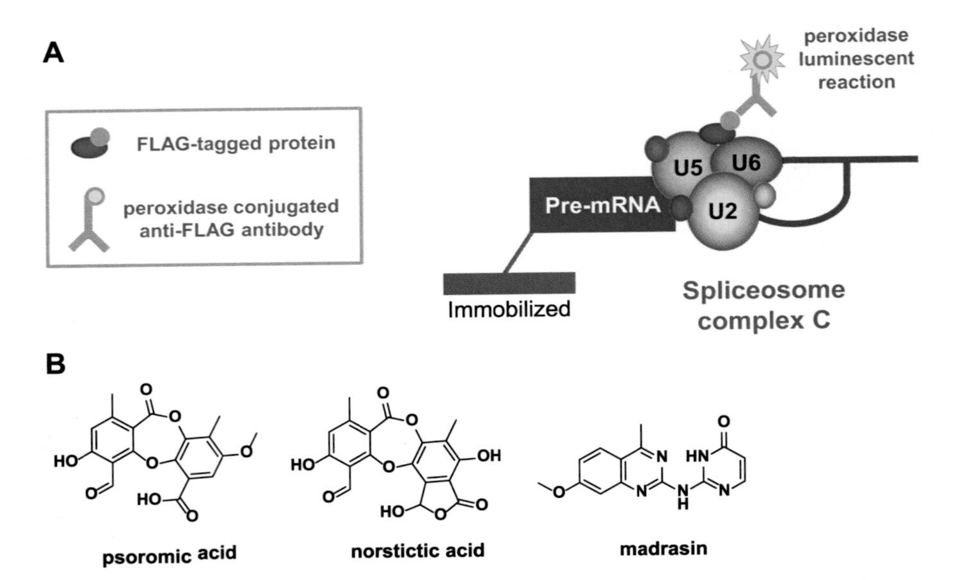

Fig. 25 (**a**) Assay designed to measure the amount of spliceosome complex-C formed nuclear extracts. (**b**) Small molecules that inhibit complex-C formation

splicing inhibitors, madrasin may have limited clinical benefit due to a lack of selectivity.

5 Conclusion

In humans the assembly of most mRNA transcripts requires the precise cleavage and formation of multiple phosphodiester bonds in nascent pre-mRNA polymers. This process of splicing is highly controlled, yet maintains sufficient flexibility to proceed down alternative paths. The combination of control and flexibility presents a unique opportunity for molecular intervention during gene expression. We have described in this chapter various chemical agents that have the capacity to alter the natural process of pre-mRNA splicing, thereby producing different levels of splicing products than found under natural conditions. This technique has powerful therapeutic application where mutation has caused certain splice variants to be under- or over-represented. Additionally, splicing manipulation may be used to bypass a deleterious section of mRNA and to reduce the expression of an undesired gene product.

Over the past several years exceptional advances have been made with SSOs and small molecules, the two leading therapeutic modalities for splicing modulation. SSOs have long been recognized for sequence selectivity, but often knocked for difficulty in administration and tissue penetration. Recent advances in delivery and penetration-enhancing agents have greatly improved the in vivo efficacy of SSOs. Small molecules, on the other hand, rely on the strength of oral delivery and broad distribution, but traditionally have suffered from the lack of selectivity in splicing biology. Recent discoveries of selective small molecule splicing modifiers in the areas of SMA and FD have begun to challenge this selectivity paradigm. We expect that additional advances in drugging pre-mRNA splicing will be achieved as our understanding of the potential targets and mechanisms increases. A combination of creative design and applied knowledge should lead to promising new drugs in this field in the future.

References

1. Sharp PA (2005) Trends Biochem Sci 30:279
2. Lee Y, Rio DC (2015) Annu Rev Biochem 84:291
3. Wang ET, Sandberg R, Luo S, Khrebtukova I, Zhang L, Mayr C, Kingsmore SF, Schroth GP, Burge CB (2008) Nature 456:470
4. Merkin J, Russell C, Chen P, Burge CB (2012) Science 338:1593
5. Barbosa-Morais NL, Irimia M, Pan Q, Xiong HY, Gueroussov S, Lee LJ, Slobodeniuc V, Kutter C, Watt S, Colak R, Kim T, Misquitta-Ali CM, Wilson MD, Kim PM, Odom DT, Frey BJ, Blencowe BJ (2012) Science 338:1587
6. Reyes A, Anders S, Weatheritt RJ, Gibson TJ, Steinmetz LM, Huber W (2013) Proc Natl Acad Sci U S A 110:15377
7. Melé M, Ferreira PG, Reverter F, DeLuca DS, Monlong J, Sammeth M, Young TR, Goldmann JM, Pervouchine DD, Sullivan TJ, Johnson R, Segrè AV, Djebali S, Niarchou A, The GTEx Consortium, Wright FA, Lappalainen T, Calvo M, Getz G, Dermitzakis ET, Ardlie KG, Guigó R (2015) Science 348:660
8. Tress ML, Abascal F, Valencia A (2017) Trends Biochem Sci 42:98
9. Mudge JM, Frankish A, Fernandez-Banet J, Alioto T, Derrien T, Howald C, Reymond A, Guigó R, Hubbard T, Harrow J (2011) Mol Biol Evol 28:2949
10. Cascino I, Fiucci G, Papoff G, Ruberti G (1995) J Immunol 154:2706
11. Medina MW, Gao F, Naidoo D, Rudel LL, Temel RE, McDaniel AL, Marshall SM, Krauss RM (2011) PLoS One 6(4):e19420
12. Lareau LF, Brenner SE (2015) Mol Biol Evol 32:1072
13. van Roon-Mom WMC, Aartsma-Rus A (2012) Methods Mol Biol 867:79
14. Havens MA, Duelli DM, Hastings ML (2013) WIREs RNA 4:247
15. Eteplirsen NDA #206466. www.fda.gov
16. Nusinersen NDA #209531. www.fda.gov
17. Lefebvre S, Bürglen L, Reboullet S, Clermont O, Burlet P, Viollet L, Benichou B, Cruaud C, Millasseau P, Zeviani M, Paslier DL, Frézal J, Cohen D, Weissenbach J, Munnich A, Melki J (1995) Cell 80:155
18. Lorson CL, Hahnen E, Androphy J, Wirth B (1999) Proc Natl Acad Sci U S A 96:6307
19. Cho S, Dreyfuss G (2010) Genes Dev 24:438
20. Burnett BG, Muñoz E, Tandon A, Kwon DY, Sumner CJ, Fischbeck KH (2009) Mol Cell Biol 29:1107
21. Baughan TD, Dickson A, Osman EY, Lorson CL (2009) Hum Mol Genet 18:1600
22. Hua Y, Vickers TA, Baker BF, Bennett CF, Krainer AR (2007) PLoS Biol 5:729
23. Hua Y, Vickers TA, Okunola HL, Bennett CF, Krainer AR (2008) Am J Hum Genet 82:834
24. Singh NK, Singh NN, Androphy EJ, Singh RN (2006) Mol Cell Biol 26:1333
25. Lim SR, Hertel KJ (2001) J Biol Chem 276:45476

26. Madocsai C, Lim SR, Geib T, Lam BJ, Hertel KJ (2005) Mol Ther 12:1013
27. Dickson A, Osman E, Lorson CL (2008) Hum Gene Ther 19:1307
28. Odermatt P, Trüb J, Furrer L, Fricker R, Marti A, Schümperli D (2016) Mol Ther 24:1797
29. Williams JH, Schray RC, Patterson CA, Ayitey SO, Tallent MK, Lutz GJ (2009) J Neurosci 29:7633
30. Hua Y, Sahashi K, Rigo F, Hung G, Horev G, Bennett CF, Krainer AR (2011) Nature 478:123
31. Porensky PN, Mitrpant C, McGovern VL, Bevan AK, Foust KD, Kaspar BK, Wilton SD, Burghes AHM (2012) Hum Mol Genet 21:1625
32. Zhou H, Janghra N, Mitrpant C, Dickinson RL, Anthony K, Price L, Eperon IC, Wilton SD, Morgan J, Muntoni F (2013) Hum Gene Ther 24:331
33. Hsieh-Li HM, Chang JG, Jong YJ, Wu MH, Wang NM, Tsai CH, Li H (2000) Nat Genet 24:66
34. Passini MA, Bu J, Richards AM, Kinnecom C, Sardi SP, Stanek LM, Hua Y, Rigo F, Matson J, Hung G, Kaye EM, Shihabuddin LS, Krainer AR, Bennett CF, Cheng SH (2011) Sci Transl Med 3:72ra18
35. Mitrpant C, Porensky P, Zhou H, Price L, Muntoni F, Fletcher S, Wilton SD, Burghes AHM (2013) PLoS One 8:e62114
36. Dervan PB, Edelson BS (2003) Curr Opin Struct Biol 13:284
37. Thomas JR, Hergenrother PJ (2008) Chem Rev 108:1171
38. Guan L, Disney MD (2012) ACS Chem Biol 7:73
39. Rzuczek SG, Colgan LA, Nakai Y, Cameron MD, Furling D, Yasuda R, Disney MD (2017) Nat Chem Biol 13:188
40. Woll MG, Qi H, Turpoff A, Zhang N, Zhang X, Chen G, Li C, Huang S, Yang T, Moon Y-C, Lee C-S, Choi S, Almstead NG, Naryshkin NA, Dakka A, Narasimhan J, Gabbeta V, Welch E, Zhao X, Risher N, Sheedy J, Weetall M, Karp GM (2016) J Med Chem 59:6070
41. Naryshkin NA, Weetall M, Dakka A, Narasimhan J, Zhao X, Feng Z, Ling KKY, Karp GM, Qi H, Woll MG, Chen G, Zhang N, Gabbeta V, Vazirani P, Bhattacharyya A, Furia B, Risher N, Sheedy J, Kong R, Ma J, Turpoff A, Lee CS, Zhang X, Moon Y-C, Trifillis P, Welch EM, Colacino JM, Babiak J, Almstead NG, Peltz SW, Eng LA, Chen KS, Mull JL, Lynes MS, Rubin LL, Fontoura P, Santarelli L, Haehnke D, McCarthy KD, Schmucki R, Ebeling M, Sivaramakrishnan M, Ko C-P, Paushkin SV, Ratni H, Gerlach I, Ghosh A, Metzger F (2014) Science 345:688
42. Ratni H, Karp GM, Weetall M, Naryshkin NA, Paushkin SV, Chen KS, McCarthy KD, Qi H, Turpoff A, Woll MG, Zhang X, Zhang N, Yang T, Dakka A, Vazirani P, Zhao X, Pinard E, Green L, David-Pierson P, Tuerck D, Poirier A, Muster W, Kirchner S, Mueller L, Gerlach I, Metzger F (2016) J Med Chem 59:6086
43. SMN2 splicing modifier RG7800 increases SMN protein in first study in SMA patients. In: World Muscle Society Congress, Brighton, UK, 30 Sept–4 Oct 2015
44. Palacino J, Swalley SE, Song C, Cheung AK, Shu L, Zhang X, Van Hoosear M, Shin Y, Chin DN, Keller CG, Beibel M, Renaud NA, Smith TA, Salcius M, Shi X, Hild M, Servais R, Jain M, Deng L, Bullock C, McLellan M, Schuierer S, Murphy L, Blommers MJJ, Blaustein C, Berenshteyn F, Lacoste A, Thomas JR, Roma G, Michaud GA, Tseng BS, Porter JA, Myer VE, Tallarico JA, Hamann LG, Curtis D, Fishman MC, Dietrich WF, Dales NA, Sivasankaran R (2015) Nat Chem Biol 11:511
46. Anderson SL, Coli R, Daly IW, Kichula EA, Rork MJ, Volpi SA, Ekstein J, Rubin BY (2001) Am J Hum Genet 68:753
47. Slaugenhaupt SA, Blumenfeld A, Gill SP, Leyne M, Mull J, Cuajungco MP, Liebert CB, Chadwick B, Idelson M, Reznik L, Robbins CM, Makalowska I, Brownstein MJ, Krappmann D, Scheidereit C, Maayan C, Axelrod FB, Gusella JF (2001) Am J Hum Genet 68:598
48. Ibrahim EC, Hims MM, Shomron N, Burge CB, Slaugenhaupt SA, Reed R (2007) Hum Mutat 28:41

49. Slaugenhaupt SA, Mull J, Leyne M, Cuajungco MP, Gill SP, Hims MM, Quintero F, Axelrod FB, Gusella JF (2004) Hum Mol Genet 13:429
50. Cuajungco MP, Leyne M, Mull J, Gill SP, Lu W, Zagzag D, Axelrod FB, Maayan C, Gusella JF, Slaugenhaupt SA (2003) Am J Hum Genet 72:749
51. Anderson SL, Qiu J, Rubin BY (2003) Biochem Biophys Res Commun 310:627
52. Keren H, Donyo M, Zeevi D, Maayan C, Pupko T, Ast G (2010) PLoS One 5:e15884
53. Liu B, Anderson SL, Qiu J, Rubin BY (2013) FEBS J 280:3632
54. Hims MM, Ibrahim EC, Leyne M, Mull J, Liu L, Lazaro C, Shetty RS, Gill S, Gusella JF, Reed R, Slaugenhaupt SA (2007) J Mol Med 85:149
55. Hims MM, Shetty RS, Pickel J, Mull J, Leyne M, Liu L, Gusella JF, Slaugenhaupt SA (2007) Genomics 90:389
56. Shetty RS, Gallagher CS, Chen YT, Hims MM, Mull J, Leyne M, Pickel J, Kwok D, Slaugenhaupt SA (2011) Hum Mol Genet 20:4093
57. Axelrod FB, Liebes L, Gold-Von Simson G, Mendoza S, Mull J, Leyne M, Norcliffe-Kaufmann L, Kaufmann H, Slaugenhaupt SA (2011) Pediatr Res 70:480
58. Slaugenhaupt SA, Johnson G, Paquette WD, Zhang W, Marugan J (2016) WO 2016115434 A1
59. Yoshida M, Kataoka N, Miyauchi K, Ohe K, Iida K, Yoshida S, Nojima T, Okuno Y, Onogi H, Usui T, Takeuchi A, Hosoya T, Suzuki T, Hagiwara M (2015) Proc Natl Acad Sci U S A 112:2764
60. Monaco AP, Neve RL, Colletti-Feener C, Bertelson CJ, Kurnit DM, Kunkel LM (1986) Nature 323:646
61. Monaco AP, Bertelson CJ, Liechti-Gallati S, Moser H, Kunkel KL (1988) Genomics 2:90
62. Aartsma-Rus A, Fokkema I, Verschuuren J, Ginjaar I, van Deutekom J, van Ommen GJ, den Dunnen JT (2009) Hum Mutat 30:293
63. Nicholson LV, Davison K, Johnson MA, Slater CR, Young C, Bhattacharya S, Gardner-Medwin D, Harris JB (1989) J Neurol Sci 94:137
64. Fanin M, Danieli GA, Vitiello L, Senter L, Angelini C (1992) Neuromuscul Disord 2:41
65. Arechavala-Gomeza V, Kinali M, Feng L, Guglieri M, Edge G, Main M, Hunt D, Lehovsky J, Straub V, Bushby K, Sewry CA, Morgan JE, Muntoni F (2010) Neuromuscul Disord 20:295
66. Pigozzo SR, Da Re L, Romualdi C, Mazzara PG, Galletta E, Fletcher S, Wilton SD, Vitiello L (2013) PLoS One 28:e72147
67. Bulfield G, Siller WG, Wight PA, Moore KJ (1984) Proc Natl Acad Sci U S A 81:1189
68. Lu QL, Morris GE, Wilton SD, Ly T, Artem'yeva OV, Strong P, Partridge TA (2000) J Cell Biol 148:985
69. Wilton SD, Dye DE, Laing NG (1997) Muscle Nerve 20:728
70. Bladen CL, Salgado D, Monges S, Foncuberta ME, Kekou K, Kosma K, Dawkins H, Lamont L, Roy AJ, Chamova T, Guergueltcheva V, Chan S, Korngut L, Campbell C, Dai Y, Wang J, Barišić N, Brabec P, Lahdetie J, Walter MC, Schreiber-Katz O, Karcagi V, Garami M, Viswanathan V, Bayat F, Buccella F, Kimura E, Koeks Z, van den Bergen JC, Rodrigues M, Roxburgh R, Lusakowska A, Kostera-Pruszczyk A, Zimowski J, Santos R, Neagu E, Artemieva S, Rasic VM, Vojinovic D, Posada M, Bloetzer C, Jeannet PY, Joncourt F, Díaz-Manera J, Gallardo E, Karaduman AA, Topaloğlu H, El Sherif R, Stringer A, Shatillo AV, Martin AS, Peay HL, Bellgard MI, Kirschner J, Flanigan KM, Straub V, Bushby K, Verschuuren J, Aartsma-Rus A, Béroud C, Lochmüller H (2015) Hum Mutat 36:395
71. Matsuo M, Masumura T, Nishio H, Nakajima T, Kitoh Y, Takumi T, Koga J, Nakamura H (1991) J Clin Invest 87:2127
72. Takeshima Y, Nishio H, Sakamoto H, Nakamura H, Matsuo M (1995) J Clin Invest 95:515
73. Pramono ZA, Takeshima Y, Alimsardjono H, Ishii A, Takeda S, Matsuo M (1996) Biochem Biophys Res Commun 226:445
74. Takeshima Y, Wada H, Yagi M, Ishikawa Y, Ishikawa Y, Minami R, Nakamura H, Matsuo M (2001) Brain Dev 23:788
75. Takeshima Y, Yagi M, Wada H, Matsuo M (2005) Brain Dev 27:488

76. Takeshima Y, Yagi M, Wada H, Ishibashi K, Nishiyama A, Kakumoto M, Sakaeda T, Saura R, Okumura K, Matsuo M (2006) Pediatr Res 59:690
77. Klein CJ, Coovert DD, Bulman DE, Ray PN, Mendell JR, Burghes AH (1992) Am J Hum Genet 50:950
78. Wallgren-Pettersson C, Jasani B, Rosser LG, Lazarou LP, Nicholson LVB, Clarke A (1993) J Neurol Sci 118:56
79. Wilton SD, Lloyd F, Carville K, Fletcher S, Honeyman K, Agrawal S (1999) Neuromuscul Disord 9:330
80. Dunckley MG, Manoharan M, Villiet P, Eperon IC, Dickson G (1998) Hum Mol Genet 7:1083
81. van Deutekom JC, Bremmer-Bout M, Janson AA, Ginjaar IB, Baas F, den Dunnen JT, van Ommen GJ (2001) Hum Mol Genet 10:1547
82. Bremmer-Bout M, Aartsma-Rus A, De Meijer EJ, Kaman WE, Janson AAM, Vossen RHAM, Van Ommen G-JB, Den Dunnen JT, Van Deutekom JCT (2004) Mol Ther 10:232
83. Aartsma-Rus A, De Winter CL, Janson AA, Kaman WE, Van Ommen GJ, Den Dunnen JT, Van Deutekom JC (2005) Oligonucleotides 15:284
84. Wilton SD, Fall AM, Harding PL, McClorey G, Coleman C, Fletcher S (2007) Mol Ther 15:1288
85. Adkin CF, Meloni PL, Fletcher S, Adams AM, Muntoni F, Wong B, Wilton SD (2012) Neuromuscul Disord 22:297
86. Aartsma-Rus A, Janson AAM, Kaman WE, Bremmer-Bout M, van Ommen G-JB, den Dunnen JT, van Deutekom JCT (2004) Am J Hum Genet 74:83
87. Mitrpant C, Fletcher S, Wilton SD (2009) Curr Mol Pharmacol 2:110
88. Jarver P, O'Donovan L, Gait MJ (2014) Nucleic Acids Ther 24:37
89. Aartsma-Rus A, Kaman WE, Bremmer-Bout M, Janson AA, den Dunnen JT, van Ommen GJ, van Deutekom JC (2004) Gene Ther 11:1391
90. Arzumanov A, Walsh AP, Rajwanshi VK, Kumar R, Wengel J, Gait MJ (2001) Biochemistry 40:14645
91. Saleh AF, Arzumanov AA, Gait MJ (2012) Methods Mol Biol 867:365
92. Fletcher S, Honeyman K, Fall AM, Harding PL, Johnsen RD, Wilton SD (2006) J Gene Med 8:207
93. Yin H, Betts C, Saleh AF, Ivanova GD, Lee H, Seow Y, Kim D, Gait MJ, Wood MJ (2010) Mol Ther 18:819
94. Goyenvalle A, Griffith G, Babbs A, El Andaloussi S, Ezzat K, Avril A, Dugovic B, Chaussenot R, Ferry A, Voit T, Amthor H, Bühr C, Schürch S, Wood MJ, Davies KE, Vaillend C, Leumann C, Garcia L (2015) Nat Med 21:270
95. Summerton J, Weller D (1997) Antisense Nucleic Acid Drug Dev 7:187
96. Popplewell LJ, Adkin C, Arechavala-Gomeza V, Aartsma-Rus A, de Winter CL, Wilton SD, Morgan JE, Muntoni F, Graham IR, Dickson G (2010) Neuromuscul Disord 20:102
97. Aartsma-Rus A (2010) RNA Biol 7:453
98. Heemskerk HA, de Winter CL, de Kimpe SJ, van Kuik-Romeijn P, Heuvelmans N, Platenburg GJ, van Ommen GJ, van Deutekom JC, Aartsma-Rus A (2009) J Gene Med 11:257
99. Vila MC, Klimek MB, Novak JS, Rayavarapu S, Uaesoontrachoon K, Boehler JF, Fiorillo AA, Hogarth MW, Zhang A, Shaughnessy C, Gordish-Dressman H, Burki U, Straub V, Lu QL, Partridge TA, Brown KJ, Hathout Y, van den Anker J, Hoffman EP, Nagaraju K (2015) Skelet Muscle 5:44
100. Alter J, Lou F, Rabinowitz A, Yin H, Rosenfeld J, Wilton SD, Partridge TA, Lu QL (2006) Nat Med 12:175
101. Yokota T, Lu QL, Partridge T, Kobayashi M, Nakamura A, Takeda S, Hoffman E (2009) Ann Neurol 65:667
102. Aoki Y, Yokota T, Nagata T, Nakamura A, Tanihata J, Saito T, Duguez SM, Nagaraju K, Hoffman EP, Partridge T, Takeda S (2012) Proc Natl Acad Sci U S A 109:13763

103. Echigoya Y, Aoki Y, Miskew B, Panesar D, Touznik A, Nagata T, Tanihata J, Nakamura A, Nagaraju K, Yokota T (2015) Mol Ther Nucleic Acids 4:e225
104. Echigoya Y, Yokota T (2014) Nucleic Acid Ther 24:57
105. Lebleu B, Moulton HM, Abes R, Ivanova GD, Abes S, Stein DA, Iversen PL, Arzumanov AA, Gait MJ (2008) Adv Drug Deliv Rev 60:517
106. Boisguérin P, Deshayes S, Gait MJ, O'Donovan L, Godfrey C, Betts CA, Wood MJ, Lebleu B (2015) Adv Drug Deliv Rev 87:52
107. Jirka SM, Heemskerk H, Tanganyika-de Winter CL, Muilwijk D, Pang KH, de Visser PC, Janson A, Karnaoukh TG, Vermue R, 't Hoen PA, van Deutekom JC, Aguilera B, Aartsma-Rus A (2014) Nucleic Acid Ther 24:25
108. Hammond SM, Hazell G, Shabanpoor F, Saleh AF, Bowerman M, Sleigh JN, Meijboom KE, Zhou H, Muntoni F, Talbot K, Gait MJ, Wood MJ (2016) Proc Natl Acad Sci U S A 113:10962
109. Fletcher S, Honeyman K, Fall AM, Harding PL, Johnsen RD, Steinhaus JP, Moulton HM, Iversen PL, Wilton SD (2007) Mol Ther 15:1587
110. Wu B, Moulton HM, Iverson PL, Jiang J, Li J, Li J, Spurney CF, Sali A, Guerron AD, Nagaraju K, Doran T, Lu P, Xiao X, Lu QL (2008) PNAS 105:14814
111. Moulton HM, Moulton JD (2010) Biochim Biophys Acta 1798:2296
112. van Westering TL, Betts CA, Wood MJ (2015) Molecules 20:8823
113. Amantana A, Moulton HM, Cate ML, Reddy MT, Whitehead T, Hassinger JN, Youngblood DS, Iversen PL (2007) Bioconjug Chem 18:1325
114. Incitti T, De Angelis FG, Cazzella V, Sthandier O, Pinnarò C, Legnini I, Bozzoni I (2010) Mol Ther 18:1675
115. Cazzella V, Martone J, Pinnarò C, Santini T, Twayana SS, Sthandier O, D'Amico A, Ricotti V, Bertini E, Muntoni F, Bozzoni I (2012) Mol Ther 20:2134
116. De Angelis FG, Sthandier O, Berarducci B, Toso S, Galluzzi G, Ricci E, Cossu G, Bozzoni I (2002) Proc Natl Acad Sci U S A 99:9456
117. Martone J, De Angelis FG, Bozzoni I (2012) Methods Mol Biol 867:239
118. Goyenvalle A, Vulin A, Fougerousse F, Leturcq F, Kaplan JC, Garcia L, Danos O (2004) Science 306:1796
119. Vulin A, Barthélémy I, Goyenvalle A, Thibaud JL, Beley C, Griffith G, Benchaouir R, le Hir M, Unterfinger Y, Lorain S, Dreyfus P, Voit T, Carlier P, Blot S, Garcia L (2012) Mol Ther 20:2120
120. Goyenvalle A (2012) Methods Mol Biol 867:259
121. Sharma M, McFarlane C, Kambadur R, Kukreti H, Bonala S, Srinivasan S (2015) IUBMB Life 67:589
122. Kang JK, Malerba A, Popplewell L, Foster K, Dickson G (2011) Mol Ther 19:159
123. Kemaladewi DU, Hoogaars WM, van Heiningen SH, Terlouw S, de Gorter DJ, den Dunnen JT, van Ommen GJ, Aartsma-Rus A, ten Dijke P, 't Hoen PA (2011) BMC Med Genomics 4:36
124. Lu-Nguyen NB, Jarmin SA, Saleh AF, Popplewell L, Gait MJ, Dickson G (2015) Mol Ther 23:1341
125. Amthor H, Macharia R, Navarrete R, Schuelke M, Brown SB, Otto A, Voit T, Muntoni F, Vrbova G, Partridge T, Zammit P, Bunger L, Patel K (2007) Proc Natl Acad Sci U S A 104:1835
126. Koo T, Wood MJ (2013) Hum Gene Ther 24:479
127. Shimizu-Motohashi Y, Miyatake S, Komaki H, Takeda S, Aoki Y (2016) Am J Transl Res 8:2471
128. Havens MA, Hastings ML (2016) Nucleic Acids Res 44:6549
129. van Deutekom JC, Janson AA, Ginjaar IB, Frankhuizen WS, Aartsma-Rus A, Bremmer-Bout M, den Dunnen JT, Koop K, van der Kooi AJ, Goemans NM, de Kimpe SJ, Ekhart PF, Venneker EH, Platenburg GJ, Verschuuren JJ, van Ommen GJ (2007) N Engl J Med 357:2677

130. Kinali M, Arechavala-Gomeza V, Feng L, Cirak S, Hunt D, Adkin C, Guglieri M, Ashton E, Abbs S, Nihoyannopoulos P, Garralda ME, Rutherford M, McCulley C, Popplewell L, Graham IR, Dickson G, Wood MJ, Wells DJ, Wilton SD, Kole R, Straub V, Bushby K, Sewry C, Morgan JE, Muntoni F (2009) Lancet Neurol 8:918
131. Muraki M, Ohkawara B, Hosoya T, Onogi H, Koizumi J, Koizumi T, Sumi K, Yomoda J, Murray MV, Kimura H, Furuichi K, Shibuya H, Krainer AR, Suzuki M, Hagiwara M (2004) J Biol Chem 279:24246
132. Nishida A, Kataoka N, Takeshima Y, Yagi M, Awano H, Ota M, Itoh K, Hagiwara M, Matsuo M (2011) Nat Commun 2:308
133. Kendall GC, Mokhonova EI, Moran M, Sejbuk NE, Wang DW, Silva O, Wang RT, Martinez L, Lu QL, Damoiseaux R, Spencer MJ, Nelson SF, Miceli MC (2012) Sci Transl Med 4:164
134. Hu Y, Wu B, Zillmer A, Lu P, Benrashid E, Wang M, Doran T, Shaban M, Wu X, Lu QL (2010) Mol Ther 18:812
135. Verhaart IE, Aartsma-Rus A (2012) PLoS Curr 4
136. Bonnal S, Vigevani L, Valcárcel J (2012) Nat Rev Drug Discov 11:847
137. Salton M, Misteli T (2016) Trends Mol Med 22:28
138. Nakajima H, Hori Y, Terano H, Okuhara M, Manda T, Matsumoto S, Shimomura K (1996) J Antibiot (Tokyo) 49:1204
139. Kaida D, Motoyoshi H, Tashiro E, Nojima T, Hagiwara M, Ishigami K, Watanabe H, Kitahara T, Yoshida T, Nakajima H, Tani T, Horinouchi S, Yoshida M (2007) Nat Chem Biol 3:576
140. Mizui Y, Sakai T, Iwata M, Uenaka T, Okamoto K, Shimizu H, Yamori T, Yoshimatsu K, Asada M (2004) J Antibiot (Tokyo) 57:188
141. Kotake Y, Sagane K, Owa T, Mimori-Kiyosue Y, Shimizu H, Uesugi M, Ishihama Y, Iwata M, Mizui Y (2007) Nat Chem Biol 3:570
142. Albert BJ, McPherson PA, O'Brien K, Czaicki NL, Destefino V, Osman S, Li M, Day BW, Grabowski PJ, Moore MJ, Vogt A, Koide K (2009) Mol Cancer Ther 8:2308
143. Fan L, Lagisetti C, Edwards CC, Webb TR, Potter PM (2011) ACS Chem Biol 6:582
144. Hasegawa M, Miura T, Kuzuya K, Inoue A, Won Ki S, Horinouchi S, Yoshida T, Kunoh T, Koseki K, Mino K, Sasaki R, Yoshida M, Mizukami T (2011) ACS Chem Biol 6:229
145. Effenberger KA, Urabe VK, Prichard BE, Ghosh AK, Jurica MS (2016) RNA 22:350
146. Roybal GA, Jurica MS (2010) Nucleic Acids Res 38:6664
147. Corrionero A, Miñana B, Valcárcel J (2011) Genes Dev 25:445
148. Effenberger KA, Urabe VK, Jurica MS (2016) WIREs RNA 8:e1381. doi:10.1002/wrna.1381
149. Iwata M, Ozawa Y, Uenaka T, Shimizu H, Niijima J, Kanada RM, Fukuda Y, Nagai M, Kotake Y, Yoshida M, Tsuchida T, Mizui Y, Yoshimatsu K, Asada M (2004) Proc Am Assoc Cancer Res 45:691
150. Hong DS, Kurzrock R, Naing A, Wheler JJ, Falchook GS, Schiffman JS, Faulkner N, Pilat MJ, O'Brien J, LoRusso P (2014) Invest New Drugs 32:436
151. O'Brien K, Matlin AJ, Lowell AM, Moore MJ (2008) J Biol Chem 283:33147
152. Yoon SO, Shin S, Lee HJ, Chun HK, Chung AS (2006) Mol Cancer Ther 5:2666
153. Samatov TR, Wolf A, Odenwälder P, Bessonov S, Deraeve C, Bon RS, Waldmann H, Lührmann R (2012) Chembiochem 13:640
154. Pawellek A, McElroy S, Samatov T, Mitchell L, Woodland A, Ryder U, Gray D, Lührmann R, Lamond AI (2014) J Biol Chem 289:34683

Top Med Chem (2018) 27: 177–206
DOI: 10.1007/7355_2016_22

Published online: 4 May 2017

Targeting RNA G-Quadruplexes for Potential Therapeutic Applications

Satyaprakash Pandey, Prachi Agarwala, and Souvik Maiti

Abstract RNA G-quadruplexes are non-canonical structures formed in G-rich regions of transcriptome. Considerable research points towards the potential role for these dynamic structures in various genes involved in multiple pathways. The ability of these structures to influence important biological processes has been demonstrated in few selected important genes and more recently in a global manner across the transcriptome. RNA G-quadruplexes are implicated in fundamental processes such as translational regulation, alternative splicing, mRNA transport and telomere homeostasis to name a few. The involvement of these structures in key biological processes makes them attractive targets for therapeutic interventions. Here we discuss the structural features of RNA G-quadruplex that make it more monomorphic to develop ligands unlike the polymorphic DNA G-quadruplexes. Furthermore, the role of RNA G-quadruplexes in important genes involved in diseases and disorders such as neurodegeneration, cancer and viral infections has been highlighted. A summary of ligands and their established roles in targeting RNA G-quadruplex structures has been discussed in context of important processes.

S. Pandey
European Research Institute for the Biology of Ageing, University Medical Center Groningen (UMCG), Groningen, The Netherlands

P. Agarwala
Chemical and Systems Biology Unit, CSIR-Institute of Genomics and Integrative Biology, Mathura Road, New Delhi 110020, India
Academy of Scientific and Innovative Research (AcSIR), Anusandhan Bhawan, 2 RafiMarg, New Delhi 110001, India

S. Maiti (✉)
Chemical and Systems Biology Unit, CSIR-Institute of Genomics and Integrative Biology, Mathura Road, New Delhi 110020, India
Academy of Scientific and Innovative Research (AcSIR), Anusandhan Bhawan, 2 RafiMarg, New Delhi 110001, India
CSIR-National Chemical Laboratory, Dr. Homi Bhabha Road, Pune 400008, India
e-mail: souvik@igib.res.in

Moreover, we highlight the need to develop ligands that are highly specific in targeting RNA G-quadruplexes, a new target to restore homeostasis in dysregulated conditions.

Keywords Cancer, Homesostasis, Ligands, Neurodegeneration, RNA G-quadruplexes, Viral infections

Contents

1 Introduction 178
2 Structural Aspects of RNA G-Quadruplexes 181
3 Biological Functions of G-Quadruplexes 184
3.1 G-Quadruplexes in Neurodegeneration 186
3.2 G-Quadruplexes in Non-coding Transcriptome 190
3.3 RNA G-Quadruplex in Virology 193
3.4 RNA G-Quadruplex in Telomere Biology 194
4 Targeting RNA G-Quadruplexes Using Ligands 195
4.1 NRAS 195
4.2 TRF2 196
4.3 p53 196
4.4 Bcl-2 197
4.5 Other Studies Utilising G-Quadruplex Ligands 198
5 Conclusions and Future Directions 198
References 199

1 Introduction

Central dogma of biology states a unidirectional flow of genetic information from DNA to proteins via RNA as the intermediate messenger molecule [1]. Conventionally, RNA is restricted to acting as blueprint for the translational machinery to produce proteins from DNA code. However, apart from being the messenger, RNA has the capacity to adopt various structures to carry out its functions. ENCODE analysis reported that 85% of human genome is pervasively transcribed but only 3% codes for proteins suggesting additional roles for RNA thereby expanding its repertoire in biological functions [2]. For any structure to be biologically functional, it requires two attributes – dynamicity and specificity [3]. Dynamicity refers to its ability to form and resolve in response to stimulus while specificity refers to its ability to distinguish the true stimulus from the false ones. RNA can undergo Watson–Crick base pairing to form hairpin structures containing double stranded stem with single stranded loop region or mismatched regions that result in bulges. Extensive functional characterisation has been carried out to examine the importance of these structures in different biological modules such as ribozymes, riboswitches, microRNAs (miRNAs) and mRNAs [3, 4–6].

Various antibiotics that are used to treat infections in humans target the translational apparatus by binding to RNA motifs [7, 8]. One of the examples, streptomycin, has been shown to act by binding to structural features of 16S rRNA thus interfering with translation and ultimately resulting in cell death [8]. These initial discoveries had highlighted the importance of structuredness in RNA and its contribution to functions of RNA. A number of techniques such as in-line probing [9], DMS footprinting [10], enzymatic footprinting [11], PARS (Parallel Analysis of RNA Structure) [12] and SHAPE (selective 2′ hydroxyl acylation analysed by primer extension) [13] have been used to understand the structuredness of single and global transcripts. Various high-end structural techniques such as NMR (nuclear magnetic resonance) and X-ray diffraction have also been employed in certain specific cases [14–16]. With the advent of next-generation sequencing technologies, a global structuredness of RNA can be studied using chemicals and enzymatic reactions on the total transcriptome of the cell. SHAPE and PARS are two of the most widely used techniques that have been used to gain insights into the structuredness of whole transcriptome [17, 18].

RNA is devoid of a complementary strand and hence has more potential to form intramolecular structures. Canonical Watson–Crick base pairing structures such as hairpin and stem loop exist and serve as interacting scaffold for different molecules [19], peptides [20] and proteins [21, 22] thus serving as potential therapeutic targets. Apart from these canonical RNA structures, certain non-canonical RNA structures also exist. Triplex RNAs which form on the existing double stranded DNA moiety by non-canonical base pairing with RNA are being explored recently and have been shown to play a role in coding and non-coding RNAs governing processes such as RNA stability and gene regulation [23–26]. One of the widely studied and characterised non-canonical RNA structures which is both dynamic and specific is G-quadruplexes.

G-quadruplexes were discovered by Davies and Gellert when they found that higher concentrations of guanylic acids form gel like structures in aqueous solutions [27]. X-ray diffraction studies later showed these mesh-like structures to be G-quadruplexes [27]. G-quadruplexes are non-canonical secondary structures found in G-rich tracts of nucleic acids both at DNA and RNA level. The basic unit consists of four guanines held together by Hoogsteen bonds wherein each guanine can act as a donor and acceptor of two hydrogen bonds. Four guanine molecules are arranged in a square, planar arrangement with a metal ion coordinated to minimise the repulsion between inwardly oriented oxygen atoms to form a G-quartet (Fig. 1) [28]. Two or more such G-quartets stack upon each other to form a three-dimensional G-quadruplex. Monovalent cations stabilise the G-quadruplex structure in the order of potassium $>$ sodium $>$ lithium [29, 30]. The intervening bases between G-quartets are termed as loops of G-quadruplex and influence the stability of G-quadruplexes. G-quadruplexes can adopt various conformations (parallel, antiparallel or mixed) and topologies which depend upon the sugar conformations, loop length and composition [30].

DNA G-quadruplexes have been ascribed regulatory roles owing to their predominant occurrence at important genomic locations such as telomeric regions and in promoters of proto-oncogenes and depletion in tumour suppressor genes [31]. Various molecules and peptides have been screened in a wake to alter these structures and interfere with their biological processing thus making them attractive therapeutic targets [32, 33]. However in the past two decades, attention has shifted from DNA G-quadruplexes to RNA G-quadruplexes.

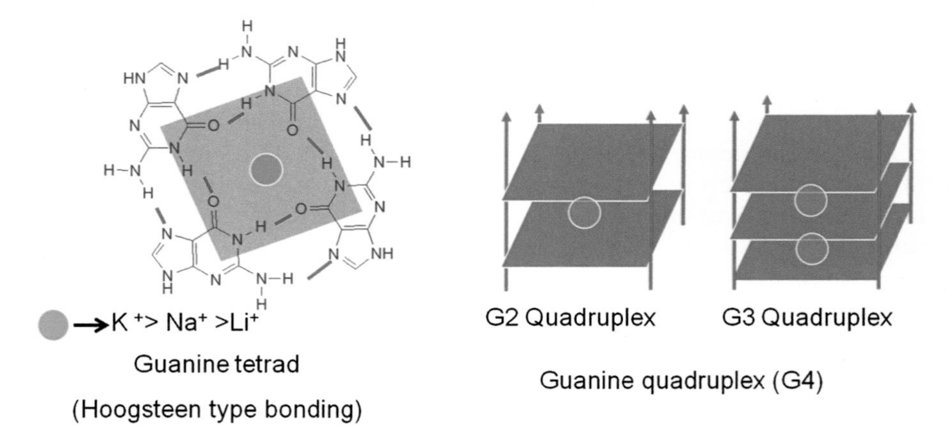

Fig. 1 G-quadruplex structure. Guanines arrange by Hoogsteen bonds (denoted in *red*) to form guanine tetrad which stacks upon each other in the presence of metal ions (denoted in *green spheres*) to form G2 or G3 quadruplexes

RNA G-quadruplexes were discovered by Kim et al. in 1991 in 3′ terminus of 5S rRNA of *Escherichia coli* [34]. RNA G-quadruplex was also found to be located in 3′ untranslated region (UTR) of insulin growth factor II (IGFII) gene downstream of an endonucleolytic cleavage site [35]. The field has grown in the subsequent years with studies of RNA G-quadruplexes located in different mRNAs, long non-coding RNAs (lncRNAs) and telomeric regions [31, 36–40]. Comparative studies have shown RNA G-quadruplexes to be more stable than their DNA counterparts suggesting them to be of potential functionality inside the cell [41, 42]. Bioinformatics studies predict RNA G-quadruplexes to occur in UTRs and telomeric transcripts substantially more than in coding regions of mRNAs [43]. Strategic location of RNA G-quadruplexes coupled with evolutionary conservation in transcriptome suggests that these structures can act as regulatory motifs whereby they influence the total output of the cell.

RNA G-quadruplexes can be viewed as regulatory modules/switches which if interfered can turn on and off the expression of desired gene. RNA G-quadruplexes indeed have been shown to be involved in a variety of biological processes such as translation, mRNA localisation, alternative splicing and telomere homeostasis to name a few [37–39, 44–46]. Balasubramanian and group for the first time demonstrated visual localisation of RNA G-quadruplex within the cytoplasm of human cells using G-quadruplex specific antibody [47]. A reverse transcriptase stalling assay has been used in the past to map the location of G-quadruplexes in a transcript by utilising the inability of reverse transcriptase enzyme to resolve these structures under favourable conditions [48]. Recently, Kwok et al. coupled reverse transcriptase stalling with next-generation sequencing to map the location of G4 structures at nucleotide resolution throughout the transcriptome [49]. A very recent study by Bartel and group however questions the existence of RNA G-quadruplexes inside the cells. The authors use DMS-seq (sequencing of RNA obtained from cells treated with Dimethyl Sulfate) and argue that there exist counter mechanisms that unfold RNA G-quadruplexes in higher eukaryotes [50]. Despite these contrasting reports, a plethora of data exists for biological relevance of RNA G-quadruplex which may form transiently in response to specific stimulus and dictate cellular processes.

RNA structures have been tested for potential targets in both prokaryotes and eukaryotes. Aptamer sequences binding to specific molecules are selected from combinatorial screening approach in an unbiased manner to develop potential drugs that can interfere with structured regions of genome and transcriptome. Short aptameric sequences binding to specific aminoglycosidic ligands are selected from random pool of aptamers. Moreover, inserting these sequences in UTR of reporter genes and targeting them with cognate small molecules established that small molecule–regulatory module interactions can control gene expression in eukaryotes [19]. Such studies established the prowess of targeting naturally occurring RNA structures with small molecules in cells. Potential RNA G-quadruplex motifs exist within UTRs of important genes such as NRAS, Zic1, MT3, VEGF and TGFβ2 to name a few. Biophysical and in cellulo studies demonstrate the influence of RNA G-quadruplex in dictating translational output of these genes [37–39, 51, 52]. Several pioneers in the field have argued that it is imperative to study the structural and thermodynamic features of RNA G-quadruplexes to better understand their functions inside the cell. This knowledge will in turn aid in designing specific ligands that can interfere and regulate naturally occurring RNA G-quadruplexes in important genes. Specific ligands that can interfere by stabilisation or destabilisation of these structures may help to restore normalcy in pathophysiological conditions. RNA G-quadruplexes occur in a wide variety of genes implicated in different processes such as telomere biology, cancer, neurodegeneration and virology and thus targeting RNA G-quadruplexes holds promise in the future. Development of ligands that can bind to specific RNA G-quadruplexes and innovative strategies that can aid in targeting these structures in important genes is constantly evolving. Here we summarise and discuss in detail the structural features and functional aspects of RNA G-quadruplexes evident in the literature. But more importantly, we provide current knowledge about the RNA G-quadruplexes in various important genes that modulate biological processes and the use of ligands against RNA G-quadruplexes highlighting them as targets for potential therapeutics in diseased conditions.

2 Structural Aspects of RNA G-Quadruplexes

Specific ligands and molecules against specific structures and scaffolds can be obtained by rational designing or by an unbiased screening in a high throughput manner. Rational design approaches require structural information at a high resolution. Although structural studies exist for G-quadruplexes, a lacuna exists in terms of the information available for all the possible conformational landscapes and predicted topologies. Structural conformation and stability of G-quadruplex is primarily governed by the number of G-quartets, loop length and composition, monovalent cations and flanking bases. Based on the orientation of the strands, ribose conformations and patterns of connecting the G tetrads, loops are classified as lateral or edge wise, diagonal and propeller type, mainly based on their appearance [30]. (a) Edge wise or lateral loops connect two adjacent strands in an antiparallel arrangement. (b) Diagonal loops connect the diagonally/oppositely oriented strands arranged in an antiparallel manner. (c) Propeller or strand reversal loops connect the top of one strand with the bottom of the other in case of parallelly oriented strands. Studies by several groups have established the loop length to be inversely correlated with the stability of G-quadruplexes both in DNA and RNA level [53–56]. Longer loop lengths tend to destabilise the three-dimensional G-quartets irrespective of their composition. Apart from length, the composition of the loop nucleotides also influences the stability of G-quadruplexes. Biophysical studies by Balasubramanian and group suggested shorter loops to be more stable for RNA G-quadruplexes [56]. However studies from our group with a more comprehensive library of different combinations of loop lengths and composition suggest that stability of RNA G-quadruplex is dependent on loop length up to a certain limit beyond which the stability of RNA G-quadruplexes is loop length independent [57]. These observations were further explored by Jean-Pierre Perreault and group where they explored the formation and biological relevance of RNA G-quadruplexes with loop lengths varying from two to 70 nucleotides in length [58]. The authors used in-line probing technique to determine the formation of G-quadruplexes and reported that RNA G-quadruplexes with a single central long loop of up to 70 nucleotides in length can form under in vitro conditions. They further validated the relevance of their in vitro results using loop lengths of 11 and 32 nucleotides under in cellulo conditions using luciferase gene reporter system. Thus these studies suggest the need to redefine the existing prediction tools of G-quadruplexes as they employ a very conservative loop length maximum limit of seven nucleotides.

Conceptually the key differences between DNA and RNA G-quadruplex are the presence of ribose sugar in RNA which is deoxygenated in DNA. Higher stability of RNA G-quadruplex is attributed to differences in hydration and increased intramolecular hydrogen bonding due to 2′ of ribose sugar [41, 59]. Although RNA G-quadruplexes are more heat stable than DNA counterparts, ribose sugar also restricts them to a parallel conformation. The most probable reason for this monomorphic nature is that the riboguanines involved in tetrad formation usually do not adopt *syn* conformation, a prerequisite for formation of an antiparallel quadruplex. The rigid *anti* conformation of the guanines is a result of 2′-OH group which imparts C3′-endo pucker conformation to the guanines involved in G-tetrad [60].

As mentioned earlier, targeting RNA G-quadruplexes for potential medicinal applications requires substantial information about the structural features of target RNA G-quadruplexes. NMR spectrum of an RNA quadruplex was first reported by Uesugi et al. for R14 sequence containing GGA repeats found in mRNAs of immunoglobin regions [61]. The sequence consisted of four GGAGG stretches separated by UUUU residues. NMR structure reported a dimeric parallel quadruplex with a propeller or strand reversal loop in which two parallel quadruplex blocks are stacked upon one another. Comparative studies by the same group further pointed towards the key differences between R-14 and D-14 stating that the same 14-mer in DNA forms an antiparallel quadruplex with diagonal loops as opposed to the propeller and parallel conformation in RNA [62].

Most of structural elucidation of G-quadruplexes has been performed on the sequences present in telomeric regions. Telomeres are the ends of chromosomes that maintain the genomic integrity of the cell. Telomeres of vertebrates contain tandem repeats of TTAGGG which can extend from 100 to 1,000 and more in length. An enzyme telomerase is responsible to maintain the telomere length which is important for cell viability [63]. Moreover, telomerase activity and telomere length have been shown to be reduced in ageing cells and significantly higher in highly dividing cancerous cells thus attracting telomerase inhibition as potential cancer target [64]. The single stranded G-rich regions of the telomeres have the potential to form G-quadruplex structures inside the cell and interfere with the telomerase activity [65]. Given the importance of telomeres in biology, these G-quadruplex regions have been probed extensively using NMR and X-ray crystallography techniques to understand the structural conformation. X-ray crystal structures of the "12mers" (TAGGGTTAGGGT) and "22mers" [AGGG(TTAGGG)] of telomeres display an all parallel structure with a propeller loop arrangement projecting away from the stacked tetrads [66, 67]. NMR and X-ray crystal structures of telomeric regions are in disagreement however. NMR spectroscopy showed telomeric region to form an antiparallel conformation in the presence of sodium ions; however, X-ray crystal reports it to be a parallel conformer in the presence of potassium ions. It is important to note that the level of potassium ions in the cell is significantly higher than sodium ions, suggesting the potassium ionic conditions to be more cellular mimicking. Zhou and group examined the telomeric G-quadruplexes under molecular mimicking conditions over a period of time [68]. The authors reported that telomeric regions initially display a mixed conformation containing both the antiparallel and parallel conformers; however, with increasing time, only the parallel conformer is observed suggesting the more stable nature of parallel G-quadruplex.

Recently the structure of telomeric repeat containing RNA (TERRA) has been elucidated by ESI-MS and crystallography. ESI-MS studies demonstrated the higher order assemblies of RNA quadruplex formed in human telomeric sequences of variable lengths (12mer, 22mer and 45mer) [69]. Studies by Martadinata and Phan have also supported the stacking nature of RNA quadruplexes by comparing it with "beads on a string" model for the quadruplex assembly via stacking of quadruplex blocks upon each other [70]. Extending the solution based studies to the crystal environment exhibited similar topologies for DNA and RNA quadruplex with differences in hydration networks of the RNA quadruplex due to 2′-OH groups and rigid C3′-endo sugar puckering of guanine residues in the tetrad [42, 67].

Another key difference between DNA and RNA is the presence of a methyl group in uracil which is absent in its counterpart thymine present in DNA. Substitution of 2′-OH by chemical analogues also has shown to destabilise G-quadruplexes further highlighting its importance in G-quadruplex structure [71]. Substitution of thymine by uracil also leads to stabilisation of G-quadruplexes suggesting the importance of loop residues in influencing the G-quadruplex stability. Thymine to uracil substitutions were carried out in thrombin binding aptamer (TBA) which possesses a two-tetrad G-quadruplex (G2 quadruplex). Uracil containing quadruplexes were found to be more stable than thymine harbouring ones [72]. This difference in stability was a result of loss of hydration which leads to stabilisation of stacking interactions due to loss of structured water molecules. Thus RNA G-quadruplexes are more stable than their DNA counterparts owing to their inherent chemistry of RNA molecule containing uracil and a 2′-OH of ribose sugar. Utilising these differences between RNA and DNA can increase the selectivity and specificity of available G-quadruplex ligands towards RNA G-quadruplex.

3 Biological Functions of G-Quadruplexes

A plethora of data exists delineating the role of DNA G-quadruplexes ranging from replication to transcription to telomere maintenance by systematic studies from several groups. However in the past few decades, interest has shifted from DNA to RNA G-quadruplexes [73, 74]. Two of the major reasons for this shift are: the absence of complementary strand in RNA that makes it more flexible to adopt various structures and the strategic occurrence of RNA G-quadruplexes in key locations of important genes making them attractive targets for therapeutic interventions. Several studies in recent past have discovered new functions for RNA G-quadruplexes extending their roles in translational repression, localisation and alternative splicing to name a few.

The first discovery concerned with a functional role of RNA G-quadruplex was made in 3′-UTR of IGF-2 mRNA. A G-quadruplex motif at the 3′-UTR of the insulin-like growth factor II mRNA folded into a structured domain under in vitro conditions. IGF2 mRNA undergoes a specific cleavage reaction in vivo at a site 5′ to the G-quadruplex motif, suggesting a role of G-quadruplex in mRNA-processing events [35]. Similar role for G-quadruplex motifs has been found for p53 mRNA wherein G-quadruplex structure interacts with hnRNP H/F and helps to maintain p53 expression by stabilising the transcript in wake of DNA damage thus leading to apoptosis of the cell [45, 75].

After the initial discovery of role of G-quadruplex in IGF2 mRNA and its effect on mRNA-processing events, the field was subjected to intense analyses and it was found that a large portion of UTRs harbour predictive G-quadruplexes in them [43]. The first experimental demonstration for role of G-quadruplexes in UTR was performed in NRAS gene by Balasubramanian group [37]. The authors used a cell free system and a reporter gene assay containing a wild type and a mutant G-quadruplex of NRAS to measure the reporter expression system. It was found that the thermodynamically stable G-quadruplex inhibited translation under the in vitro conditions. Although this experiment cemented the role of G-quadruplex in translation, it did not reflect the behaviour of G-quadruplex structures inside the cell. The first demonstration of G-quadruplex mediated translational repression in UTRs came from in cellulo studies conducted on Zic-1 (a zinc finger protein) UTR [38]. The study showed that G-quadruplexes inhibited translation inside the cell using a luciferase reporter system assay. More importantly, the inhibition was at the level of translation and not transcription thus suggesting a translational regulation role of G-quadruplexes in UTRs of mRNAs. Several studies followed up these observations in different mRNAs involved in different biological processes ending with similar conclusions for most if not all. Systematic studies were conducted by Hartig and group to understand the role of G-quadruplexes as RNA thermometers inside the cell using bacterial system [76]. G-quadruplexes with different lengths and stabilities were introduced in a GFP reporter system to measure the effect of these structures. It was found that thermodynamically stable G-quadruplexes inhibited translation in a stability dependent manner. Although most of the initial studies pointed towards an inhibitory role for these structures, a seminal study by Basu and group showed a contrasting role (translational activation) of G-quadruplex in translational regulation [51]. Basu and group studied a G2-quadruplex (G-quadruplex with two tetrads) in vascular endothelial growth factor (VEGF) UTR. Biophysical and biochemical studies were carried out to establish the existence and stability of the structure in VEGF UTR under in vitro conditions. However reporter system assays in human cell lines revealed that the G-quadruplex motif increased translation rather than diminishing it. It was recently demonstrated that these structures might serve as an independent recruiting sites similar to Internal Ribosome Entry Site (IRES) motifs thereby attracting the translational machinery and hence increasing the translation [77]. Several studies including a few from our group have been conducted with an aim to explore the structural diversity and understand the role of G-quadruplexes on translation in a stability and location dependent manner [39, 78–81]. Many rules have been laid out that follow inhibitory role of RNA G-quadruplexes. However few examples deviate from the conventionally established principles such as 5′-UTR of Transforming Growth factor β2 (TGFβ2) [52]. TGFβ2 contains a G3 quadruplex (G-quadruplex with three stacks) in its 5′-UTR as predicted bioinformatically using available algorithms. Biophysical and biochemical analyses with the G-quadruplex motif clearly demonstrated its existence and stability pattern. Reporter system assays using only the G-quadruplex domain displayed a decreased translation for the luciferase enzyme inside the cell. However when whole UTR of TGFβ2 was inserted in the reporter system, contrasting results were obtained. It was found that G-quadruplex of TGFβ2 increased translation when present along with flanking bases of UTR but inhibited translation without flanking bases. Further analyses suggested the presence of a G2 quadruplex in its UTR along with the conventional G3 quadruplex. The synergistic effect of G3 and G2 quadruplex is somehow responsible for the increased translation but its detailed mechanism still remains elusive. Thus contrasting roles of G-quadruplex exists as translational regulators thus making it difficult to predict the functions of newly found G-quadruplexes.

Very few studies have explored the role of G-quadruplexes at 3′-end of mRNAs. Arora et al. used the above described luciferase reporter system to understand the functionality of G-quadruplex in proto-oncogene serine/threonine-protein kinase 1 (PIM1) [82]. A translational suppressive role was demonstrated for PIM1 G-quadruplex but nevertheless without a detailed mechanism.

G-quadruplexes can be hypothesised as regulatory switches or interaction domains for different proteins and hence could play an important role in cellular processes. Their role in various diseases has been discussed elsewhere [83]. Targeting these structures for potential therapeutic applications are discussed below.

3.1 G-Quadruplexes in Neurodegeneration

3.1.1 Fragile X Mental Retardation Protein

Fragile X mental retardation protein (FMRP) is deregulated in mental disease conditions – fragile X mental retardation syndrome (FXS) and FRAXE-associated mental retardation syndrome (FRAXE) [84]. FXS is the most common form of inherited intellectual disability with an incidence of one in 5,160 males who develop childhood seizures, autistic behaviour and immature dendritic spines [85]. FMRP is a protein which is absent in patients with FXS pathology. FMRP is an RNA binding protein containing two ribonucleoprotein K homology domain (KH domain) and RGG domains, both characteristic of RNA binding proteins. In FXS patients, expansion of CGG repeats is observed in FMR1 that codes for FMRP protein. This expansion leads to silencing of FMRP expression and ultimately absence of FMRP protein inside the cells [84, 86]. FMRP is an RNA binding protein which is involved in dynamics of mRNA shuttling between nucleus and cytoplasm [87]. FMRP RGG domains recognise its targets but it has been shown that binding of FMRP is a combination of both structure and sequence. Extensive studies have been carried out to understand the RNA binding activity of FMRP. FMRP contains RGG (Arginine–Glycine–Glycine) box which serves as the binding domain to its various targets [88]. In an effort to identify the transcripts that are bound by FMRP, extensive pull-down studies have been carried out followed by sequencing to reveal the identity of the interacting RNAs [89, 90]. Initial studies by Darnell and group suggested that most of the RNA targets of FMRP harbour G-quadruplex motifs in their 5′-UTR, coding regions and 3′-UTR regions. Many of these mRNA targets have been shown to harbour G-quadruplex structures in their 5′-UTR, CDS and 3′-UTR, although only a subset of them has been experimentally validated. *Catalytic subunit of Protein Phosphatase 2A* (*PP2Ac*) and *Microtubules Associated Protein 1b* (*MAP1B*) are FMRP targets harbouring one or more G-quadruplex structures in their 5′ UTR. FMRP represses the translation of MAP1B and PP2Ac by blocking translational machinery. In the absence of FMRP, MAP1B and PP2Ac both lead to changes in cytoskeletal filaments. In the absence of FMRP, MAP1B levels increase leading to abnormal microtubule stability and in turn affecting the formation of dendritic spines in neurons [91]. Similarly, the absence of FMRP increases the level of PP2Ac leading to altered actin remodelling in fibroblast cells [92].

Expression of the G-quadruplex disrupting proteins hnRNP A2 and CBF-A was shown to alleviate the translational block [93]. Interestingly, FMRP was found to regulate its own activity by binding to G-quadruplex present in coding region of its mRNA and inhibiting its translation by a negative feedback loop [94]. Likewise, FMRP binds to G-quadruplex in coding region of Amyloid Precursor Protein (APP) and regulates its translation [95]. A comprehensive analysis of four major datasets containing mRNAs associated with FMRP revealed significant overlap with each other [96]. Most of the associated mRNAs contained a WGGA motif (where W = A or U) and these motifs were shown to be highly clustered. The authors hypothesise that WGGA motifs might form G-quadruplexes which serve as the recognition motifs for FMRP. FMRP binds to

4% of the RNAs present in the mammalian brain and majorly functions as a translational regulator governing the level of translation inside the cell. Evidence of G-quadruplex motifs in FMRP bound RNAs points towards a significant role of G-quadruplexes inside the cell and strengthens the role of G-quadruplex as sequestrating agents for proteins.

FMRP is postulated to stabilise these RNA molecules, as in the absence of FMRP variety of genes are deregulated. Ligands that can compensate for FMRP loss by stabilising G4 structures may restore normal metabolism of RNA targets of FMRP. If cells lacking FMRP respond to such ligands, they will provide new unheralded roles of FMRP biology. Moreover understanding the disease progression will be the key to designing therapeutics. If G4 ligands can compensate for FMRP function, then they possibly open up new avenues for pathologic diseases linked with FMRP loss. It will be of significance to ascertain the biological role of G-quadruplex and ligands in the context of diseases linked with deregulated G-quadruplex binding proteins. A global stabilisation of G4 motifs could increase the stability of mRNA targets and hence help to repair the deregulated pathways in pathological conditions.

3.1.2 C9orf72 Gene in Amyotrophic Lateral Sclerosis and Fronto-Temporal Dementia

RNA G-quadruplexes have also been shown to be involved in neurodegenerative disorders. Amyotrophic lateral sclerosis (ALS) results from degeneration of motor neurons in the brain leading to dysregulated motor skills and paralysis and is fatal. Fronto-temporal dementia (FTD) is a result of degenerated neurons in frontal and temporal lobes and leads to altered behaviour and language impairment [97]. Both ALS and FTD are considered to be part of the same disease. Neurodegenerative diseases such as FTD and ALS are a result of multiple defects in cellular machinery of neurons. But more importantly, expanded repeats expansion of GGGGCC $(G_4C_2)_n$ has been observed in first intron of C9orf72 gene in patients with FTD and ALS [98, 99]. Normal individuals contain <30 repeats while patients with ALS and FTD contain ~500 to few thousand repeats in their genome. NMR and CD experiments have demonstrated that these repeats are able to form RNA G-quadruplexes structures [100]. It was hypothesised that G-quadruplex structures may bind and sequester important proteins during the pathogenesis of ALS and FTD. Haeusler et al. demonstrated the functional consequence of these repeats in C9orf72 gene [101]. They showed that these repeats lead to abortive transcription by RNA polymerase. The rate of abortive transcription events was found to be directly proportional to number of GGGGCC repeats. These repeats thus stall the movement of RNA polymerase leading to incomplete transcription events thereby decreasing pool of full-length transcripts. The authors further went on to determine the binding partners for abortive transcripts using RNA pull-down assays followed by quantitative mass spectrometry using SILAC. Among many proteins identified, nucleolin was found to be a robust binding partner to $(G_4C_2)_n$ repeat containing RNA. Furthermore evidence of nucleolar stress was attributed to dispersed localisation of nucleolin in patient cells. Nucleolin was found to be more dispersed throughout the nucleus under such conditions. Nucleolus is the site of rRNA maturation and deregulated nucleolin levels correlate with decreased rRNA levels in the patient cells. The authors were also able to recapitulate the findings from the patient cells by expressing abortive transcripts in normal cells.

RNA in situ hybridisation experiments further confirmed the interaction of $(G_4C_2)_n$ RNA with nucleolin inside the patient cells. Thus the authors demonstrated that patients with repeat expansions in C9orf72 gene showcased abortive transcription, nucleolar stress, decreased rRNA maturation and more sensitivity to proteotoxic stress. These deregulated conditions alter homeostasis and the ability of these patients to handle stresses declines rapidly making them more vulnerable to proteotoxic stresses with increased age. Interfering with G-quadruplex structures in such patients may help to relieve some of the stresses manifested in these patient cells. Different groups have tried to identify the exact causes of ALS and FTD and postulate that the effects can be seen by RNA mediated toxicity and/or peptide mediated toxicity or a combination of both resulting in compromised nucleocytoplasmic transport mechanism [102]. Genome wide screens in yeast showed the effect to be caused by dipeptide toxicity alone [103]. Dipeptide chains can be translated from expanded repeats in a repeat associated non-AUG translation. Compelling evidence exists supporting each of the hypotheses in different model organisms. Nevertheless, combination of multiple deregulated pathways could also cause the disease and needs to be tested in future. Zamiri et al. utilised G4 ligand molecules to disrupt the secondary structures adopted by expanded repeats in C9orf72 gene [104]. They tested cationic porphyrin (5,10,15,20-tetra(N-methyl-4-pyridyl) porphyrin ($T_mP_yP_4$)) and found it to actively distort the G-quadruplexes adopted by $(G_4C_2)_n$ repeats in C9orf72 transcripts. The authors also tested the ability of TmPyP4 to interfere with interaction between repeats and other proteins inside the cells. They found that the presence of $T_mP_yP_4$ interfered with binding of ASF/SF2 and hnRNP A1 binding to $(G_4C_2)_n$ repeats suggesting that G4 structure in neurological diseases and their protein interaction can be disrupted using specific small molecules. Similar strategy was also utilised by Su et al. wherein they screened bioactive small molecules that can target expanded repeats of C9orf72 $(G_4C_2)_n$ repeats in cells isolated from patients with ALS/FTD [105]. These studies indicate that small molecules targeting $(G_4C_2)_n$ repeats present a viable therapeutic strategy in ALS/FTD patients.

3.1.3 G-Quadruplexes in Alzheimer's Disease

Alzheimer's disease is caused due to accumulation of amyloid plaques and neurofibrillary tangles in neuronal cells leading to severe form of dementia in the patients. Amyloid plaques are a result of accumulation of amyloid β peptide (Aβ), a proteolytic product of the APP [106]. Both under- and overexpression of APP protein have been linked to dysregulated neuronal structure and function [107, 108]. Early onset of Alzheimer's is a result of high accumulation of amyloid β peptide (Aβ) that is a result of elevated levels whereas reduced levels of APP are linked to impaired learning and immature dendritic spines [106]. Crenshaw et al. demonstrated the presence of a G-quadruplex in 3′-UTR of APP gene that controls its expression inside the cell [109]. G-quadruplex in 3′-UTR of APP does not alter the expression levels of APP mRNA but regulates its translational output. Using luciferase studies, it was demonstrated that mutated G-quadruplex in 3′-UTR of APP increases the translational output of APP. Thus 3′-UTR G4 in APP negatively regulates its translational levels thereby keeping APP expression in check as any deviation in APP levels is deleterious for the cell. A wide variety of proteins interact with G-quadruplexes and APP levels may also be subjected to expression of these interacting proteins and need to be probed further. On contrary, mutations disrupting APP 3′-UTR G-quadruplex could also increase APP levels and lead to pathological features associated with Alzheimer's disease. APP 3′-UTR G-quadruplex can help understand the progression of Alzheimer's disease and can be employed diagnostically to reveal mutations or disruptions in regulation of APP levels inside the cell. Moreover, in diseases with lower levels of APP, G-quadruplex ligands can be employed to increase the levels of APP by destabilising APP 3′-UTR quadruplex and restore APP levels. But more importantly, APP levels are maintained in a critical range and any deviation could give rise to dysregulated neuronal biology and hence needs to be considered for future therapeutic interventions. Role of G-quadruplexes in Alzheimer's disease can also be at the level of APP cleavage by alpha and beta secretase enzymes both of which are themselves under G-quadruplex mediated regulation. Two important genes BACE1 and ADAM10 are involved in APP proteolytic cleavage and both of these genes have been shown to be regulated by G-quadruplex structures. ADAM10 belongs to a family of a disintegrin and metalloproteinase family which contribute to anti-amyloidogenic activity [110]. ADAM10 contains an RNA G-quadruplex in its 5′-UTR that represses the translation of ADAM10. Moderate overexpression of ADAM10 can decrease amyloid β production inside the cells. β-site APP cleaving enzyme 1 (BACE1) is responsible for cleaving APP to amyloid β peptides. BACE1 is subjected to G-quadruplex regulation during its alternative splicing [111]. Mutation in G-quadruplex motif decreases full-length active BACE1 by increasing the short inactive BACE1 protein. Thus, Alzheimer's pathology is subjected to G-quadruplex mediated regulation at various key steps and alteration or interference with G-quadruplex motifs could prove to be a valuable intervention step in understanding the progression of disease and ultimately leading to potential therapeutic targets. Rescuing the negative translational control by destabilising G-quadruplex in ADAM10 and by decreasing full-length BACE1 may alter the balance towards more normal physiology of neurons thus counteracting the production of amyloid β peptides and decreasing neurofibrillary tangles associated with Alzheimer's diseases. However developing molecules that can target G-quadruplex ligands specifically by crossing blood–brain barrier is a challenging task but nevertheless is a worthwhile pursuit in future.

3.1.4 G-Quadruplex as Zipcodes in Neurons

An intriguing case of G-quadruplex involvement in neuronal RNA metabolism was published by Moine et al. Moine et al. predicted that 30% of dendritic mRNAs harbour potential quadruplex motifs in their 3′-UTRs raising the possibility that G4 motifs may function as cis-acting elements dictating the localisation of dendritic mRNAs. Using Lambda reporter GFP system and mutational studies, they showed that G-quadruplex motifs in UTRs of PSD-95 and CamKIIA – two of well-known dendritic mRNAs – function as zipcodes for neurite targeting. They also demonstrated that the localisation is dependent on the ability of mRNA to form G-quadruplex structures as mutated G-quadruplex structures did not function as neurite signalling modules for mRNAs. FMRP has been shown to be involved in variety of mRNA localisation in neurons and the authors claimed that the absence of FMRP did not alter the basal transport of G-quadruplex containing dendritic mRNAs suggesting that these modules function as cis-acting elements and may respond to different gradients in cationic compositions to execute its function. These G-quadruplex sequences can help to target mRNAs to different locations of cells. Transport of proteins is an energy expensive task but transport of mRNAs to desired locations inside the cells helps to spontaneously increase protein production inside the cells (Fig. 2).

3.2 G-Quadruplexes in Non-coding Transcriptome

With the advent of next-generation sequencing technologies, it was realised that a major portion of genome is transcribed (~85%) but less than 3% of it codes for protein coding genes [112, 113]. The remaining portion of transcripts was classified as non-coding RNAs which were further sub-divided into two major classes – lncRNAs and small non-coding RNAs (sRNAs). The distinction between lncRNAs and sRNAs is based on arbitrary cut-off of length 200 nucleotides. Both classes of RNAs have been shown to be involved in dictating biological output of the cell and have been sought after as potential therapeutic targets in various pathological conditions [113]. Enrichment of G-quadruplex in untranslated sequences prompted us to speculate that G-quadruplexes may have a role in non-coding transcriptome. Our group have successfully predicted and tested the role of G-quadruplexes in lncRNA and miRNAs thus opening up the field to explore the function of G-quadruplexes in non-coding transcriptome.

3.2.1 G-Quadruplexes in Long Non-coding RNAs

lncRNAs are transcripts greater than 200 nucleotides that do not contain an ORF sequence. lncRNAs have been implicated to have roles in development and differentiation, and their deregulation is linked to diseases in various organisms [113]. A major subset of lncRNAs is enriched in nucleus and influences gene expression by binding to various proteins involved in chromatin remodelling [114]. Binding to different interactors by lncRNAs is based on their structural features which acts as scaffold for assembly of protein complexes to execute their functions. G-quadruplexes may act as scaffold structures in lncRNAs and aid in binding of different proteins thereby guiding the assembly of proteins at their preferred location or altering the localisation of proteins leading to pathological conditions. Our group performed bioinformatics prediction of

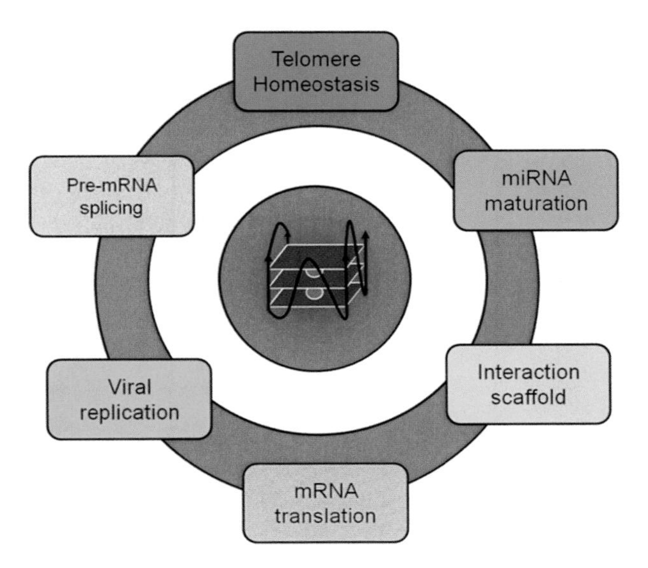

Fig. 2 Biological functions of RNA G-quadruplexes

126,406 annotated lncRNAs mapping 749 lncRNA to harbour at least one G4 motif in their sequence [40]. Further analysis demonstrated G-quadruplex motifs to harbour single and dinucleotide loops suggesting them to be of highly stable nature and hinting towards role of G-quadruplexes in lncRNA biology. Less abundance of cytosines in the loop nucleotides further provided impetus that G-quadruplexes do not compete for GC rich structures in lncRNAs and thus may form more efficiently. LncRNA biology is a promising field with many different facets still unexplored. Contribution of G-quadruplex structures to functionality of lncRNA is a promising avenue and needs to be tested in future. We are currently exploring the role of G4 structures in lncRNA functions using candidate lncRNAs and modulating its function with mutations and state-of-the-art transcriptomics and proteomics approaches. Development of specific G-quadruplex ligands which can be tagged with sequences complementary to desired lncRNA can help to target lncRNAs specifically and interfere with their function especially under diseased conditions.

3.2.2 G-Quadruplexes in MicroRNA Biology

Another important role of G-quadruplex which is of potential therapeutic value is in the field of miRNA biology. miRNAs are the most abundant class of small non-coding RNAs regulating gene expression and homeostasis of the cell [115]. Deregulated miRNA levels are linked to various processes that cause disturbance to normal physiology of cells leading to pathological conditions [116–118]. Various approaches ranging from antisense therapeutics to small molecules have been developed to target miRNA involved in pathological conditions [119–122]. Screening of small molecules to decrease oncogenic miRNA or upregulate tumour suppressor miRNA is prompting the field to explore various methods that can aid in regulating miRNA levels. miRNAs are 22-nucleotide long molecules developed in a series of concerted cleavage and transport reactions from nucleus to cytoplasm. Primary miRNAs (pri-miRNAs) are transcribed to several kilobases in nucleus by RNA polymerase II and processed by Drosha (RNAse II enzyme) to yield precursor miRNAs (pre-miRNAs) that are exported to cytoplasm [123]. Pre-miRNAs are further cleaved by Dicer enzyme into 22 nucleotides long double stranded RNA molecules that bind to their cognate target mRNAs and silence their expression either by translational suppression or destabilisation of mRNA [115]. Processing of pre-miRNA to mature miRNA involves a catalytic nuclease enzyme–Dicer. Dicer acts a molecular ruler recognising stem-loop structures in pre-miRNA and cleaving them to produce mature miRNA molecules. Our group speculated the role of putative G-quadruplex structures in pre-miRNA and its effect on mature miRNA production [124]. We hypothesised that Dicer mediated activity on pre-miRNAs will be hampered by the ability of pre-miRNA to form a G-quadruplex. Using in vitro Dicer and in cellulo models, we established that there exists an RNA G-quadruplex to stem-loop equilibrium that dictates the production of mature miRNA molecule inside the cells. Both G2 and G3 quadruplexes were tested for their ability to modulate miRNA levels in synthetic and naturally occurring pre-miRNAs and found to be effective in influencing miRNA production inside the cells. Bioinformatics predictions suggested that 13% of pre-miRNAs possess potential G-quadruplex motifs in their natural sequence. Using a high throughput qRT-PCR method, we tested cationic porphyrin $T_mP_yP_4$ to establish the role for G-quadruplex structure in dictating mature miRNA production. Cationic porphyrin $T_mP_yP_4$ has been shown to interfere with G-quadruplex formation in RNA by destabilising them [104]. Our analysis indicated that a major subset of pre-miRNAs is affected by $T_mP_yP_4$ as evident by deregulated mature miRNA production. $T_mP_yP_4$ is a highly non-specific binder to RNA G-quadruplex and hence more in-depth studies will be required to evaluate its effect on miRNA biology. Thus both metastable G2 quadruplexes and G3 quadruplexes can act as structural interference to Dicer processing leading to deregulated miRNA levels. Interference with miRNA processing is an attractive field and employing G-quadruplex ligands to shift the equilibrium between G-quadruplex and stem-loop structures could provide another dimension to target miRNAs involved in pathophysiological conditions. Role of G-quadruplex structures in the regulation of miR-92b and miR-125a has also been demonstrated by different groups. Arachchilage et al. demonstrated that production of miR-92b is dictated by a G-quadruplex structure in its precursor form and was shown to be dependent on different cations [125]. This study was the first report establishing the in vivo role of G-quadruplex structure in miRNA biology.

Another study demonstrated the role of G-quadruplex in miRNA-125a targeting its cognate mRNA [126]. MiR-125a regulates the levels of postsynaptic density protein 95 (PSD-95), a cognate FMRP binding mRNA. FMRP, Argonaute2 (Ago2), forms a silencing complex on the 3′-UTR of PSD-95 thereby repressing the translation of PSD-95. FMRP when phosphorylated binds to 3′-UTR of PSD-95 via a potential G-quadruplex structure and decreases translation of PSD-95. Under specific stimuli, FMRP is dephosphorylated and Ago2 is released from PSD-95 UTR thereby activating translation of PSD-95. The target site of miR-125a is hidden or enclosed in a G-rich region postulated to be bound by FMRP. Multiple G-quadruplex conformers can form in G-rich region and miR-125a can access its target site only in the presence of one G-quadruplex conformer. This study provided a novel role for alternative G-quadruplexes that can mask the miRNA target sites and regulate miRNA mediated targeting inside the cells by mechanism involving G-quadruplex binding proteins, messenger RNAs and miRNAs.

3.3 RNA G-Quadruplex in Virology

Many different viral genomes have been shown to utilise G-quadruplex structures to control the expression of key viral proteins thereby mounting an escape mechanism from host immune response.

Epstein–Barr virus (EBV) belongs to gamma herpesvirus family of viruses involved in various lymphoblastoid malignancies [127]. EBV integrates within B-lymphocytes establishing a long-term latent infection by episome formation within the cells [128]. EBV nuclear antigen 1 (EBNA1) is the gene responsible for latent infection and maintenance of viral episome in the cells [129]. EBNA1 contains distinct RGG motifs which bind to G-rich regions with propensity to from G-quadruplex structures in RNA [130]. EBNA1 recruits origin of replication complexes (ORC) to critically aid in viral replication and maintenance [129]. Several studies have tried to understand the role of this interaction in viral replication and whether G-quadruplex binding compounds can interfere with this interaction ultimately inhibiting viral replication. Lieberman and group tested various quadruplex ligands such compounds $T_mP_yP_3$, $T_mP_yP_4$ and BRACO-19 interfered with EBNA1 recruitment of ORC [131]. Long-term treatment with BRACO-19 decreased cell viability and increased sensitivity of EBV positive cells to BRACO-19. Small molecules that can specifically target EBV viral RNA G4s could help in designing drugs that can target EBV positive cells.

A recent paper demonstrates the presence of RNA G-quadruplex in Ebola virus [132]. Ebola virus disease (EVD) is caused by infection with Zaire ebolavirus (EBOV). The genome of Ebola virus is a 19 kb long negative strand RNA molecule with a potential to form a myriad of structures that can be used as targets to inhibit viral replication. EBOV L gene contained a putative G-quadruplex in its UTR. Using established biophysical methods in the field of G4 biology, the authors demonstrated the ability of this sequence to form a stable G4 structure in RNA. Not many drugs exist for treatment of Ebola virus with high efficiency. The authors tested $T_mP_yP_4$ based on the assumption that $T_mP_yP_4$ will interfere with G-quadruplex in L1 gene and thereby decrease viral load inside the cells. Using mini replicon assays, the authors went onto demonstrate that viral replication is dramatically reduced (up to 9.6-fold) upon treatment with $T_mP_yP_4$ in wild type G-quadruplex containing mini genome as compared to mutant G-quadruplex containing mini genome. G-quadruplex targeting ligands was found to be more effective than ribavirin (antiviral drug) suggesting they can be used to target Ebola virus replication lifecycle but still effect of such ligands on the host cells needs to be determined. Nevertheless targeting EBOV L gene G-quadruplex structure with small molecule opens up new class of drugs that can be screened against these deadly pathogens.

3.4 RNA G-Quadruplex in Telomere Biology

One of the most widely studied functions of G-quadruplexes is its role in the telomeric regions. Telomeres are the ends of chromosomes that protect the chromosomal integrity and have been implicated in ageing and cancer. Telomeric regions are extended by an enzyme termed telomerase that extends TTAGGG repeats and carries out its function. Telomeric regions had been considered transcriptionally silent for a long time until a seminal study demonstrated that telomeres are actively transcribed to form long transcripts ranging from 100 nucleotides to 9 kb [133]. TERRA repeats contain UUAGGG repeats which have been predicted to form G-quadruplexes. Studies by various groups have attempted to delineate the role of these structures in TERRA transcripts, making it one of the most actively studied G-quadruplexes harbouring transcript [134, 135]. Mass spectrometric studies have been carried out to catalogue TERRA interacting proteins [136]. A high number of hnRNP proteins were found to be enriched in TERRA bound fraction. An example hnRNP A1 has already been shown to be a G-quadruplex binder in other studies further suggesting that these structures might act as hooks or sequestrating agents to control the localisation and function of different proteins that bind to them. From the above studies, G-quadruplexes were hypothesised to act as interaction domains for different proteins.

RNA being inherently prone to forming structures requires equilibrium between different structures to execute its function. Not all the conformations adopted by single mRNA molecule may be of biological significance while in some cases, propensity to form multiple structures may serve as regulatory roles by structural interference mechanisms. Human telomerase RNA forms an important helical structure termed as P1 helix [44]. P1 helix determines the template boundary for telomerase reverse transcription activity limiting it to six nucleotides. P1 helix is preceded by a 5′ tail which has the propensity to from G-quadruplexes. It was found that G-quadruplex formation interferes with P1 helix formation. Ligand 360A shifted the equilibrium more towards G-quadruplex formation [44]. Initial in vitro biophysical studies paved the way to further understand this structural interference by G-quadruplex motifs in human telomerase RNA. It was found that RHAU a G-quadruplex resolvase protein binds and resolves it to favour P1 helix formation inside the cells [137]. Telomerase is expressed at high levels in various cancers. A preliminary analysis suggests that RHAU levels are also increased in various cancers. This parallel suggests that RHAU may be involved in telomere maintenance in cancer cells thus contributing to their proliferative capacity inside the cells. Various G-quadruplex targeting ligands have been used against therapeutically important genes (Fig. 3).

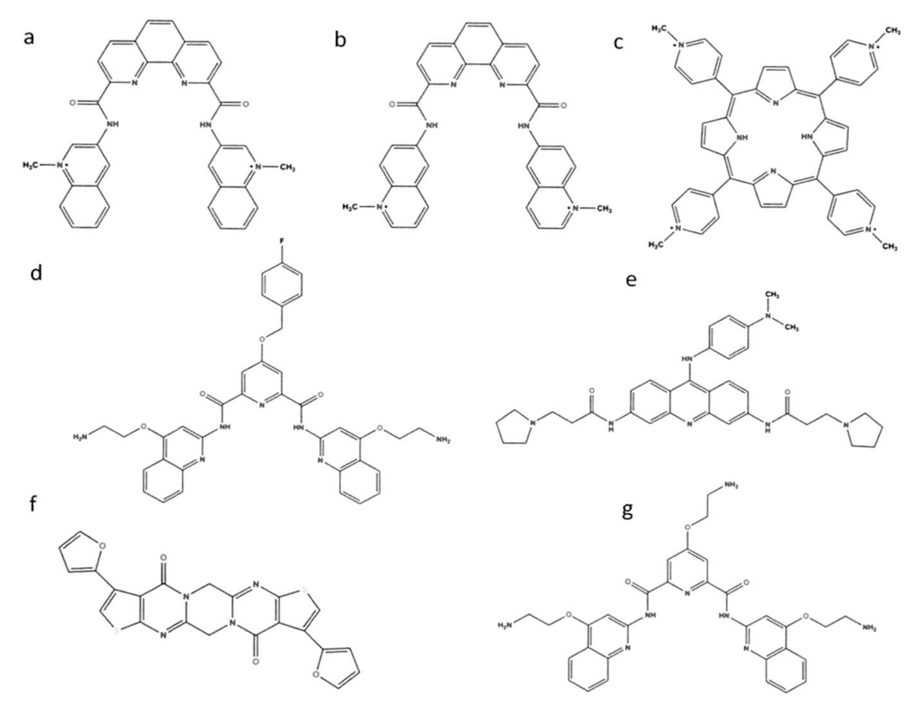

Fig. 3 G-quadruplex ligands. Different ligands employed against RNA G-quadruplexes. (**a**) Phen-DC3, (**b**) Phen DC-6, (**c**) $T_mP_yP_4$, (**d**) RR110, (**e**) BRACO-19, (**f**) RGB-1 and (**g**) Pyridostatin

4 Targeting RNA G-Quadruplexes Using Ligands

4.1 NRAS

First demonstration of targeting naturally occurring RNA G-quadruplex in UTR with a small molecule was from Balasubramanian group [138]. The authors tested pyridine-2,6-bis-quinolino-dicarboxamide derivative (RR82) in an in cellulo translational assay system utilising NRAS gene cloned upstream of luciferase. Pyridine-2,6-bis-quinolino-dicarboxamide derivative RR82 decreased translation of the reporter system by 50% at 1.25 μM concentration in an in cellulo assay. The authors further in their wake to improve selectivity derivatised RR82 by replacing a positively charged side chain with a more apolar fluorophenyl group and created RR110. RR110 significantly downregulated the reporter gene expression (twofold) without affecting the non-G4 control demonstrating its more selective recognition of the target RNA G-quadruplex. This study paved the way to develop ligands highly specific for RNA G-quadruplex of interest thus providing an early insight in the optimisation of RNA G-quadruplex ligand [138].

4.2 TRF2

As pointed earlier, most of the initial studies performed for RNA G-quadruplex and small molecule targeting focused on telomere regulation. Telomere repeat-binding factor 2 (TRF2) is an important component of shelterin that protects the telomeric ends from being recognised as double stranded breaks. Absence of TRF2 is associated with growth arrest and senescence owing to shorter telomeres in human cells [139]. Targeting TRF2 expression in glioblastoma cells led to decreased tumorigenesis in mice models using viral mediated shRNA delivery targeting TRF2 gene [140]. Gomez et al. demonstrated the role of a potential G-quadruplex located 19 nucleotides upstream of translation initiation site in TRF2 gene [141]. The authors used a GFP reporter assay and cloned TRF2 quadruplex and its mutant counterpart to ascertain the role for the structure inside the cell. They found 2.8-fold decrease in GFP expression of G-quadruplex containing plasmid as compared to its mutant counterpart suggesting that G-quadruplex structures regulate the expression of TRF2 in a cellular context. No change in mRNA levels further suggested a translational suppressor mechanism by RNA G-quadruplex. Potent G4 binders – 360A molecule, a pyridine dicarboxamide derivate, Phen-DC3 and Phen-DC6, and two bisquinolinium compounds – were further tested in FRET based experiments and displayed higher affinity for TRF2 RNA G-quadruplex than its mutant counterpart. Using an in vitro translation system, they further showed that all three molecules decrease translation but Phen-DC3 and Phen-DC6 were more potent than 360A molecule. Phen-DC3 decreased GFP expression by fourfold as compared to the mutant suggesting its selective nature. Taken together, these observations indicate that G-quadruplex ligands inhibit protein synthesis proving further proof of principle for ligand mediated regulation of gene expression by targeting natural RNA G-quadruplexes motifs. Furthermore, TRF2 was shown to bind simultaneously to TERRA and telomeric duplex or G-quadruplex DNA making TERRA G-quadruplex an important part of telomere structure [142].

4.3 p53

p53 is the holy grail of oncogenes highly involved in cancer progression and is one of the most sought out cancer therapeutic targets. P53 is a tumour suppressor gene which is mutated in various cancer cells [143]. TP53 gene exists in multiple isoforms at mRNA levels [144, 145]. Many of these isoforms show lack of N-terminal regions. N-terminal region of p53 contains a transactivation domain and a binding site for Hdm2 that targets p53 to proteasome machinery [146]. An alternate isoform p53I2 gives rise to truncated p53, Δ40p53, which lacks 39 residues at N-terminal region. Δ40p53 counteracts the growth suppression effect of p53 by negatively affecting the expression of p53 target genes [144]. Differential splicing in intron 3 leads to equilibrium shift between Δ40p53 and full-length p53. A G-quadruplex motif in intron 3 was found to drive this equilibrium between expressed transcripts of p53 and Δ40p53 [45]. Both in mice and human models, it has been suggested that small variations in levels of Δ40p53 could counteract growth suppression effect of p53 [143, 147]. Stable G-quadruplex structures thus direct proper maturation of p53 mRNA and expression of p53 inside the cells. The authors also tested 360A ligand to stabilise G-quadruplex structures and examine the effect on p53 expression [45]. 360A increased full-length isoform levels by fourfold and decreased the expression of p53I2 that codes for Δ40p53. These observations suggest that stabilising G-quadruplexes can be a potential therapeutic to shift the equilibrium between full-length p53 and its alternate isoforms. More importantly, the authors also mapped mutations affecting the G-quadruplex region and risk of cancer. They found out that common polymorphism *TP53* PIN3 leads to increased risk of various cancers. Furthermore, an independent epidemiological study quantified the expression levels of p53 RNA and found it to be reduced in TP53 PIN allele [148]. Mutant p53, a gain of function, is found in 74% of TP53 mutated tumours [143]. Mutant p53 has been shown to bind to G-rich regions with propensity to form G-quadruplex structures [149]. These observations further highlight the role of G-quadruplex structures in governing the p53 network. Various groups are in pursuit of repairing the mutated p53 proteins to counteract the proliferative potential of cancer cells. Similarly, screening of G-quadruplex ligands which can shift the equilibrium between different isoforms of p53 would serve as a potent method to target cancer.

4.4 *Bcl-2*

Cancer is an abnormal growth of cells due to disturbances in the signals that govern survival and death of cells. Several important oncogenes belong to the family of apoptotic genes wherein important anti-apoptotic regulators are overexpressed thus leading to proliferative potential of tumours. Bcl-2 gene is an important regulator in apoptosis pathway and considered to be an oncogene. Bcl-2 blocks the pro-apoptotic proteins Bax and Bcl-wl and prolongs the survival properties of cancer cells. Different strategies have been conducted in wake to knock down bcl-2 using antisense [150] and small molecules [151] and many of these have shown promising results. 5′ UTR of bcl-2 gene was reported to contain a potential G-quadruplex motif which under in cellulo conditions downregulates the bcl-2 expression [80]. Targeting bcl-2-RNA G-quadruplex thereby presents us with another promising alternative for downregulation of bcl-2 in wake to target cancer cells. Bcl-2 levels are significantly upregulated in cancer cells as opposed to normal cells providing them oncogenesis potential and resistance to chemotherapy [152]. Combinatorial drug therapies targeting different cellular components are an effective approach albeit with side effects. Bcl-2 gene also possesses a promoter G4 [153] thus providing opportunity to target the genes at both its transcription and translational levels by DNA and RNA G-quadruplex binding molecules.

4.5 Other Studies Utilising G-Quadruplex Ligands

Twenty-four-membered pyridine ring containing polyoxazole macrocyclic compounds was tested for anticancer activity and one of the compounds exhibited highly selective binding to cell cycle check point dependent kinase Aurora A further demonstrating that specific ligands targeting specific RNA G-quadruplexes can be obtained. Compound displayed antitumour activity in metastatic cancers in mice model further proving that RNA G-quadruplexes are highly promising therapeutic targets for pathophysiological diseases such as cancer [154].

Recently, a new molecule RGB1 has been obtained by Uesugi with more specific RNA G-quadruplex binding than reported elsewhere [155]. Using an in vitro screening of 8,000 molecules, the authors searched for RNA G-quadruplex stabilisers that can stall reverse transcriptase mediated extension. RGB-1 was further shown to bind to TERRA G-quadruplexes suggesting its binding to different RNA G-quadruplex sequences. Comparison of RGB1 with $T_mP_yP_4$ further demonstrated its ability to specifically target RNA G-quadruplex whereas $T_mP_yP_4$ was found to affect both G-quadruplex and its mutant counterpart in a luciferase assay. Although the binding strength of RGB-1 is lesser than currently used ligands such as pyridostatin, its ability to distinguish between RNA and DNA G-quadruplex makes it a highly promising molecule to exploit RNA G-quadruplex targeting in future. More in-depth structural analysis with complex of RNA G-quadruplex with RGB-1 will pave the way to design ligands specific for RNA G-quadruplex.

5 Conclusions and Future Directions

RNA secondary structures provide chemical biologists with an opportunity to target genes in a specific manner. Global downregulation of genes harbouring similar structure or specific deregulation by specific ligands targeting a desired gene of choice both are appealing to biology and clinicians. RNA G-quadruplexes are one of the potential targets that can influence gene expression globally and individually. DNA G-quadruplexes can exist as parallel, antiparallel or mixed topology depending on its sequence, cations, loop length and composition. RNA G-quadruplexes exist in a single parallel topology due to its C3′-endo conformation that makes it monomorphic. Currently the field suffers from lack of ligands that can distinguish between RNA and DNA G-quadruplexes. Recent high throughput screening studies do provide candidates that may be able to bind to RNA G-quadruplexes without recognising DNA G-quadruplexes. More such approaches could open the field for specific targeting of RNA G-quadruplexes harboured by genes involved in important biological processes. It is also important to distinguish between functional and inert G-quadruplexes. Not all the predicted G-quadruplex motifs would adopt structure or be formed inside the cells. Innovative approaches that can categorise functional G-quadruplexes will provide researchers with a curated G-quadruplex list that needs to be interfered to restore homeostasis in pathological cells. Genome editing tools such as ZFNs, TALENS and CRISPR-Cas have provided us with tools which can be employed to substitute G-quadruplex structures in cells with mutated sequences. Such studies will help to understand the regulatory potential of these structures inside the cells in both normal and abnormal conditions. Identification of mechanism and biomolecules that interfere with dynamics of G-quadruplex formation and their deregulation in adverse conditions will open up new therapeutic targets in future.

Acknowledgements The authors acknowledge Manoj Teltumbade for help with chemical structures and valuable inputs. This work was financially supported by the BSC0123 project (Project: Genome Dynamics in Cellular Organization, Differentiation and Enantiostasis) from the Council of Scientific and Industrial Research (CSIR), Government of India.

References

1. Crick F (1970) Central dogma of molecular biology. Nature 227:561–563
2. Hangauer MJ, Vaughn IW, McManus MT (2013) Pervasive transcription of the human genome produces thousands of previously unidentified long intergenic noncoding RNAs. PLoS Genet 9:e1003569
3. Wan Y, Kertesz M, Spitale RC, Segal E, Chang HY (2011) Understanding the transcriptome through RNA structure. Nat Rev Genet 12:641–655
4. Doudna JA, Cech TR (2002) The chemical repertoire of natural ribozymes. Nature 418:222–228
5. Mandal M, Breaker RR (2004) Gene regulation by riboswitches. Nat Rev Mol Cell Biol 5:451–463
6. Bartel DP (2004) MicroRNAs: genomics, biogenesis, mechanism, and function. Cell 116:281–297
7. Carter AP, Clemons WM, Brodersen DE, Morgan-Warren RJ, Wimberly BT, Ramakrishnan V (2000) Functional insights from the structure of the 30S ribosomal subunit and its interactions with antibiotics. Nature 407:340–348
8. Brodersen DE, Clemons WM Jr, Carter AP, Morgan-Warren RJ, Wimberly BT, Ramakrishnan V (2000) The structural basis for the action of the antibiotics tetracycline, pactamycin, and hygromycin B on the 30S ribosomal subunit. Cell 103:1143–1154
9. Regulski EE, Breaker RR (2008) In-line probing analysis of riboswitches. Methods Mol Biol 419:53–67
10. Tijerina P, Mohr S, Russell R (2007) DMS footprinting of structured RNAs and RNA-protein complexes. Nat Protoc 2:2608–2623
11. Kolupaeva VG, Pestova TV, Hellen CU (2000) An enzymatic footprinting analysis of the interaction of 40S ribosomal subunits with the internal ribosomal entry site of hepatitis C virus. J Virol 74:6242–6250
12. Kertesz M, Wan Y, Mazor E, Rinn JL, Nutter RC, Chang HY, Segal E (2010) Genome-wide measurement of RNA secondary structure in yeast. Nature 467:103–107
13. Lucks JB, Mortimer SA, Trapnell C, Luo S, Aviran S, Schroth GP, Pachter L, Doudna JA, Arkin AP (2011) Multiplexed RNA structure characterization with selective 2'-hydroxyl acylation analyzed by primer extension sequencing (SHAPE-Seq). Proc Natl Acad Sci U S A 108:11063–11068
14. Puglisi JD, Chen L, Frankel AD, Williamson JR (1993) Role of RNA structure in arginine recognition of TAR RNA. Proc Natl Acad Sci U S A 90:3680–3684
15. Laughlan G, Murchie AI, Norman DG, Moore MH, Moody PC, Lilley DM, Luisi B (1994) The high-resolution crystal structure of a parallel-stranded guanine tetraplex. Science 265:520–524

16. Phillips K, Dauter Z, Murchie AI, Lilley DM, Luisi B (1997) The crystal structure of a parallel-stranded guanine tetraplex at 0.95 A resolution. J Mol Biol 273:171–182
17. Mortimer SA, Trapnell C, Aviran S, Pachter L, Lucks JB (2012) SHAPE-Seq: high-throughput RNA structure analysis. Curr Protoc Chem Biol 4:275–297
18. Wan Y, Qu K, Ouyang Z, Kertesz M, Li J, Tibshirani R, Makino DL, Nutter RC, Segal E, Chang HY (2012) Genome-wide measurement of RNA folding energies. Mol Cell 48:169–181
19. Werstuck G, Green MR (1998) Controlling gene expression in living cells through small molecule-RNA interactions. Science 282:296–298
20. Weeks KM, Crothers DM (1991) RNA recognition by Tat-derived peptides: interaction in the major groove? Cell 66:577–588
21. Zalfa F, Giorgi M, Primerano B, Moro A, Di Penta A, Reis S, Oostra B, Bagni C (2003) The fragile X syndrome protein FMRP associates with BC1 RNA and regulates the translation of specific mRNAs at synapses. Cell 112:317–327
22. Kedde M, Strasser MJ, Boldajipour B, Oude Vrielink JA, Slanchev K, le Sage C, Nagel R, Voorhoeve PM, van Duijse J, Orom UA, Lund AH, Perrakis A, Raz E, Agami R (2007) RNA-binding protein Dnd1 inhibits microRNA access to target mRNA. Cell 131:1273–1286
23. Martianov I, Ramadass A, Serra Barros A, Chow N, Akoulitchev A (2007) Repression of the human dihydrofolate reductase gene by a non-coding interfering transcript. Nature 445:666–670
24. Jain A, Wang G, Vasquez KM (2008) DNA triple helices: biological consequences and therapeutic potential. Biochimie 90:1117–1130
25. Buske FA, Mattick JS, Bailey TL (2011) Potential in vivo roles of nucleic acid triple-helices. RNA Biol 8:427–439
26. Mondal T, Subhash S, Vaid R, Enroth S, Uday S, Reinius B, Mitra S, Mohammed A, James AR, Hoberg E, Moustakas A, Gyllensten U, Jones SJ, Gustafsson CM, Sims AH, Westerlund F, Gorab E, Kanduri C (2015) MEG3 long noncoding RNA regulates the TGF-beta pathway genes through formation of RNA-DNA triplex structures. Nat Commun 6:7743
27. Gellert M, Lipsett MN, Davies DR (1962) Helix formation by guanylic acid. Proc Natl Acad Sci U S A 48:2013–2018
28. Smith FW, Feigon J (1992) Quadruplex structure of Oxytricha telomeric DNA oligonucleotides. Nature 356:164–168
29. Hardin CC, Watson T, Corregan M, Bailey C (1992) Cation-dependent transition between the quadruplex and Watson-Crick hairpin forms of d(CGCG3GCG). Biochemistry 31:833–841
30. Burge S, Parkinson GN, Hazel P, Todd AK, Neidle S (2006) Quadruplex DNA: sequence, topology and structure. Nucleic Acids Res 34:5402–5415
31. Eddy J, Maizels N (2006) Gene function correlates with potential for G4 DNA formation in the human genome. Nucleic Acids Res 34:3887–3896
32. Gomez D, Lemarteleur T, Lacroix L, Mailliet P, Mergny JL, Riou JF (2004) Telomerase downregulation induced by the G-quadruplex ligand 12459 in A549 cells is mediated by hTERT RNA alternative splicing. Nucleic Acids Res 32:371–379
33. Monchaud D, Teulade-Fichou MP (2008) A hitchhiker's guide to G-quadruplex ligands. Org Biomol Chem 6:627–636
34. Kim J, Cheong C, Moore PB (1991) Tetramerization of an RNA oligonucleotide containing a GGGG sequence. Nature 351:331–332
35. Christiansen J, Kofod M, Nielsen FC (1994) A guanosine quadruplex and two stable hairpins flank a major cleavage site in insulin-like growth factor II mRNA. Nucleic Acids Res 22:5709–5716
36. Patel DJ, Phan AT, Kuryavyi V (2007) Human telomere, oncogenic promoter and 5'-UTR G-quadruplexes: diverse higher order DNA and RNA targets for cancer therapeutics. Nucleic Acids Res 35:7429–7455
37. Kumari S, Bugaut A, Huppert JL, Balasubramanian S (2007) An RNA G-quadruplex in the 5' UTR of the NRAS proto-oncogene modulates translation. Nat Chem Biol 3:218–221
38. Arora A, Dutkiewicz M, Scaria V, Hariharan M, Maiti S, Kurreck J (2008) Inhibition of translation in living eukaryotic cells by an RNA G-quadruplex motif. RNA 14:1290–1296

39. Morris MJ, Basu S (2009) An unusually stable G-quadruplex within the 5'-UTR of the MT3 matrix metalloproteinase mRNA represses translation in eukaryotic cells. Biochemistry 48:5313–5319
40. Jayaraj GG, Pandey S, Scaria V, Maiti S (2012) Potential G-quadruplexes in the human long non-coding transcriptome. RNA Biol 9:81–86
41. Arora A, Maiti S (2009) Differential biophysical behavior of human telomeric RNA and DNA quadruplex. J Phys Chem B 113:10515–10520
42. Zhang DH, Fujimoto T, Saxena S, Yu HQ, Miyoshi D, Sugimoto N (2010) Monomorphic RNA G-quadruplex and polymorphic DNA G-quadruplex structures responding to cellular environmental factors. Biochemistry 49:4554–4563
43. Huppert JL, Bugaut A, Kumari S, Balasubramanian S (2008) G-quadruplexes: the beginning and end of UTRs. Nucleic Acids Res 36:6260–6268
44. Gros J, Guedin A, Mergny JL, Lacroix L (2008) G-Quadruplex formation interferes with P1 helix formation in the RNA component of telomerase hTERC. Chembiochem 9:2075–2079
45. Marcel V, Tran PL, Sagne C, Martel-Planche G, Vaslin L, Teulade-Fichou MP, Hall J, Mergny JL, Hainaut P, Van Dyck E (2011) G-quadruplex structures in TP53 intron 3: role in alternative splicing and in production of p53 mRNA isoforms. Carcinogenesis 32:271–278
46. Subramanian M, Rage F, Tabet R, Flatter E, Mandel JL, Moine H (2011) G-quadruplex RNA structure as a signal for neurite mRNA targeting. EMBO Rep 12:697–704
47. Biffi G, Di Antonio M, Tannahill D, Balasubramanian S (2014) Visualization and selective chemical targeting of RNA G-quadruplex structures in the cytoplasm of human cells. Nat Chem 6:75–80
48. Hagihara M, Yoneda K, Yabuuchi H, Okuno Y, Nakatani K (2010) A reverse transcriptase stop assay revealed diverse quadruplex formations in UTRs in mRNA. Bioorg Med Chem Lett 20:2350–2353
49. Kwok CK, Marsico G, Sahakyan AB, Chambers VS, Balasubramanian S (2016) rG4-seq reveals widespread formation of G-quadruplex structures in the human transcriptome. Nat Methods 13:841–844
50. Guo JU, Bartel DP (2016) RNA G-quadruplexes are globally unfolded in eukaryotic cells and depleted in bacteria. Science 353
51. Morris MJ, Negishi Y, Pazsint C, Schonhoft JD, Basu S (2010) An RNA G-quadruplex is essential for cap-independent translation initiation in human VEGF IRES. J Am Chem Soc 132:17831–17839
52. Agarwala P, Pandey S, Mapa K, Maiti S (2013) The G-quadruplex augments translation in the 5' untranslated region of transforming growth factor beta2. Biochemistry 52:1528–1538
53. Hazel P, Huppert J, Balasubramanian S, Neidle S (2004) Loop-length-dependent folding of G-quadruplexes. J Am Chem Soc 126:16405–16415
54. Bugaut A, Balasubramanian S (2008) A sequence-independent study of the influence of short loop lengths on the stability and topology of intramolecular DNA G-quadruplexes. Biochemistry 47:689–697
55. Guedin A, Gros J, Alberti P, Mergny JL (2010) How long is too long? Effects of loop size on G-quadruplex stability. Nucleic Acids Res 38:7858–7868
56. Zhang AY, Bugaut A, Balasubramanian S (2011) A sequence-independent analysis of the loop length dependence of intramolecular RNA G-quadruplex stability and topology. Biochemistry 50:7251–7258
57. Pandey S, Agarwala P, Maiti S (2013) Effect of loops and G-quartets on the stability of RNA G-quadruplexes. J Phys Chem B 117:6896–6905
58. Jodoin R, Bauer L, Garant JM, Mahdi Laaref A, Phaneuf F, Perreault JP (2014) The folding of 5'-UTR human G-quadruplexes possessing a long central loop. RNA 20:1129–1141
59. Joachimi A, Benz A, Hartig JS (2009) A comparison of DNA and RNA quadruplex structures and stabilities. Bioorg Med Chem 17:6811–6815

60. Tang CF, Shafer RH (2006) Engineering the quadruplex fold: nucleoside conformation determines both folding topology and molecularity in guanine quadruplexes. J Am Chem Soc 128:5966–5973
61. Liu H, Kanagawa M, Matsugami A, Tanaka Y, Katahira M, Uesugi S (2000) NMR study of a novel RNA quadruplex structure. Nucleic Acids Symp Ser:65–66
62. Liu H, Kugimiya A, Sakurai T, Katahira M, Uesugi S (2002) A comparison of the properties and the solution structure for RNA and DNA quadruplexes which contain two GGAGG sequences joined with a tetranucleotide linker. Nucleosides Nucleotides Nucleic Acids 21:785–801
63. Greider CW, Blackburn EH (1989) A telomeric sequence in the RNA of Tetrahymena telomerase required for telomere repeat synthesis. Nature 337:331–337
64. Shay JW, Wright WE (2011) Role of telomeres and telomerase in cancer. Semin Cancer Biol 21:349–353
65. Wang Q, Liu JQ, Chen Z, Zheng KW, Chen CY, Hao YH, Tan Z (2011) G-quadruplex formation at the 3' end of telomere DNA inhibits its extension by telomerase, polymerase and unwinding by helicase. Nucleic Acids Res 39:6229–6237
66. Parkinson GN, Lee MP, Neidle S (2002) Crystal structure of parallel quadruplexes from human telomeric DNA. Nature 417:876–880
67. Collie GW, Haider SM, Neidle S, Parkinson GN (2010) A crystallographic and modelling study of a human telomeric RNA (TERRA) quadruplex. Nucleic Acids Res 38:5569–5580
68. Xu L, Feng S, Zhou X (2011) Human telomeric G-quadruplexes undergo dynamic conversion in a molecular crowding environment. Chem Commun (Camb) 47:3517–3519
69. Collie GW, Parkinson GN, Neidle S, Rosu F, De Pauw E, Gabelica V (2010) Electrospray mass spectrometry of telomeric RNA (TERRA) reveals the formation of stable multimeric G-quadruplex structures. J Am Chem Soc 132:9328–9334
70. Martadinata H, Phan AT (2009) Structure of propeller-type parallel-stranded RNA G-quadruplexes, formed by human telomeric RNA sequences in K+ solution. J Am Chem Soc 131:2570–2578
71. Sacca B, Lacroix L, Mergny JL (2005) The effect of chemical modifications on the thermal stability of different G-quadruplex-forming oligonucleotides. Nucleic Acids Res 33:1182–1192
72. Olsen CM, Marky LA (2009) Energetic and hydration contributions of the removal of methyl groups from thymine to form uracil in G-quadruplexes. J Phys Chem B 113:9–11
73. Ji X, Sun H, Zhou H, Xiang J, Tang Y, Zhao C (2011) Research progress of RNA quadruplex. Nucleic Acid Ther 21:185–200
74. Agarwal T, Jayaraj G, Pandey SP, Agarwala P, Maiti S (2012) RNA G-quadruplexes: G-quadruplexes with "U" turns. Curr Pharm Des 18:2102–2111
75. Decorsiere A, Cayrel A, Vagner S, Millevoi S (2011) Essential role for the interaction between hnRNP H/F and a G quadruplex in maintaining p53 pre-mRNA 3'-end processing and function during DNA damage. Genes Dev 25:220–225
76. Wieland M, Hartig JS (2007) RNA quadruplex-based modulation of gene expression. Chem Biol 14:757–763
77. Bhattacharyya D, Diamond P, Basu S (2015) An independently folding RNA G-quadruplex domain directly recruits the 40S ribosomal subunit. Biochemistry 54:1879–1885
78. Bonnal S, Schaeffer C, Creancier L, Clamens S, Moine H, Prats AC, Vagner S (2003) A single internal ribosome entry site containing a G quartet RNA structure drives fibroblast growth factor 2 gene expression at four alternative translation initiation codons. J Biol Chem 278:39330–39336
79. Halder K, Wieland M, Hartig JS (2009) Predictable suppression of gene expression by 5'-UTR-based RNA quadruplexes. Nucleic Acids Res 37:6811–6817
80. Shahid R, Bugaut A, Balasubramanian S (2010) The BCL-2 5' untranslated region contains an RNA G-quadruplex-forming motif that modulates protein expression. Biochemistry 49:8300–8306
81. Endoh T, Kawasaki Y, Sugimoto N (2013) Suppression of gene expression by G-quadruplexes in open reading frames depends on G-quadruplex stability. Angew Chem Int Ed Engl 52:5522–5526

82. Arora A, Suess B (2011) An RNA G-quadruplex in the 3' UTR of the proto-oncogene PIM1 represses translation. RNA Biol 8:802–805
83. Maizels N (2015) G4-associated human diseases. EMBO Rep 16:910–922
84. Pieretti M, Zhang FP, Fu YH, Warren ST, Oostra BA, Caskey CT, Nelson DL (1991) Absence of expression of the FMR-1 gene in fragile X syndrome. Cell 66:817–822
85. Coffee B, Keith K, Albizua I, Malone T, Mowrey J, Sherman SL, Warren ST (2009) Incidence of fragile X syndrome by newborn screening for methylated FMR1 DNA. Am J Hum Genet 85:503–514
86. Sutcliffe JS, Nelson DL, Zhang F, Pieretti M, Caskey CT, Saxe D, Warren ST (1992) DNA methylation represses FMR-1 transcription in fragile X syndrome. Hum Mol Genet 1:397–400
87. Feng Y, Gutekunst CA, Eberhart DE, Yi H, Warren ST, Hersch SM (1997) Fragile X mental retardation protein: nucleocytoplasmic shuttling and association with somatodendritic ribosomes. J Neurosci 17:1539–1547
88. Blackwell E, Zhang X, Ceman S (2010) Arginines of the RGG box regulate FMRP association with polyribosomes and mRNA. Hum Mol Genet 19:1314–1323
89. Brown V, Jin P, Ceman S, Darnell JC, O'Donnell WT, Tenenbaum SA, Jin X, Feng Y, Wilkinson KD, Keene JD, Darnell RB, Warren ST (2001) Microarray identification of FMRP-associated brain mRNAs and altered mRNA translational profiles in fragile X syndrome. Cell 107:477–487
90. Darnell JC, Jensen KB, Jin P, Brown V, Warren ST, Darnell RB (2001) Fragile X mental retardation protein targets G quartet mRNAs important for neuronal function. Cell 107:489–499
91. Lu R, Wang H, Liang Z, Ku L, O'Donnell WT, Li W, Warren ST, Feng Y (2004) The fragile X protein controls microtubule-associated protein 1B translation and microtubule stability in brain neuron development. Proc Natl Acad Sci U S A 101:15201–15206
92. Castets M, Schaeffer C, Bechara E, Schenck A, Khandjian EW, Luche S, Moine H, Rabilloud T, Mandel JL, Bardoni B (2005) FMRP interferes with the Rac1 pathway and controls actin cytoskeleton dynamics in murine fibroblasts. Hum Mol Genet 14:835–844
93. Khateb S, Weisman-Shomer P, Hershco-Shani I, Ludwig AL, Fry M (2007) The tetraplex (CGG)n destabilizing proteins hnRNP A2 and CBF-A enhance the in vivo translation of fragile X premutation mRNA. Nucleic Acids Res 35:5775–5788
94. Schaeffer C, Bardoni B, Mandel JL, Ehresmann B, Ehresmann C, Moine H (2001) The fragile X mental retardation protein binds specifically to its mRNA via a purine quartet motif. EMBO J 20:4803–4813
95. Westmark CJ, Malter JS (2007) FMRP mediates mGluR5-dependent translation of amyloid precursor protein. PLoS Biol 5:e52
96. Suhl JA, Chopra P, Anderson BR, Bassell GJ, Warren ST (2014) Analysis of FMRP mRNA target datasets reveals highly associated mRNAs mediated by G-quadruplex structures formed via clustered WGGA sequences. Hum Mol Genet 23:5479–5491
97. Ling SC, Polymenidou M, Cleveland DW (2013) Converging mechanisms in ALS and FTD: disrupted RNA and protein homeostasis. Neuron 79:416–438
98. Renton AE, Majounie E, Waite A, Simon-Sanchez J, Rollinson S, Gibbs JR, Schymick JC, Laaksovirta H, van Swieten JC, Myllykangas L, Kalimo H, Paetau A, Abramzon Y, Remes AM, Kaganovich A, Scholz SW, Duckworth J, Ding J, Harmer DW, Hernandez DG, Johnson JO, Mok K, Ryten M, Trabzuni D, Guerreiro RJ, Orrell RW, Neal J, Murray A, Pearson J, Jansen IE, Sondervan D, Seelaar H, Blake D, Young K, Halliwell N, Callister JB, Toulson G, Richardson A, Gerhard A, Snowden J, Mann D, Neary D, Nalls MA, Peuralinna T, Jansson L, Isoviita VM, Kaivorinne AL, Holtta-Vuori M, Ikonen E, Sulkava R, Benatar M, Wuu J, Chio A, Restagno G, Borghero G, Sabatelli M, Heckerman D, Rogaeva E, Zinman L, Rothstein JD, Sendtner M, Drepper C, Eichler EE, Alkan C, Abdullaev Z, Pack SD, Dutra A, Pak E, Hardy J, Singleton A, Williams NM, Heutink P, Pickering-Brown S, Morris HR, Tienari PJ, Traynor BJ (2011) A hexanucleotide repeat expansion in C9ORF72 is the cause of chromosome 9p21-linked ALS-FTD. Neuron 72:257–268

99. DeJesus-Hernandez M, Mackenzie IR, Boeve BF, Boxer AL, Baker M, Rutherford NJ, Nicholson AM, Finch NA, Flynn H, Adamson J, Kouri N, Wojtas A, Sengdy P, Hsiung GY, Karydas A, Seeley WW, Josephs KA, Coppola G, Geschwind DH, Wszolek ZK, Feldman H, Knopman DS, Petersen RC, Miller BL, Dickson DW, Boylan KB, Graff-Radford NR, Rademakers R (2011) Expanded GGGGCC hexanucleotide repeat in noncoding region of C9ORF72 causes chromosome 9p-linked FTD and ALS. Neuron 72:245–256
100. Fratta P, Mizielinska S, Nicoll AJ, Zloh M, Fisher EM, Parkinson G, Isaacs AM (2012) C9orf72 hexanucleotide repeat associated with amyotrophic lateral sclerosis and frontotemporal dementia forms RNA G-quadruplexes. Sci Rep 2:1016
101. Haeusler AR, Donnelly CJ, Periz G, Simko EA, Shaw PG, Kim MS, Maragakis NJ, Troncoso JC, Pandey A, Sattler R, Rothstein JD, Wang J (2014) C9orf72 nucleotide repeat structures initiate molecular cascades of disease. Nature 507:195–200
102. Freibaum BD, Lu Y, Lopez-Gonzalez R, Kim NC, Almeida S, Lee KH, Badders N, Valentine M, Miller BL, Wong PC, Petrucelli L, Kim HJ, Gao FB, Taylor JP (2015) GGGGCC repeat expansion in C9orf72 compromises nucleocytoplasmic transport. Nature 525:129–133
103. Jovicic A, Mertens J, Boeynaems S, Bogaert E, Chai N, Yamada SB, Paul JW 3rd, Sun S, Herdy JR, Bieri G, Kramer NJ, Gage FH, Van Den Bosch L, Robberecht W, Gitler AD (2015) Modifiers of C9orf72 dipeptide repeat toxicity connect nucleocytoplasmic transport defects to FTD/ALS. Nat Neurosci 18:1226–1229
104. Zamiri B, Reddy K, Macgregor RB Jr, Pearson CE (2014) TMPyP4 porphyrin distorts RNA G-quadruplex structures of the disease-associated r(GGGGCC)n repeat of the C9orf72 gene and blocks interaction of RNA-binding proteins. J Biol Chem 289:4653–4659
105. Su Z, Zhang Y, Gendron TF, Bauer PO, Chew J, Yang WY, Fostvedt E, Jansen-West K, Belzil VV, Desaro P, Johnston A, Overstreet K, Oh SY, Todd PK, Berry JD, Cudkowicz ME, Boeve BF, Dickson D, Floeter MK, Traynor BJ, Morelli C, Ratti A, Silani V, Rademakers R, Brown RH, Rothstein JD, Boylan KB, Petrucelli L, Disney MD (2014) Discovery of a biomarker and lead small molecules to target r(GGGGCC)-associated defects in c9FTD/ALS. Neuron 83:1043–1050
106. Nunan J, Small DH (2000) Regulation of APP cleavage by alpha-, beta- and gamma-secretases. FEBS Lett 483:6–10
107. Tanzi RE (2005) The synaptic Abeta hypothesis of Alzheimer disease. Nat Neurosci 8:977–979
108. Weyer SW, Zagrebelsky M, Herrmann U, Hick M, Ganss L, Gobbert J, Gruber M, Altmann C, Korte M, Deller T, Muller UC (2014) Comparative analysis of single and combined APP/APLP knockouts reveals reduced spine density in APP-KO mice that is prevented by APPsalpha expression. Acta Neuropathol Commun 2:36
109. Crenshaw E, Leung BP, Kwok CK, Sharoni M, Olson K, Sebastian NP, Ansaloni S, Schweitzer-Stenner R, Akins MR, Bevilacqua PC, Saunders AJ (2015) Amyloid precursor protein translation is regulated by a 3'UTR guanine quadruplex. PLoS One 10:e0143160
110. Lammich S, Kamp F, Wagner J, Nuscher B, Zilow S, Ludwig AK, Willem M, Haass C (2011) Translational repression of the disintegrin and metalloprotease ADAM10 by a stable G-quadruplex secondary structure in its 5'-untranslated region. J Biol Chem 286:45063–45072
111. Fisette JF, Montagna DR, Mihailescu MR, Wolfe MS (2012) A G-rich element forms a G-quadruplex and regulates BACE1 mRNA alternative splicing. J Neurochem 121:763–773
112. Kapranov P, Cawley SE, Drenkow J, Bekiranov S, Strausberg RL, Fodor SP, Gingeras TR (2002) Large-scale transcriptional activity in chromosomes 21 and 22. Science 296:916–919
113. Rinn JL, Chang HY (2012) Genome regulation by long noncoding RNAs. Annu Rev Biochem 81:145–166
114. Derrien T, Johnson R, Bussotti G, Tanzer A, Djebali S, Tilgner H, Guernec G, Martin D, Merkel A, Knowles DG, Lagarde J, Veeravalli L, Ruan X, Ruan Y, Lassmann T, Carninci P, Brown JB, Lipovich L, Gonzalez JM, Thomas M, Davis CA, Shiekhattar R, Gingeras TR, Hubbard TJ, Notredame C, Harrow J, Guigo R (2012) The GENCODE v7 catalog of human

long noncoding RNAs: analysis of their gene structure, evolution, and expression. Genome Res 22:1775–1789
115. Bartel DP (2009) MicroRNAs: target recognition and regulatory functions. Cell 136:215–233
116. Lujambio A, Lowe SW (2012) The microcosmos of cancer. Nature 482:347–355
117. Mendell JT, Olson EN (2012) MicroRNAs in stress signaling and human disease. Cell 148:1172–1187
118. Zampetaki A, Mayr M (2012) MicroRNAs in vascular and metabolic disease. Circ Res 110:508–522
119. Esau CC (2008) Inhibition of microRNA with antisense oligonucleotides. Methods 44:55–60
120. Young DD, Connelly CM, Grohmann C, Deiters A (2010) Small molecule modifiers of microRNA miR-122 function for the treatment of hepatitis C virus infection and hepatocellular carcinoma. J Am Chem Soc 132:7976–7981
121. Deiters A (2010) Small molecule modifiers of the microRNA and RNA interference pathway. AAPS J 12:51–60
122. Bose D, Jayaraj G, Suryawanshi H, Agarwala P, Pore SK, Banerjee R, Maiti S (2012) The tuberculosis drug streptomycin as a potential cancer therapeutic: inhibition of miR-21 function by directly targeting its precursor. Angew Chem Int Ed Engl 51:1019–1023
123. Kim VN, Han J, Siomi MC (2009) Biogenesis of small RNAs in animals. Nat Rev Mol Cell Biol 10:126–139
124. Pandey S, Agarwala P, Jayaraj GG, Gargallo R, Maiti S (2015) The RNA stem-loop to G-quadruplex equilibrium controls mature microRNA production inside the cell. Biochemistry 54:7067–7078
125. Mirihana Arachchilage G, Dassanayake AC, Basu S (2015) A potassium ion-dependent RNA structural switch regulates human pre-miRNA 92b maturation. Chem Biol 22:262–272
126. Stefanovic S, Bassell GJ, Mihailescu MR (2015) G quadruplex RNA structures in PSD-95 mRNA: potential regulators of miR-125a seed binding site accessibility. RNA 21:48–60
127. Young LS, Rickinson AB (2004) Epstein-Barr virus: 40 years on. Nat Rev Cancer 4:757–768
128. Kanda T, Kamiya M, Maruo S, Iwakiri D, Takada K (2007) Symmetrical localization of extrachromosomally replicating viral genomes on sister chromatids. J Cell Sci 120:1529–1539
129. Yates JL, Warren N, Sugden B (1985) Stable replication of plasmids derived from Epstein-Barr virus in various mammalian cells. Nature 313:812–815
130. Snudden DK, Hearing J, Smith PR, Grasser FA, Griffin BE (1994) EBNA-1, the major nuclear antigen of Epstein-Barr virus, resembles 'RGG' RNA binding proteins. EMBO J 13:4840–4847
131. Norseen J, Johnson FB, Lieberman PM (2009) Role for G-quadruplex RNA binding by Epstein-Barr virus nuclear antigen 1 in DNA replication and metaphase chromosome attachment. J Virol 83:10336–10346
132. Wang SR, Zhang QY, Wang JQ, Ge XY, Song YY, Wang YF, Li XD, Fu BS, Xu GH, Shu B, Gong P, Zhang B, Tian T, Zhou X (2016) Chemical targeting of a G-quadruplex RNA in the Ebola Virus L gene. Cell Chem Biol 23:1113–1122
133. Azzalin CM, Reichenbach P, Khoriauli L, Giulotto E, Lingner J (2007) Telomeric repeat containing RNA and RNA surveillance factors at mammalian chromosome ends. Science 318:798–801
134. Luke B, Lingner J (2009) TERRA: telomeric repeat-containing RNA. EMBO J 28:2503–2510
135. Xu Y, Suzuki Y, Ito K, Komiyama M (2010) Telomeric repeat-containing RNA structure in living cells. Proc Natl Acad Sci U S A 107:14579–14584
136. Lopez de Silanes I, Stagno d'Alcontres M, Blasco MA (2010) TERRA transcripts are bound by a complex array of RNA-binding proteins. Nat Commun 1:33
137. Booy EP, Meier M, Okun N, Novakowski SK, Xiong S, Stetefeld J, McKenna SA (2012) The RNA helicase RHAU (DHX36) unwinds a G4-quadruplex in human telomerase RNA and promotes the formation of the P1 helix template boundary. Nucleic Acids Res 40:4110–4124

138. Bugaut A, Rodriguez R, Kumari S, Hsu ST, Balasubramanian S (2010) Small molecule-mediated inhibition of translation by targeting a native RNA G-quadruplex. Org Biomol Chem 8:2771–2776
139. van Steensel B, Smogorzewska A, de Lange T (1998) TRF2 protects human telomeres from end-to-end fusions. Cell 92:401–413
140. Bai Y, Lathia JD, Zhang P, Flavahan W, Rich JN, Mattson MP (2014) Molecular targeting of TRF2 suppresses the growth and tumorigenesis of glioblastoma stem cells. Glia 62:1687–1698
141. Gomez D, Guedin A, Mergny JL, Salles B, Riou JF, Teulade-Fichou MP, Calsou P (2010) A G-quadruplex structure within the 5'-UTR of TRF2 mRNA represses translation in human cells. Nucleic Acids Res 38:7187–7198
142. Biffi G, Tannahill D, Balasubramanian S (2012) An intramolecular G-quadruplex structure is required for binding of telomeric repeat-containing RNA to the telomeric protein TRF2. J Am Chem Soc 134:11974–11976
143. Olivier M, Hollstein M, Hainaut P (2010) TP53 mutations in human cancers: origins, consequences, and clinical use. Cold Spring Harb Perspect Biol 2:a001008
144. Courtois S, Verhaegh G, North S, Luciani MG, Lassus P, Hibner U, Oren M, Hainaut P (2002) DeltaN-p53, a natural isoform of p53 lacking the first transactivation domain, counteracts growth suppression by wild-type p53. Oncogene 21:6722–6728
145. Bourdon JC, Fernandes K, Murray-Zmijewski F, Liu G, Diot A, Xirodimas DP, Saville MK, Lane DP (2005) p53 isoforms can regulate p53 transcriptional activity. Genes Dev 19:2122–2137
146. Vousden KH, Lane DP (2007) p53 in health and disease. Nat Rev Mol Cell Biol 8:275–283
147. Pehar M, O'Riordan KJ, Burns-Cusato M, Andrzejewski ME, del Alcazar CG, Burger C, Scrable H, Puglielli L (2010) Altered longevity-assurance activity of p53:p44 in the mouse causes memory loss, neurodegeneration and premature death. Aging Cell 9:174–190
148. Gemignani F, Moreno V, Landi S, Moullan N, Chabrier A, Gutierrez-Enriquez S, Hall J, Guino E, Peinado MA, Capella G, Canzian F (2004) A TP53 polymorphism is associated with increased risk of colorectal cancer and with reduced levels of TP53 mRNA. Oncogene 23:1954–1956
149. Quante T, Otto B, Brazdova M, Kejnovska I, Deppert W, Tolstonog GV (2012) Mutant p53 is a transcriptional co-factor that binds to G-rich regulatory regions of active genes and generates transcriptional plasticity. Cell Cycle 11:3290–3303
150. Dias N, Stein CA (2002) Potential roles of antisense oligonucleotides in cancer therapy. The example of Bcl-2 antisense oligonucleotides. Eur J Pharm Biopharm 54:263–269
151. Kang MH, Reynolds CP (2009) Bcl-2 inhibitors: targeting mitochondrial apoptotic pathways in cancer therapy. Clin Cancer Res 15:1126–1132
152. Sartorius UA, Krammer PH (2002) Upregulation of Bcl-2 is involved in the mediation of chemotherapy resistance in human small cell lung cancer cell lines. Int J Cancer 97:584–592
153. Dexheimer TS, Sun D, Hurley LH (2006) Deconvoluting the structural and drug-recognition complexity of the G-quadruplex-forming region upstream of the bcl-2 P1 promoter. J Am Chem Soc 128:5404–5415
154. Rzuczek SG, Pilch DS, Liu A, Liu L, LaVoie EJ, Rice JE (2010) Macrocyclic pyridyl polyoxazoles: selective RNA and DNA G-quadruplex ligands as antitumor agents. J Med Chem 53:3632–3644
155. Katsuda Y, Sato S, Asano L, Morimura Y, Furuta T, Sugiyama H, Hagihara M, Uesugi M (2016) A small molecule that represses translation of G-quadruplex-containing mRNA. J Am Chem Soc 138:9037–9040

Top Med Chem (2018) 27: 207–236
DOI: 10.1007/7355_2017_25

Published online: 25 July 2017

The Therapeutic Targeting of Long Noncoding RNA

Caroline J. Woo

Abstract While only 1–2% of the human genome is dedicated to protein-coding genes, much of the genome is actively transcribed. Long noncoding RNAs (lncRNAs) are a subset of noncoding RNAs that arise from this "dark matter." They are involved in nearly every aspect of cellular biology, and in particular, transcriptional regulation through epigenetic protein complexes. Using an oligonucleotide-based approach, which can afford specificity, lncRNAs serve as potential therapeutic targets in many disease areas. This chapter discusses their biogenesis, biological roles, and considerations for drug development.

Keywords Epigenetics, Long non-coding RNA, Oligomers, Upregulation

Contents

1 Introduction 208
2 Definition 208
3 Discovery 209
4 Functions 211
5 *Cis*-Acting and *Trans*-Acting Nuclear LncRNAs 212
6 Epigenetic Modifications of Histones and DNA 212
7 Repetitive Elements 215
8 Specificity 216
9 Oligonucleotide-Based Therapeutics 218
10 Target Identification and Validation 220
11 Animal Models 221
12 Long Noncoding RNA as Therapeutic Targets 222
13 Selected Examples of lncRNA Preclinical Studies 223
  13.1 Example 1: ciRS-7 as a Target for Cancer Therapy 223
  13.2 Example 2: APOA1-AS as a Target for Hypercholesterolemia 225
  13.3 Example 3: SMN-FL as a Target for SMA 226

C.J. Woo (✉)
RaNA Therapeutics, 200 Sidney Street, Cambridge, MA 02139, USA
e-mail: cwoo@ranarx.com

13.4 Example 4: DBE-T as a Target for FSHD 227
14 Discussion 229
References 230

1 Introduction

The discovery of long noncoding RNAs (lncRNAs) has greatly increased the scope of potential therapeutic drug targets over those afforded by directly targeting proteins or messenger RNAs (mRNAs). Of the estimated 20,000 proteins in the human body, less than 2% have been successfully targeted with approved small molecule drugs or antibodies [1]. In contrast, more than 30,000 unique lncRNAs have been identified [2] and are potential drug targets. Importantly, as synthetic oligonucleotides (oligomers) have recently been used to target mRNAs, oligomers may also be developed by rational drug design to target lncRNAs. Many diseases posing a therapeutic challenge in the past are now understood to be caused by genetic and/or epigenetic changes outside of protein-coding sequences. Therapies for such diseases may be contemplated by targeting the RNAs arising from the noncoding regions of the genome.

The upregulation of protein targets by traditional therapeutic modalities, such as small molecules and antibodies, has been an elusive goal. However, lncRNA antagonism affords an effective way to upregulate proteins since a subset of noncoding RNAs (ncRNAs) function to repress protein transcription/translation by localizing silencing chromatin modifiers. The recruitment of complexes that modify chromatin architecture could then change the expression of a protein-coding gene(s). Consequently, inherited disorders with low protein expression or the expression of proteins with impaired function may be treatable by identifying and antagonizing lncRNA activity. Many common life-threatening diseases such as cancer, diabetes type II, Alzheimer's disease, and Parkinson's disease appear to have common dysregulated epigenetic patterns and may therefore be amenable to therapies directed against noncoding RNAs [3, 4]. The functions of most lncRNAs are still being evaluated, but a wealth of genome-wide information is publicly available in databases to facilitate their identification, functional characterization, and target validation [5]. Thus, therapeutic approaches based upon lncRNAs hold promise for future transformative therapies of such clinically challenging maladies.

2 Definition

The three main classes of ncRNAs include: housekeeping RNAs (e.g., ribosomal RNAs, transfer RNAs, small nuclear RNAs and small nucleolar RNAs), short noncoding RNAs [e.g., microRNAs (miRNAs)], and lncRNAs. LncRNAs are broadly classified as those greater than 200 nucleotides in length. Unlike the first two classes of ncRNAs, whose functions are well-defined, lncRNAs have diverse roles and are potentially involved in nearly every aspect of gene expression. Since the discovery and annotation of lncRNAs are occurring faster than their functional characterizations, lncRNAs are primarily characterized by their genomic location relative to protein-coding genes. Long intergenic RNAs (lincRNAs) are those transcribed from genomic regions between protein-coding genes. They may arise from intergenic enhancer regions and are then called enhancer RNAs (eRNAs) [6, 7]. There are several types of lncRNAs that overlap with protein-coding regions [6]. *Cis* natural antisense RNAs (NATs) are transcripts that partially or wholly overlap with genes and are transcribed on the opposite strand from the same genomic locus [8, 9]. In general, NATs partially overlap a protein-coding gene and are believed to be involved in the regulation of the expression of that gene. Sense-overlapping intronic lncRNAs are transcribed from within the protein-coding gene on the same strand as the gene and their functions are unclear [10, 11].

Most lncRNAs are processed similarly to mRNAs, including transcription by RNA Polymerase II, capping at the 5′ end, splicing by conserved splice-signals and polyadenylation [12]. They are similar in length to protein-coding RNAs, although they generally have fewer exons. Additionally, lncRNA genes harbor many of the histone modifications that correlate with expression states along protein-coding genes. However, meaningful differences do exist as lncRNAs are mostly enriched in the nucleus, have often tissue- or temporally restricted expression, and some can be processed further into smaller ncRNAs [12]. Interestingly, in one study examining several cell lines, promoters of tissue-specific lincRNAs were found to be depleted of all histone marks except H3K9me3, a histone mark typically associated with silenced genes [13]. The association of H3K9me3 with these promoters was suggested to be involved in tissue-restricted expression. LncRNAs generally do not code for proteins or peptides, and have low cellular abundance. Even so, lncRNAs make up approximately 64% (excluding ribosomal, mitochondrial, and repeat sequences) of all human cellular transcripts [14].

3 Discovery

Nearly two decades ago, the FANTOM (Functional Annotation of Mammalian) genome consortium began classifying all expressed mammalian RNAs by sequencing cDNA clones. It was through this international collaboration that thousands of new ncRNAs were identified. Cap analysis gene expression (CAGE) technology was developed that could determine which RNAs were expressed in a cell by sequencing the 5′ ends of mRNAs [15]. Thus, it was revealed that much of the genome is transcribed even outside of the protein-coding genes [16, 17]. Originally focused on mouse cDNAs, FANTOM soon included human data as well. With the advent of next-generation sequencing technology, entire transcriptomes (total cellular RNA) could be sequenced in parallel by RNA-sequencing (RNA-seq) and this greatly expanded the number of new ncRNA discoveries. The Encyclopedia of DNA elements (ENCODE) Project, which began in 2003, employed these techniques and genome-wide computational methods with the goal of identifying and determining functional elements of the entire human genome outside of protein-coding genes. It became clear that, though at least 85% of the genome is transcribed, only 1–2% of it comprises protein-coding genes [18–20].

Sequencing-based methodologies have produced genome-wide transcriptomes from multiple species, cell types, and disease-states. Given the potential importance of lncRNAs, extensive ongoing research efforts are focused on their functional roles. Databases from large consortia exist such as FANTOM and GENCODE/ENCODE which catalog genome-wide data, including lncRNAs [21]. Other databases, specific to lncRNAs, exist as well, with varying degrees of validation [5].

One of the first described lncRNAs was *XIST*, a well-characterized and highly conserved lncRNA. It binds to Polycomb Repressive Complex 2 (PRC2) and spreads to multiple sites on one X-chromosome in placental female mammals to silence most genes, resulting in an inactive X chromosome [22–24]. *XIST* was discovered because it was only expressed on the inactive X chromosome and lacked an open reading frame. *XIST* was observed to coat the entire inactive X chromosome in a unique nuclear compartment resulting in a condensed chromatin structure that is now known to be associated with gene silencing. Deletion of *XIST* resulted in failed X-chromosome inactivation, whereas forced expression of *XIST* on X-chromosomes (females) or autosomes (males) triggered gene silencing [23–25]. These studies demonstrated that *XIST* was a key component required for X-chromosome inactivation. It further demonstrated that an RNA that did not code for any protein was an important modulator of chromosome function and participated in events that positioned the chromatin to modulate the expression of protein-coding genes. On a molecular level, *XIST* interacts with various protein complexes, and these interactions occur with specific regions of the *XIST* sequence [26–28]. For example, transcriptional silencing requires the repeat A domain [29], which recruits PRC2 [24], while a different region known as the repeat C region is required for *XIST* binding to chromatin. The deletion of one region did not affect the association of proteins with the other regions and therefore their domains were modular.

HOX transcript antisense RNA (*HOTAIR*) is another well-studied lncRNA, which also contains distinct domains that associate with PRC2 and a histone deacetylase complex, LSD1/CoREST/REST. It was first discovered in foreskin fibroblasts using microarray technology that probed RNA expression from the HOX gene clusters containing both the HOX genes and their intergenic regions [30]. The basic principles discovered by studying *XIST* and *HOTAIR* can potentially be applied generally across many classes of lncRNA.

4 Functions

As more lncRNAs were discovered, it became apparent that they work through diverse mechanisms and that cellular localization is essential to their function. LncRNAs exist both in the nucleus and the cytoplasm, although they are mostly nuclear. Enriched in the nucleus, some lncRNAs can modulate the chromatin landscape to suppress or activate gene function [31, 32]. LncRNAs may direct regulatory complexes to promoter regions to fine-tune the expression of single genes, gene clusters and, as was described for *XIST*, an entire chromosome [33]. It has been demonstrated that they can act as molecular scaffolds with discrete modular binding regions that recruit regulatory protein complexes to promoter sites [33, 34]. LncRNAs are also necessary for imprinting (the silencing of a cluster of genes from one parental germline as determined by the sex of the parent chromosome) because their expression from one parental allele will result in the silencing of the genes from that same allele through the recruitment of chromatin modifiers. The classic example of imprinting is the expression of XIST for X inactivation. It is estimated that approximately 30% of lncRNAs associate with distinct protein complexes that regulate chromatin. In addition to localizing proteins to their sites of action, some lncRNAs can activate or inactivate transcriptional modulators by removing them from their promoter regulation sites [35]. LncRNAs also function in sub-nuclear compartments such as paraspeckles (cytoplasmic transport of mRNA), permissive inter-chromatin granules (pre-mRNA splicing), repressive Polycomb bodies (gene silencing/repression), telomeres (cell division and senescence), and the mammalian-specific ribonucleoprotein bodies (stress responses, processing/storage of mRNA, autocrine/paracrine exosomes). In addition, several lncRNAs have been identified where the process of their transcription alone changes the chromatin configuration to allow the transcription of neighboring genes [36, 37].

LncRNAs also exist in the cytoplasm where they serve many functions. One example is growth arrest-specific 5 (*GAS5*) lncRNA, which folds into a secondary structure that that acts as a decoy by binding to the DNA binding domain of the glucocorticoid receptor to prevent signal transduction [38]. Cytoplasmic lncRNAs also regulate the expression of mRNAs post-transcriptionally. They have been shown to interact with mRNAs to recruit regulatory proteins that accelerate its degradation or serve as molecular sponges to neutralize specific miRNAs that negatively modulate the translation of various mRNA targets [39–41]. MiRNAs, are a class of small noncoding RNAs, which suppress gene transcription by binding to the 3′ untranslated regions (UTRs) of their target mRNAs or to DNA promoter regions [35, 42, 43]. By acting as molecular sponges, lncRNAs inhibit miRNA function, resulting in increased expression of their target mRNAs [42]. Other lncRNA functions include the modulation of gene splicing, altering the stability of mRNA, modifying the stability and conformation of proteins, and nuclear export of mRNAs to the cytoplasm for translation [31]. Thus, lncRNAs can affect most cellular processes involved in protein expression.

LncRNAs also circulate in exosomes and can be detected in blood and urine [44]. The function of these exosomal lncRNAs is not known but it is thought that they may play a role in autocrine or paracrine cell signaling. Since various lncRNAs are sometimes differentially expressed in patients compared with healthy individuals, exosomal RNA profiles have diagnostic potential for classifying diseases and stratifying disease states. The ability to do so may allow for the optimization of therapeutic strategies tailored to individual patients in a personalized medicine paradigm. LncRNAs are involved in a wide range of regulatory structures and mechanisms that are ripe for therapeutic targeting [11, 32, 45].

5 *Cis*-Acting and *Trans*-Acting Nuclear LncRNAs

A lncRNA is *cis*-acting when it regulates the transcription of neighboring genes on the same allele from which it is transcribed. An lncRNA is *trans*-acting when it regulates genes on a different allele or non-neighboring genes on the same allele. The spectrum of lncRNA functions tends to balance stability, abundance, and target proximity to optimally affect their specific targets. Newly transcribed *cis*-acting lncRNAs (and some *trans*-acting lncRNAs) act immediately at their proximal targets, though they are often expressed at very low copy numbers (roughly 1–10 copies per cell). *Cis*-acting lncRNAs affect their function at their target genes because they are present at appropriate concentrations. ANRIL is an example of an lncRNA that functions in *cis* by recruiting PRC2 to silence the *Ink4A/ARF* locus, which encode for tumor suppressor proteins p15, p14, and p16 [46].

Most *trans*-acting lncRNAs act distally in a diffusion-mediated manner and are often characterized by a long half-life and relatively high cellular abundance to achieve sufficient concentration at their target sites. For example, metastasis associated lung adenocarcinoma transcript 1 (*MALAT1*) is present at approximately 3,000 copies per cell and localizes primarily by diffusion from its site of transcription to numerous distal gene targets on other chromosomes. When *MALAT1* is expressed ectopically, it can still affect its various gene targets. Due to the three-dimensional organization of chromosomes within the nucleus, some *trans*-acting lncRNAs interact with targets in close spatial proximity even though they are located on a different chromosome. For example, the lncRNA *CISTR-ACT*, which is transcribed on chromosome 12 and modulates the functions on the proximally positioned chromosome 17 *SOX9* locus [47].

6 Epigenetic Modifications of Histones and DNA

The human genome is comprised of approximately three billion basepairs, containing both protein-coding and noncoding DNA, that is organized and packaged within a nucleus that confers both structure and functionality. The foundation of this organization is built by a set of proteins called histones. The histones combine to form an octamer composed of four dimeric subunits (H2A, H2B, H3, and H4) around which approximately 146 basepairs of DNA are wrapped to form a nucleosome unit. Collectively, the DNA and histone octamers are called chromatin. The dynamic association between histones and DNA allows changes in their conformation to mask or expose promoter regions of DNA to transcriptional machinery in order to control transcription [48].

Regulatory proteins, in conjunction with chromatin, orchestrate transcription to create unique transcriptome profiles in differentiated cells. Regions of the genome that do not contribute to the appropriate cellular function must be silenced since each cell contains a full complement of DNA, while other functionally relevant regions of the genome must be transcriptionally active. Histones play a key role in this selectivity, since they are covalently modified by regulatory protein complexes to modulate the transcription of their associated genes. Different mechanisms exist by which protein complexes localize to specific regions of the genome. One way in which the regulatory proteins can be recruited to chromatin to create (write), remove (erase), or recognize (read) marks on histone tails is through lncRNA-mediated mechanisms [49]. Histone writers include lysine acetyltransferases, arginine/lysine methyltransferases, and serine/threonine/tyrosine phosphorylases. These marks are covalent modifications on specific amino acids catalyzed by enzyme complexes that are localized to their sites of action by associating with lncRNAs, the result of which is gene expression, repression, or silencing [49]. Marks catalyzed by writers form a pattern that can predict silencing, repression (partial silencing), or enhanced active transcription. For example, histone acetyltransferases are writers that acetylate lysines along histones and histone deacetylases are erasers that remove acetyl groups. Together, they create a dynamic regulation of gene expression driven, in part, by specific histone acetylation. Other enzymes known as readers bind to the histone marks along with effector proteins to initiate the modulation of gene transcription unique to that locus.

Along with histone lysine acetylation, the best-characterized histone marks include lysine and arginine methylation. Lysine acetylation on histone tails is associated with transcriptional activation. Neutralizing the positive lysine side chain charge by acetylation decreases the interaction with the negatively charged DNA to unwind from histones, thereby exposing transcription sites. Although methylation does not change the charge of lysine, it serves as a recognition site for many chromatin effector complexes. The trimethyl states of histone H3 at lysine 4 (designated as H3K4me3) and lysine 36 (H3K36me3) by histone methyltransferase complexes are associated with active transcription. In contrast, the trimethylation of lysine 27 (H3K27me3) by PRC2 serves as the recognition site for PRC1 resulting in more stable gene silencing through chromatin compaction [50].

Chromatin- and DNA-modifying protein complexes modulate and maintain chromatin structures at chromatin regulatory sites with activities that may include ATP-dependent nucleosome remodeling, histone modification, DNA modification,

DNA/protein binding regions, and often lncRNAs. For example, various lncRNAs have been shown to interact with PRC2 to recruit their gene silencing function to a specific gene. *HOTAIR* is a *trans*-acting lncRNA is transcribed from the HOXC gene cluster on chromosome 12 and participates in the silencing of genes from the HOXD cluster on chromosome 2 as well as thousands of other sites throughout the genome by localizing PRC2 and H3K27me3 deposition. As such, knockdown of *HOTAIR* by RNA interference (RNAi) leads to de-repression of HOXD genes and decrease of PRC2 occupancy and H3K27 methylation of the locus [30]. The aberrant overexpression of *HOTAIR* has been associated with nearly 20 different cancer types by repressing critical tumor suppressor genes.

DNA is also a target for reader, writer, and eraser enzymes through recruitment by lncRNAs. DNA is methylated by the dinucleotide methyltransferases (DNMT) enzyme complex family that includes DNMT1 and DNMT3A/3B. DNMT1 is associated with *S*-phase cell replication and acts primarily as a maintenance methyltransferase, while DNMT3A/3B are essential for de novo methylation of DNA during embryogenesis. DNMTs transfer a methyl group from *S*-adenosylmethionine to cytosines immediately followed by guanine (CpG), thereby forming 5-methylcytosine. CpG islands – stretches of CpG, spanning 300–3,000 basepairs – are found at many promoter regions. A highly methylated state of a CpG island is associated with gene repression. In humans, 70% of all promoter regions contain CpG islands and nearly 70–80% of cytosines in CpG islands are methylated. CpG islands are often located near H3K27me3 marks and it has been shown that the EZH2 subunit of PRC2 interacts directly with DNMT complexes to guide DNA methylation of CpG islands [51]. DNMTs are involved in the regulation of gene expression from early embryogenesis throughout cell differentiation and the dysregulation of these processes may play a major role in carcinogenesis [50–52]. A body of emerging evidence supports the hypothesis that lncRNAs play key roles in the recruitment and scaffolding of DMNT writers that mark the CpG regions of DNA [51]. When comparing various cancerous versus adjacent normal tissue, the promoter regions of several tumor suppressor genes become hypomethylated resulting in their transcriptional suppression. Using RNA immunoprecipitation followed by next-generation sequencing, 148 lncRNAs were found to be associated with DNMT-1 in an Hg19 colon cancer cell line [52]. One of these was a highly colon-tissue specific lncRNA named DMNT1-associated Colon Cancer Repressed lncRNA1 (*DACOR1*) that was downregulated in 22 colon cancer samples relative to their adjacent non-cancer tissues. The lentiviral transduction of DACOR1 into two cancer cell lines demonstrated that *DACOR1* enhanced methylation at multiple genetic loci without affecting DMNT1 protein levels [51]. DACOR1 induction resulted in the downregulation of a group of oncogenes including TGFß/BMP and PHGDH, PSAT1, CBS, and ASNS leading to an inhibition of cell growth [53–55]. Thus, the expression patterns of lncRNAs could serve as biomarkers of disease progression and potential anti-neoplastic therapeutic strategies.

7 Repetitive Elements

Noncoding regions of the genome have low evolutionary constraint relative to protein- coding exons and are therefore less conserved. Gene duplication, repetitive elements, and transposable elements have all contributed to the rapid evolution of noncoding RNAs. More than half of all human DNA is made up of repetitive elements or duplicated sequences often present as tandem repeats. Transposable elements (TEs) are a type of repetitive element that can change positions within the genome – resulting in the creation of polymorphisms and new mutations – and can create additional repetitive elements in the genome. It has recently been hypothesized that TEs significantly shape the noncoding transcriptome by modifying the function, sequence, and structure of existing lincRNAs and by establishing new lncRNA loci through introduction of DNA sequences that can confer transcriptional activity by the insertion of promoter-like sequences [53, 54].

A growing body of evidence demonstrates a close association of repetitive elements with the functions of lncRNAs. Repetitive elements are found in roughly 85% of annotated lncRNAs in primates [55]. For example, the Alu element is a member of the short interspersed nuclear elements (SINE) family and is the most abundant repetitive element in primates, making up 10% of the primate genome [56]. Since lncRNAs fold in complex tertiary structures to create their functional binding regions, there is more evolutionary pressure to conserve their three-dimensional structures than their specific nucleotide sequences [56]. Alu elements contained within lncRNAs can contribute to the tertiary structures necessary for function. For example, some mRNAs contain Alu sequences in their 3′-untranslated regions (3′-UTRs) that interact with Alu repeat elements in lncRNAs. The intermolecular stem structures formed by these Alu/Alu base pairings provide the binding site for STAU1 to initiate mRNA decay as is the case for plasminogen activator inhibitor type 1 (SERPINE1) [57]. Additionally, inverted Alu elements are found in many human mRNA sequences that appear to affect translational accuracy, DNA repair, alternative splicing, RNA editing, translation regulation, and heat shock induced gene repression [58, 59]. Rodents also harbor repetitive elements, known as B-elements, in mRNA regulation. B-elements can be found at the 3′-UTR of mRNA that bind to lncRNAs with murine-specific repeats that are functionally similar to the Alu repeats found in primates. In fact, both ALU and B-repeats arose from the 7SL RNA gene more than 80 million years ago before the primate-rodent evolutionary split [56, 60–63]. Thus, in mice, B1 repeat sequences are utilized in a similar fashion to Alu repeats in primates, and murine lncRNA homologs may exist that could aid in the development of studies for proof of concept in murine models of disease.

The repetitive elements found within lncRNAs serve many other functions. They are components of NATs that contain multiple binding sites to sequester miRNAs from their target mRNAs. For example, ciRS-7, a NAT that acts as a sponge to potentially sequester up to 74 miR-7 microRNAs from its target tumor suppressor genes, is downregulated in cancer (see Sect. 10). In other cases, the repetitive elements in lncRNA domains enable high-affinity or multivalent interactions with specific regulatory protein complexes. The lncRNA *Firre* (functional intergenic repeating element) is encoded on the X chromosome but forms a punctate sphere of active sites on multiple alleles to interact with the nuclear matrix protein hnRNPU [64]. *Firre* includes 12 repetitive elements, each of which interacts with scaffold protein attachment factor A, a protein that has both DNA and RNA binding sites and is responsible for the nuclear retention of many RNAs [65]. Thus, repetitive elements may serve as the sites of interconnection of *Firre* RNA molecules at multiple genomic DNA sites [10, 66].

Genomic regions containing repetitive elements are sometimes unstable, leading to the contraction or expansion in the number of repeat units and resulting in disease. Until recently, treatment paradigms for such diseases have remained elusive although the significance of variations in repetitive elements for the manifestation of disease pathology has long been appreciated. In the case of facioscapulohumeral disease (FSHD), contraction of the number of tandem repeats in chromosome 4q35 is responsible for FSHD pathology. The significance of the repetitive sequences will be further described in the examples section. Additional examples of genetic diseases that are associated with contractions or expansions in repeat elements include fragile X syndrome, myotonic muscular dystrophy, and Charcot-Marie-Tooth type A. In some cases, a specific lncRNA is upregulated leading to pathological manifestations of the disease. Accordingly, that specific lncRNA represents a rational therapeutic target.

8 Specificity

Nearly 80% of lncRNAs in primates are specific to cell type, tissue type, cellular differentiation state or disease-state [67]. The targeting of these lncRNAs to block their functions, therefore, affords unique therapeutic specificity compared to conventional drugs that target proteins. The regulatory proteins that complex with lncRNAs are often ubiquitously expressed across cell types and function at many genes, such as PRC2, LSD1/coREST/REST, and DNMT1, which have thousands of gene targets. The lncRNAs can, in part, afford specificity of transcriptional regulators. Thus, targeting lncRNAs using rationally designed oligomers [68] or possibly small molecule inhibitors [2] may be able to provide therapeutic benefit with decreased side effects (both on-target and off-target related). Oligomers targeting lncRNAs will be discussed in the next section of this chapter. Using small molecules to modulate RNAs, whether it be at the level of splicing mRNAs or at disrupting lncRNA function, will be discussed elsewhere. Small molecule approaches to specifically target a subspecies within a class of closely related family members may pose challenges because a druggable secondary or tertiary along the lncRNA may be too conserved for their selective inhibition [69]. However, new methodologies to better characterize subtle differences in RNA structure, such as SHAPE (selective 2′-hydroxyl acylation and primer extension) [70], may open the doors to rationally designing small molecules to modulate lncRNA function.

Because of the ubiquitous expression of regulatory proteins that interact with lncRNAs, direct targeting of those proteins may lead to target-related side effects and a narrow therapeutic window, even with a subclass-specific small molecule drug. For example, histone deacetylase (HDAC) inhibitors – first discovered in small molecule screens conducted to identify compounds that induced the apoptosis of cancer cells [71] – inhibit the catalytic subunit of corepressor complexes that remove acetyl groups from histones. Doing so results in the opening of chromatin and subsequent transcriptional de-repression of hundreds of genes. As a class, small molecule HDAC inhibitors have limitations due to both target-related and off-target related side effects that hinder their clinical utility and limit dose levels that can safely be administered to patients [72]. In addition, HDAC inhibitors generally inhibit all mammalian HDAC isotypes because they generally interact with the druggable Zn^{2+} binding pocket common to all HDAC inhibitors [71, 73–75]. The similarity of this binding pocket across isoforms has stymied efforts to develop selective HDAC inhibitors for specific isoforms and therefore it is not known which isoform is predominantly responsible for the toxic effects observed in clinical trials [76, 77]. Thus, HDAC inhibitors are generally pan-inhibitors that interact with most isoforms across the HDAC class such that the FDA has approved only three drugs from more than 500 clinical trial investigations [77].

An alternative method to block aberrant histone deacetylation patterns is by targeting lncRNAs involved in their recruitment. Since HDAC corepressor complexes can be recruited by lncRNAs, this approach can potentially block HDAC activity at specific genomic loci. In some cases, the expression of various lncRNAs associated with HDACs is cancer-specific. C-terminal binding protein 1-antisense (*CTBP1-AS*) is an lncRNA that is transcribed antisense to the *CTBP1* gene and upregulated in prostate cancer [78]. *CTBP1-AS* acts in both in *cis* and in *trans*. *CTBP1-AS* recruits HDAC-Sin3A and repressor phosphotyrosine-binding–associated splicing factor (PSF) to the promoter region of *CTBP1*, resulting in histone deacetylation, condensation of local chromatin, and decreased expression of *CTBP1*. Low levels of CTPB1 promote the proliferation of cancer cells and are correlated with poor prognosis and survival in patients with prostate cancer. In addition, *CTBP1-AS* acts in *trans* by recruiting HDAC/PSF complexes to silence many genes, some of which are tumor suppressors [79, 80]. Thus, by targeting a single lncRNA, the prognosis of both androgen-dependent and androgen-independent prostate cancer patients could potentially be improved by a mechanism involving HDAC inhibition but with more selectivity and specificity than with currently available small molecule HDAC inhibitors.

9 Oligonucleotide-Based Therapeutics

The field of synthetic oligonucleotides (oligomers) has evolved rapidly over the last 25 years and the advances made can be directly leveraged to target lncRNAs. Oligomers can bind to endogenous targeted RNA sequences by complementary base pairing. The most common mechanistic approaches for lncRNA targeting are to either induce degradation of the target RNA (e.g., via gapmers or RNAi) or sterically block its interaction with regulatory protein complexes using chemistries that do not invoke RNA degradation (e.g., via uniform chemically modified mixmers or morpholinos) [81–84]. In this section, we will describe some common oligonucleotide chemistries. Other advances and chemistries not covered have been extensively reviewed elsewhere [83, 85, 86].

Oligomers are chemically modified synthetic DNA and RNA analogs comprising 7–25 bases with unique chemical backbone linkages to improve biostability, biodistribution, and/or binding affinity. The oligonucleotide backbone can be modified with phosphorothioate (PS), which involves the replacement of the phosphate oxygen with a sulfur in an oligomer, leading to a backbone with increased biostability but a small decrease in binding affinity compared with the native phosphodiester linkage. This allows the oligomer to be less prone to nuclease degradation. The 2′-O-methyl (2′OMe), locked nucleic acid (LNA), and 2′-O-(2-methoxyethyl) (MOE) modifications are common replacements for the ribose ring that confer increased biostability and affinity for target DNA and RNA. The 2′OMe modification is a naturally occurring modification of the ribose ring that confers increased stability against endonuclease-mediated degradation and slightly increases the affinity for binding to RNA. The LNA modification is the result of a bridging methyl group between the 2′ ribose hydroxyl and 4′ ribose ring carbon forcing the sugar to pucker into a high C3′-endo conformation (RNA-like) that results in strong binding to the complementary DNA and RNA targets. The MOE modification is a substitution of the 2′ ribose that confers greater stability against nuclease digestion and forms a C3′-endo conformation to favor more stable binding to RNA targets. Other modifications include a phosphorodiamidate morpholino (PMO) in which the deoxyribose ring is replaced with a morpholino and the backbone is replaced with phosphorodiamidate resulting in an uncharged oligomer. LNA, 2′OMe, MOE, and PMO based oligomers possess superior biological stability and binding affinity than DNA oligomers.

Oligomers that are used to target RNAs for RNAse-H mediated degradation are known as gapmers because of the specific arrangement of the chemically modified nucleotides. Gapmers are designed to have a central RNase H-sensitive DNA segment comprised of optimally 8–10 nucleotides and flanked by 2–5 chemically modified analogs of LNA, MOE, or 2′OMe base units [87, 88]. Gapmers activate RNase H, an endogenous enzyme that recognizes natural DNA/RNA hybrids formed by the gapmer base-pairing with an endogenous RNA, including lncRNAs, which leads to RNA degradation [59]. Gapmers are effective in both the nucleus and cytoplasm though they tend to be slightly more efficacious for nuclear targets

[89]. Short interfering RNA (siRNA) approaches can more efficiently target cytoplasmic RNAs for degradation via the RISC complex, but have also shown nuclear activity against more abundant lncRNAs [39, 90] though there is some evidence that siRNA may also function against nuclear targets [91]. Like native DNA, the PS modification supports RNase H activity that degrades double-stranded DNA/RNA duplexes. However, the LNA, 2′OMe, MOE, and PMO modifications do not support RNase-H activity. The flanking LNA, MOE, or 2′OMe oligonucleotides are still critical because they improve the affinity of the gapmer for its target RNA and increase the biostability of the hybrid which facilitates RNase H activity and knockdown of selected RNA targets.

Oligomers which contain these chemical modifications in a non-gap context do not degrade their RNA targets, but rather antagonize the binding of lncRNAs to their partners, and are referred to as mixmers [92]. They may bind to the lncRNA to either sterically block an interaction or to alter the secondary or tertiary structure of the lncRNA to be unrecognizable. As a result, using mixmers can alter gene regulation through suppression, enhanced transcription (de-repression), alternative splicing, and transcript stability in all cellular compartments [93, 94]. One example of the potential use of mixmers is to block the binding of the suppressive PRC2 complex to lncRNAs resulting in decreased PRC2 localization to promoter regions and subsequent transcriptional upregulation of associated protein-coding genes [87]. LNA, 2′OMe, MOE, or other chemical modifications are systematically placed at different loci of the oligomer sequence to balance specificity and binding affinity. For an optimal oligomer, whether it is a gapmer or mixmer, a balance between on-target high-affinity binding with minimal off-target binding is the optimal goal.

The oligomers described above generally possess similar pharmacokinetic properties. After systemic administration, they clear from the blood stream with a circulating half-life of several hours. Oligomers are presumed to be taken up into cells through endosomal transport that allow them access to the cytoplasm and nucleus. Overall, phosphorothioated oligomers have an intracellular half-life of at least several weeks and thus, their duration of action after systemic administration is prolonged once they reach their cellular target. Such relatively longer intracellular half-life is particularly attractive for CNS applications [95] requiring intrathecal delivery [96], as is demonstrated by nusinersin, a uniform MOE-based oligomer recently approved for the treatment of spinal muscular atrophy, which is administered intrathecally once every 3 months.

Oligomers primarily distribute to the kidney and liver after systemic administration. However, targeting to other tissue types to alter the biodistribution of the oligomers is an active field of research. Prior to 2015, only two oligomers were approved by the FDA – fomivirsen for cytomegalovirus retinitis [69, 97] and mipomersen for homozygous familial hypercholesterolemia [69, 86, 98]. Recently, however, the FDA approval of a splice-switching morpholino for treating Duchenne muscular dystrophy demonstrates that therapeutic levels of oligomers can reach the nuclear compartment of skeletal muscle myotubes, a more challenging in vivo target than the liver [99]. Furthermore, oligomers directly administered to the CNS have the potential to treat neurological disorders. Many lncRNAs are expressed specifically in the brain and may be amenable to oligomer modulation. This strategy of blocking chromatin writers/erasers from their lncRNA binding sites with oligomers may be sufficient to alter transcriptional patterns for therapeutic interventions.

10 Target Identification and Validation

The discovery of lncRNAs with therapeutic potential is facilitated by the availability of public databases that contain transcriptomes of human, mouse, and primates. Some databases also contain genetic and epigenetic information distinguishing between diseased and healthy tissues, including cancers and metabolic disorders. The ability to mine these constantly expanding databases is instrumental to identifying and validating lncRNAs as therapeutic targets.

The rate of lncRNA discovery has surpassed the progress to delineate their functions. The comparison of lncRNA expression patterns across tissues and cells can aid in the correlation of lncRNAs and their biological functions. This approach has been used to ascribe the functions of lncRNAs involved in cell-cycle regulation, stem cell pluripotency, inflammation, and immunity. The identification of a therapeutic target can also be inferred by divergent histone/DNA methylation patterns in cells/tissues from patients and healthy individuals. LncRNAs can be studied either in the context of a healthy or diseased tissue, depending on the therapeutic strategy and the disease setting.

Knock-down experiments with siRNAs, short hairpin RNAs (shRNAs) or gapmers as well as overexpression experiments targeting the lncRNA are used to validate the functional significance of the observed correlation. However, attributing function to lncRNAs can be complex. For example, lncRNAs such as nuclear paraspeckle assembly transcript 1 (*NEAT1*) have been identified where the process of transcription alone leads to the epigenetic modification of neighboring promoter regions, without physical association of *NEAT1*, to affect the expression of those genes. NEAT1 itself locates to paraspeckles and is required for paraspeckle formation. In such circumstances where the transcription of the lncRNA itself causes chromatin changes at neighboring genes, ectopic expression of *cis*-acting lncRNAs may not always recapitulate its effects.

Abnormal DNA methylation patterns around CpG islands can be discovered via database mining or empirically between healthy and diseased patient cells. In still other instances, mutations in protein-coding genes might be the causative event for a rare genetic disease where low concentrations of functional protein are produced. Increased transcription or corrected splicing of these genes by targeting modulatory lncRNAs is a viable approach for the upregulation of enough functional protein to provide potentially transformative benefit.

11 Animal Models

Drug efficacy is classically determined in murine models of disease for proof of concept (POC) studies. Although this is not an absolute regulatory requirement preceding first-in man studies, it provides confidence for regulatory agencies and has a strong probability of predicting therapeutic benefit. POC studies in small mammals for lncRNA targeting drugs, however, are challenged by the lack of sequence conservation of many lncRNAs across species. Protein-coding genes are generally highly conserved across species, highlighting their functional importance. This lack of sequence conservation for many lncRNAs should not discount their biological significance. An lncRNA may tolerate sequence differences between species, so long as their secondary or tertiary structures important for function remain conserved. Nonetheless, because of differences arising in many lncRNA sequences, most studies are conducted in primary human patient cells or in human cell lines that express the lncRNA target. However, in some cases lncRNAs have been observed to function differently in cell culture than in tissues [100] so it is preferable to demonstrate that the disruption of their function will produce the desired effect in vivo.

Though the primary sequence may be unique to a species, it is conceivable that preclinical studies could be conducted to target a murine homolog of a primate lncRNA that has a similar function in both species. In this instance, POC studies could be conducted with an oligomer designed to target the mouse sequence in a disease model. Studies comparing the function of human and mouse lncRNAs might be required in this case. Similar approaches have been used to demonstrate POC with murine and human antibodies directed against the same target [101].

POC studies can be conducted in primates, particularly for the upregulation/de-repression of protein-coding genes. Gapmers have been used to degrade APOA1-AS lncRNA to upregulate APOA1 in green monkeys [102]. Since this approach was used to upregulate genes primarily expressed in the liver, a 1.7-fold upregulation in of APOA1 mRNA could be assessed from liver biopsies, thereby not requiring terminal studies. Additionally, a 30% increase in the levels of circulating APOA1 and its related homologs could be measured from blood samples [102]. Various studies have suggested that upregulation of circulating APOA1 is beneficial in the treatment of cardiovascular disease [103]. Taken together such studies conducted in healthy non-human primates provide POC, if the gapmer is proven to be safe and specific, for the conduct of formal preclinical and subsequent clinical studies. In situations where an animal disease model is crucial, it may be possible to create a primate model of the disease in small non-human primates such as marmosets.

The introduction of human tumor xenografts into immune compromised mice is a practice that has been widely utilized for cancer studies. However, it is possible to graft non-tumor tissues in immune deficient mice as well. An example of such an approach is given in the facioscapulohumeral muscular dystrophy (FSHD) studies. Since the genetic locus responsible for FSHD is present in mice, human muscle cells were expanded from a biopsy and implanted into murine quadriceps muscle to conduct POC studies. Another approach is to create a humanized model of disease as was done for *SMN* for the treatment of spinal muscular atrophy (SMA). Humans are unique for the SMN locus, where two copies, SMN1 and SMN2, exist. All other species harbor only the *SMN1* ortholog. *SMN1* and *SMN2* are nearly identical in sequence with the exception of a critical C-to-T transition in SMN2 that causes the mRNA splicing machinery to skip exon 7. The protein arising from SMN2 is truncated and rapidly degraded. A severe model of SMN deficiency was created by knocking out the mouse SMN gene and knocking in the human gene cluster containing both the human SMN locus and SMN-antisense 1 (SMN-AS1) [104].

12 Long Noncoding RNA as Therapeutic Targets

In addition to their associations with multiple types of cancer, various studies have identified lncRNAs that are involved in the pathophysiology of disease in other tissues and organs. Differential expression of lncRNAs can be found in ocular diseases including glaucoma, proliferative vitreoretinopathy, and diabetic retinopathy [105]. A class of lncRNAs has been identified that serve as nuclear co-receptors for steroid hormones, such as *SRA*, and have differential expression patterns between breast cancer and normal cell lines suggesting a role in tumorigenesis [106–108]. Additionally, *SRA* co-activates perixosome proliferator-activated receptor δ (PPARδ), a nuclear receptor that regulates adipogenesis. SRA knock-out mice are protected against obesity and glucose tolerance with a possibility playing a role in diabetes. Prader Willi syndrome, a genetic disorder causing obesity, mental retardation and insatiable appetite has been linked to a loss of 116HG lncRNA expression in the CNS due to the loss of a paternally imprinted gene cluster [109, 110]. *LincRNA-COX2*, *Lethe* and *THRIL* are examples of lncRNAs involved in inflammatory responses [111, 112]. A group of lncRNAs including *BACE1-AS* are upregulated in Alzheimer's disease and regulate the transcription of proteins such as beta-amyloid involved in the formation of amyloid plaques [113]. Another antisense lncRNA, *BDNF-AS* is implicated in neurological disorders including Huntington's disease and schizophrenia [114, 115]. The lncRNA fragile X mental retardation 1 (*FMR1*) plays a role in the pathology of the trinucleotide repeat disorder such as fragile X syndrome [116, 117]. Other lncRNAs have been linked to various cardiac and muscle diseases. Thus, lncRNA have diagnostic and therapeutic potential for many diseases in nearly all tissue types that are currently unmet medical needs.

13 Selected Examples of lncRNA Preclinical Studies

In this section, the preclinical studies conducted to identify and demonstrate proof of concept have been summarized for four potential therapeutic lncRNA targets. The first is a circular RNA designated ciRS-7 that is downregulated in many types of cancer and acts as a NAT sponge for the tumor suppressor microRNA, miR-7. The second is an lncRNA that negatively modulates *APOA1* transcription whose antagonism may be an effective therapy in the treatment of cardiovascular disease. The third lncRNA, SMN-AS1, is an antisense RNA target negatively regulates SMN2 gene expression. The fourth featured lncRNA target may provide a potential therapy for facioscapulohumeral muscular dystrophy (FSHD).

13.1 Example 1: ciRS-7 as a Target for Cancer Therapy

A handful of exonic circular RNAs (circRNAs) were identified several decades ago but, until recently, were thought to be extremely rare. In 2013, Jeck et al. enriched for circRNA species in the Hs68 human fibroblast cell line using RNase R to degrade the linear species and sequenced the remaining non-linear RNA population [118]. They concluded that approximately 14% of the transcribed genes identified in the Hs68 human fibroblast cell line code for circRNA. Nearly 2000 circular RNAs have been found in humans to date [119, 120]. CircRNAs are primarily located in the cytoplasm but some appear to be enriched in the nucleus [118]. They have high biostability (approximately 48 h) relative to other ncRNAs, due to their circular structure that is resistant to exonuclease-mediated degradation. Alu elements are significantly over-represented in exons flanking the circRNAs, often in a complementary inverted orientation [118, 119]. Notably, the sequences of many circRNAs are well conserved across zebrafish, mouse, and primate, underscoring the importance of their biological roles [121]. In addition, compared with lncRNA classes that are poorly conserved across species, high conservation in circRNAs facilitates preclinical development in terms of relevant animal models.

The recently discovered circRNA, ciRS-7 (also known as *CDR1-NAT*, *CDRI-AS* and *cir-ITCH*) is transcribed from several exons of the cerebellar degeneration related protein 1 (*CDR1*) gene. The primary function of ciRS-7 is thought to be the negative modulation of miR-7 by serving as a NAT sponge to decrease its active concentration. In healthy adult mammals, ciRS-7 is expressed primarily in the CNS and to a lesser extent in peripheral tissues such as pancreatic islet cells. The expression of ciRS-7 and miR-7 is highly correlated across tissues suggesting that most miR-7 exists in a ciRS-7/miR-7 neutralized complex. Knock-out studies conducted in fruit flies suggest that miR-7 may participate in stress responses. In a model of neuroinflammation, depriving cultured astrocytes of oxygen and glucose led to increased miR-7 levels and a consequent attenuation of inflammation [122]. In another study, miR-7 overexpression in human neuroblastoma SH-SY5Y cells inhibited mitochondrial fragmentation, mitochondrial depolarization, cytochrome c release, reactive oxygen species generation, and release of mitochondrial calcium and thereby inhibited apoptosis [123]. Thus, it is possible that ciRS-7 may function to maintain a reservoir of neutralized miR-7 that can be rapidly deployed during environmental stress.

CiRS-7 is overexpressed in various cancers, including breast, hepatocellular, lung, cervical, Schwannoma, tongue, colorectal, and gastric cancers, and it is positively correlated with poor prognosis and survival [40, 124]. CiRS-7 sequesters several miRNAs including miR-7, a powerful tumor suppressor. Although ciRS-7 has other functions, its overexpression is one of its key oncogenic mechanisms that has been widely investigated across different cancers. In support of this mechanism, the activity of miR-7 inversely correlates with the expression of ciRS-7. A single copy of ciRS-7 can simultaneously chelate up to 70 units of miR-7 since each ciRS-7 has 74 potential binding sites for miR-7 [119, 125]. Additional studies conducted in cancer cell lines and tissues collectively demonstrate that ciR-7 acts as a molecular sponge for up to four other tumor-suppressing miRNAs [124]. Though the specific miRNAs sequestered by ciRS-7 may vary across cancer classes, they always include miR-7.

Several lines of evidence indicate the role of the ciRS-7/miR-7 axis in disease pathology. In the CNS, increased miR-7 activity is associated with Parkinson's via the negative modulation of alpha synuclein mRNA levels. In peripheral tissues, dysregulated miR-7 activity is implicated in diabetes through its negative modulation of the mTOR pathway. Many of the target genes of miR-7 include oncogenes and tumor suppressor genes that promote transformation and cancer progression. Dysregulation of miR-7 results in the aberrant expression of such genes including *EGFR*, *IRS-1/2*, *Raf1*, *Ack1*, *Pad1*, *PIK3CD*, *XIAP*, *KLF4*, *SNCA*, *4EBP1*, *ACK1*, *AKT*, *FAK HNF4*, *IGF1R*, *NOTCH1*, *p70S6K*, *PA28c*, *PAK1*, *mTOR*, and *DVl2*. Thus, increased ciRS-7 expression results in a powerful promotion of tumorigenesis with roles in transformation, progression, and metastatic potential, with much of its activity due to its negative regulation of free cytoplasmic miR-7 levels. One of the challenges of targeted cancer treatment is that a symphony of genes is affected (frequently silenced) so it is challenging to choose which specific protein/gene to target. However, miR-7 and other miRNAs target multiple genes involved in transformation/cancer progression and may therefore represent a more effective treatment paradigm.

Evidence to date indicates that an attractive option for cancer therapy is to target ciRS-7 for degradation with the goal of upregulating miR-7. Since both ciR-7 and miR-7 are well conserved across species, relevant animal models for preclinical development may be easy to establish.

13.2 Example 2: APOA1-AS as a Target for Hypercholesterolemia

Apolipoprotein A1, the major transport protein for high density lipoprotein (HDL) in blood circulation, functions as a co-factor for an enzyme important for the efflux of cholesterol from tissues. APOA1 is of key importance for reducing the lipid/cholesterol burden of white blood cells in arteries that are the major cause of plaque-formation in coronary artery disease. Therefore, a viable therapeutic intervention for coronary artery disease is to increase the circulating levels of APOA1 protein in patients having uncontrolled high cholesterol.

A *cis*-acting NAT sequence was recently identified leading to the discovery of APOA1-AS, an lncRNA that negatively modulates *APOA1* transcription at the APO gene cluster [102]. Other members of this cluster include *APOC3*, *APOC4*, *APOA5*, and inosine-guanosine kinase. APOA1-AS is transcribed convergently with *APOA1* and overlaps with 123 nucleotides of its exon 4. SiRNAs were used to knockdown APOA1-AS in the human liver cancer cell line HepG2. It was demonstrated that a 65% knockdown of APOA1-AS led to a threefold upregulation of *APOA1* mRNA. Other members of the APO gene cluster including APOC3 and APOC4 were also upregulated three- to fourfold. This upregulation was specific for the *APOA1*, *C3*, and *C4* genes since other genes in the locus, including inosine-guanosine kinase and *APOA5*, were not affected. Chromatin immunoprecipitation (ChIP) was used to measure the levels of active chromatin mark (H3K4me3), and H3K9me3 or H3K27me3 (repressive chromatin marks) in siRNA-treated and untreated cells [102]. No changes in H3K9me3 were observed, but H3K27me3 was significantly reduced in the promoter regions of the *APO* gene cluster. These results are consistent with a mechanism where *APO* transcription is repressed by APOA1-AS. Additional ChIP experiments were carried out using antibodies against lysine-specific demethylase (LSD1) and the PRC2 subunit SUZ12 to demonstrate that the binding of those proteins to the *APOA1* region was reduced along with the changes in histone marks. Taken together, these data suggest that APOA1-AS repressed the expression of APOA1 and related genes by recruiting the PRC2 complex to the promoter region of the *APOA1* gene.

After screening a panel of gapmers that spanned the entire APOA1-AS sequence, several were identified that upregulated APOA1 expression ($\approx$1.7-fold upregulation) in HepG2 cells. Lead optimization using different chemistries were performed before LNA gapmers were chosen for in vivo studies in African green monkeys. Here, primates were used for in vivo POC because APOA1-AS is not conserved in lower mammals.

The LNA gapmers were delivered intravenously at a dose of 10 mg/kg $\times$ 3 days and the comparison of liver biopsies on day -30 (baseline) and 72 h post last injection demonstrated a 1.7-fold elevation in *APOA1* transcripts in the liver and circulating APOA1 protein levels. Circulating APOA1 protein levels remained elevated for 18 days post last injection. Scrambled gapmer sequences produced no elevation in mRNA or protein. These studies provide POC for an LNA gapmer that could be nominated for preclinical development and subsequent clinical studies.

13.3 Example 3: SMN-FL as a Target for SMA

Spinal muscular atrophy (SMA), an autosomal recessive disease, is the leading genetic cause of infant mortality worldwide. It is caused by mutations in the survival motor neuron 1 gene (*SMN1*) and results in the premature death of motor neurons. Humans possess an evolutionary duplicate of the *SMN1* gene (*SMN2*) which is identical to *SMN1* except for a C to T point mutation that results in the skipping of exon 7 in approximately 90% of transcriptional events. The protein produced from this 'delta 7' transcript is largely non-functional and rapidly degraded. Up to 10% of the *SMN2* gene is transcribed correctly to include exon 7, producing a full length SMN protein (SMN-FL) that is identical to the protein produced from *SMN1*. SMN2 expression is polymorphic in humans, with copy numbers of SMN2 ranging from between 0 to 6. In healthy individuals, *SMN2* copy number does not dictate any phenotype. However, the number of copies of *SMN2* generally correlates inversely with the severity of SMA in patients since the only source of SMN protein is from the *SMN2* gene. Patients with a higher copy number of *SMN2* produce more functional full length SMN-FL. While other genetic factors can contribute to determining disease severity, the main cause of SMA is the homozygous loss of *SMN1*. Interestingly, *SMN1* heterozygotes are asymptomatic even though they produce only 50% wild type SMN1 protein compared to healthy individuals. Therefore, a modest increase in the SMN protein is expected to yield significant therapeutic benefit supporting the therapeutic utility of approaches to increase SMN2 expression.

ChIP sequencing data from the NIH Roadmap Epigenome Consortium showed that H3K27me3 marks (indicative of PRC2 gene suppression) were associated with *SMN2* in human fetal brain. Knocking down the PRC2 subunit EZH2 in an SMA fibroblast cell line decreased the H3K27me3 marks and resulted in upregulation of SMN-FL, suggesting that PRC2/EZH2 repressed the transcription of *SMN2*. Indications of an antisense transcript were first detected by immunoprecipitating RNAs associated with PRC2 [126]. Further studies characterized this lncRNA, SMN-AS1, which was shown to localize to the SMN locus, potentially have a *cis*-regulatory role in SMN2 expression [127]. Mixmers were designed to sterically block the interaction between SMN-AS1 and PRC2. In SMA fibroblasts, ex vivo cortical neurons as well as patient iPS-derived motor neuron cultures, the mixmers increased SMN expression as a result of inhibiting the interaction of SMN-AS1 with PRC2. Together, these studies identified and validated an lncRNA which represses SMN2 transcription as a strong target for the potential treatment of SMA patients.

A second lncRNA at the SMN locus was found to be associated with PRC2, which was also named SMN-AS1 [127]. Gapmer-mediated degradation of this lncRNA also resulted in the increase of SMN mRNA levels. Various experiments conducted in HeLa cells, SMA fibroblast cell lines, neuroblastoma cell lines, and primary mouse neurons suggested that PRC2 is recruited to the SMN promoter by SMA-AS1, resulting in repression of SMN by H3K27me3. In a humanized mouse model of SMA having the human *SMN2* locus, SMN-AS1 was found to be expressed specifically in the CNS. Gapmers were administered to the humanized mice, which resulted in a decrease in SMN-AS1 levels in the brain and spinal cord [128]. SMN2 pre-mRNA, FL-SMN2 mRNA, and SMN protein levels increased in a manner inversely proportional to SMN-AS1 knockdown. The murine phenotype, however, was unaffected. An additional study was conducted to test whether the co-administration of a low dose of splice-correcting ASO plus the gapmer targeting SMN-AS1 would improve the phenotype of this model. Mice co-administered with the splice-switching ASO and a gapmer targeting SMN-AS1 survived longer than the control groups or those treated with the splice-switching oligo alone. These studies suggest a clinical benefit of co-administration of SMN-AS1-interfering ASO and splice-switching ASO for the treatment of SMA.

13.4 Example 4: DBE-T as a Target for FSHD

Facioscapulohumeral muscular dystrophy (FSHD) is an autosomal-dominant disease causing weakness/atrophy of the face, shoulder, upper arm muscles in its early stages, followed by weakness/atrophy of the trunk and leg muscles in the later disease-state resulting in loss of ambulation. FSHD is caused by 3.3 kb long repeat elements (REs), known as D4Z4, that are located in the highly polymorphic region 4q35 in 95% of patients and are ordered in a head to tail orientation upstream from the DUX4 gene promoter. In healthy individuals, the number of D4Z4 repeats ranges from 11 to 150. In patients with FSHD, however, the number of D4Z4 repeats is contracted and ranges between 1 to 10 units. This contraction is associated with the upregulation of the DUX4 transcription factor and several of its neighboring genes (FRG1/2 and ANT1). There has been some debate about which upregulated gene is an appropriate therapeutic target, as murine models over-expressing either FRG1 or DUX4 result in a dystrophic phenotype. Recent studies, however, suggest an epigenetic disease mechanism since FSHD is associated with an alteration of epigenetic marks on DNA and histones, as well as a loosened state of the chromatin associated with de-repression of DUX4 gene transcription and its associated genes.

The region occupied by D4Z4 in healthy subjects is one of the largest CpG islands in the human genome. Three-dimensional fluorescence in situ hybridization studies showed that H3K27me3 at these CpG islands was diminished in FSHD patients relative to healthy controls. In addition, the methyltransferase subunit of PRC2, EZH2, was enriched at the promoters of *DUX4*, *FRG1*, *FRG2*, and *ANT1* in healthy subjects. These results suggested that the loss of PRC2 silencing was responsible for the increased expression of D4Z4 associated genes.

A unique lncRNA, D4Z4 Binding Element Transcript (DBE-T), was found in FSHD biopsy samples, but not in healthy controls [129]. Depletion of PRC2 by RNAi or treatment with 5-Aza-2′-deoxycytosine (5-AzadCyD), a DNA methyltransferase inhibitor, and Trichostatin acid (TSA), a histone deacetylase inhibitor, resulted in both de-repression of 4q35 associated genes and concentration-dependent elevations of DBE-T without effecting the number of D4Z4 repeats, CpG binding, or H3K27me3 enrichment. ShRNA-mediated knockdown of DBE-T in chr4/CHO cells prevented de-repression of 4q35 associated genes, establishing DBE-T as a positive regulator of 4q35 gene expression. Transient ectopic expression of *DBE-T* in chr4/CHO cells was unable to de-repress the D4Z4 associated genes, demonstrating that *DBE-T* could not act in *trans*. It was also observed that DBE-T knockdown resulted in the closed chromatin state observed in healthy cells compared with the open chromatin structure characteristic of FSHD cells. *DBE-T* was observed by FISH to be enriched in the nucleus and localized at D4Z4 loci. Taken together these results showed that *DBE-T* is an lncRNA that could de-repress the genes from the D4Z4 locus responsible for FSHD.

The recruitment of ASH1L, a histone-lysine *N*-methyltransferase enzyme, to the D4Z4 locus in chromosome 4 was observed when cells were treated with 5-AzadCyD + TSA, further elucidating the mechanism of dysregulation in FSHD [129]. In ChIP studies, ASH1L was also observed to be associated with D4Z4 in FSHD muscle biopsy samples but not in healthy controls. UV-RNA immunoprecipitation (UV-RIP) using anti-ASH1L antibodies showed enrichment of DBE-T compared to GAPDH controls, suggesting that DBE-T interacted directly with ASH1L. This was confirmed using purified recombinant ASH1L and in vitro transcribed DBE-T. ChIP-qPCR was used to show that histone mark H3K36Me2, associated with ASH1L, was increased in the FSHD locus and could be reversed by DBE-T knockdown. These results demonstrate that DBE-T aberrantly recruits ASH1L to the FSHD locus resulting in the activation of the FSHD genes.

This series of experiments indicates that the disease pathology in FSHD patients is epigenetically driven. In healthy individuals, 11 or more D4Z4 repeats result in increased CpG methylation, a closed chromatin state due to histone deacetylation, and transcriptional silencing of *DUX4*, *FRG1/2,* and *ANT1*. Fewer than 11 D4Z4 repeats signals a permissive chromatin state, resulting in DBE-T transcription, recruitment of ASH1L to the promoter regions of the *DUX4* locus and de-repression of the FSHD-associated genes.

The development of an animal model for FSHD has been challenging due to the lack of a conserved D4Z4-DUX4 region in non-primate species. Without a predictive murine model for POC studies, the development of *DBE-T*-targeting technologies will be challenging. Although this hypothesis has yet to be tested in an animal model, another group has tested a PMO for DUX4 knockdown using human muscle engrafted onto nude mice [130]. Interestingly, this group observed that knockdown of DUX4 also led to significant diminution of the downstream genes FRG1 and ANT1, suggesting that DUX4 protein may be required for activating their transcription. Further studies will be necessary to understand the interplay between *DBE-T*, *DUX4* and the neighboring genes.

14 Discussion

LncRNA targeting affords exquisite target specificity that far exceeds the ability to target a specific protein isoform. This is because lncRNAs are often expressed in a cell and/or disease-specific manner where they modulate gene expression at specific loci by attracting relatively ubiquitous regulatory protein complexes, such as HDACs and PRC2. The comparison of small molecule HDAC inhibitors with the potential inhibition of uniquely sequenced lncRNAs in specific gene loci regulated by HDACs illustrates the relative specificity that can be leveraged for diseases where HDAC inhibition has potential therapeutic utility. It is possible that each of the 18 HDAC enzyme subclasses may be directed to target promoter regions by specific *cis*-acting lncRNAs. Such specific targeting is advantageous as HDACs regulate multiple genes across different cell types, many of which are unrelated to disease phenotype.

The discovery of new lncRNAs and their mechanism of action is rapidly growing, resulting in an explosion of new therapeutic applications for a wide range of regulatory targets. The four examples provided in this chapter illustrate the potential utility of directly targeting lncRNA to treat rare genetic diseases (SMA and FSHD) and diseases with high prevalence (cancer and cardiovascular disease). Targeting lncRNAs to increase or decrease protein levels may be a superior alternative to the direct targeting of proteins in many cases.

The ability to upregulate proteins (potentially to therapeutic levels) is another clear advantage of lncRNA targeting. Most traditional modalities must find targets whose downregulation can improve disease conditions, even though the upregulation of a protein would be more beneficial. Accordingly, targeting repressive lncRNAs may allow previously un-druggable proteins to be upregulated to provide therapeutic benefit. As the four examples provided in this chapter suggest, either targeting the lncRNA for degradation or interrupting their binding to its interaction partners may be applicable to modulate target genes with specificity and robustness.

Numerous publicly available databases exist that greatly aid in the collaborative discovery and validation of lncRNA targets. Although too numerous to list here, these databases feature genome sequences across many species and include: the DNA methylome (methylation patterns in normal and disease states); orphan lncRNA sequences, whose functions have yet to be determined, and their genomic location; and microRNAs that are regulated by lncRNAs. These databases can be leveraged to expedite discovery and validation of therapeutic lncRNA targets in a process tailored for each target as exemplified in the examples provided above.

The targeting of lncRNA for therapeutic applications represents a brave new world for addressing unmet medical needs – from rare genetic diseases to high prevalence diseases such as cardiovascular disease, diabetes type II, Parkinson's disease, and cancer. The examples provided in this chapter illustrate the vast potential of lncRNAs as drug targets. Importantly, their large numbers and their molecular and cellular specificity opens the door for lncRNAs to be a rich field for research and drug development to explore.

References

1. Hopkins AL, Groom CR (2002) The druggable genome. Nat Rev Drug Discov 1(9):727–730
2. Iyer MK et al (2015) The landscape of long noncoding RNAs in the human transcriptome. Nat Genet 47(3):199–208
3. Clark SJ (2007) Action at a distance: epigenetic silencing of large chromosomal regions in carcinogenesis. Hum Mol Genet 16(1):R88–R95
4. Karpova NN, Sales AJ, Joca SR (2017) Epigenetic basis of neuronal and synaptic plasticity. Curr Top Med Chem 17(7):771–793
5. Fritah S, Niclou SP, Azuaje F (2014) Databases for lncRNAs: a comparative evaluation of emerging tools. RNA 20(11):1655–1665
6. Lai F et al (2015) Integrator mediates the biogenesis of enhancer RNAs. Nature 525 (7569):399–403
7. Rothschild G, Basu U (2017) Lingering questions about enhancer RNA and enhancer transcription-coupled genomic instability. Trends Genet 33(2):143–154
8. Werner A, Berdal A (2005) Natural antisense transcripts: sound or silence? Physiol Genomics 23(2):125–131
9. Zhang X et al (2014) The role of antisense long noncoding RNA in small RNA-triggered gene activation. RNA 20(12):1916–1928
10. St Laurent G, Wahlestedt C, Kapranov P (2015) The landscape of long noncoding RNA classification. Trends Genet 31(5):239–251
11. Wilusz JE (2016) Long noncoding RNAs: re-writing dogmas of RNA processing and stability. Biochim Biophys Acta 1859(1):128–138
12. Schlackow M et al (2017) Distinctive patterns of transcription and RNA processing for human lincRNAs. Mol Cell 65(1):25–38
13. Mele M et al (2017) Chromatin environment, transcriptional regulation, and splicing distinguish lincRNAs and mRNAs. Genome Res 27(1):27–37
14. Kapranov P et al (2010) The majority of total nuclear-encoded non-ribosomal RNA in a human cell is 'dark matter' un-annotated RNA. BMC Biol 8:149
15. Shiraki T et al (2003) Cap analysis gene expression for high-throughput analysis of transcriptional starting point and identification of promoter usage. Proc Natl Acad Sci U S A 100 (26):15776–15781
16. Carninci P et al (2005) The transcriptional landscape of the mammalian genome. Science 309 (5740):1559–1563
17. Katayama S et al (2005) Antisense transcription in the mammalian transcriptome. Science 309(5740):1564–1566
18. Birney E et al (2007) Identification and analysis of functional elements in 1% of the human genome by the ENCODE pilot project. Nature 447(7146):799–816
19. Kapranov P et al (2007) RNA maps reveal new RNA classes and a possible function for pervasive transcription. Science 316(5830):1484–1488
20. Mercer TR, Dinger ME, Mattick JS (2009) Long non-coding RNAs: insights into functions. Nat Rev Genet 10(3):155–159
21. Derrien T et al (2012) The GENCODE v7 catalog of human long noncoding RNAs: analysis of their gene structure, evolution, and expression. Genome Res 22(9):1775–1789

22. Plath K et al (2003) Role of histone H3 lysine 27 methylation in X inactivation. Science 300 (5616):131–135
23. Silva J et al (2003) Establishment of histone h3 methylation on the inactive X chromosome requires transient recruitment of Eed-Enx1 polycomb group complexes. Dev Cell 4 (4):481–495
24. Zhao J et al (2008) Polycomb proteins targeted by a short repeat RNA to the mouse X chromosome. Science 322(5902):750–756
25. Avner P, Heard E (2001) X-chromosome inactivation: counting, choice and initiation. Nat Rev Genet 2(1):59–67
26. Agrelo R et al (2009) SATB1 defines the developmental context for gene silencing by Xist in lymphoma and embryonic cells. Dev Cell 16(4):507–516
27. Hasegawa Y et al (2010) The matrix protein hnRNP U is required for chromosomal localization of Xist RNA. Dev Cell 19(3):469–476
28. Pullirsch D et al (2010) The Trithorax group protein Ash2l and Saf-A are recruited to the inactive X chromosome at the onset of stable X inactivation. Development 137(6):935–943
29. Wutz A, Rasmussen TP, Jaenisch R (2002) Chromosomal silencing and localization are mediated by different domains of Xist RNA. Nat Genet 30(2):167–174
30. Rinn JL et al (2007) Functional demarcation of active and silent chromatin domains in human HOX loci by noncoding RNAs. Cell 129(7):1311–1323
31. Engreitz JM, Ollikainen N, Guttman M (2016) Long non-coding RNAs: spatial amplifiers that control nuclear structure and gene expression. Nat Rev Mol Cell Biol 17(12):756–770
32. Khorkova O, Hsiao J, Wahlestedt C (2015) Basic biology and therapeutic implications of lncRNA. Adv Drug Deliv Rev 87:15–24
33. Guttman M, Rinn JL (2012) Modular regulatory principles of large non-coding RNAs. Nature 482(7385):339–346
34. Davidovich C, Cech TR (2015) The recruitment of chromatin modifiers by long noncoding RNAs: lessons from PRC2. RNA 21(12):2007–2022
35. Dai Q et al (2015) Competing endogenous RNA: a novel posttranscriptional regulatory dimension associated with the progression of cancer. Oncol Lett 10(5):2683–2690
36. Martens JA, Laprade L, Winston F (2004) Intergenic transcription is required to repress the Saccharomyces cerevisiae SER3 gene. Nature 429(6991):571–574
37. Schmitt S, Prestel M, Paro R (2005) Intergenic transcription through a polycomb group response element counteracts silencing. Genes Dev 19(6):697–708
38. Lucafo M et al (2015) Long noncoding RNA GAS5: a novel marker involved in glucocorticoid response. Curr Mol Med 15(1):94–99
39. Valencia-Sanchez MA et al (2006) Control of translation and mRNA degradation by miRNAs and siRNAs. Genes Dev 20(5):515–524
40. Weng W et al. (2017) Circular RNA ciRS-7 – a promising prognostic biomarker and a potential therapeutic target in colorectal cancer. Clin Cancer Res. epub ahead of print April 26, 2017
41. Kosik KS (2013) Molecular biology: circles reshape the RNA world. Nature 495 (7441):322–324
42. Ebbesen KK, Kjems J, Hansen TB (2016) Circular RNAs: identification, biogenesis and function. Biochim Biophys Acta 1859(1):163–168
43. Yamazaki T, Hirose T (2015) The building process of the functional paraspeckle with long non-coding RNAs. Front Biosci (Elite Ed) 7:1–41
44. Kim KM et al (2017) RNA in extracellular vesicles. Wiley Interdiscip Rev RNA 8(4). doi:10.1002/wrna.1413
45. Sun M, Kraus WL (2015) From discovery to function: the expanding roles of long noncoding RNAs in physiology and disease. Endocr Rev 36(1):25–64
46. Kotake Y et al (2011) Long non-coding RNA ANRIL is required for the PRC2 recruitment to and silencing of p15(INK4B) tumor suppressor gene. Oncogene 30(16):1956–1962
47. Smyk M et al (2013) Chromosome conformation capture-on-chip analysis of long-range cis-interactions of the SOX9 promoter. Chromosom Res 21(8):781–788

48. Bradbury EM, Van Holde KE (1988) Chromatin. Series in molecular biology. Springer-Verlag, New York, 530 pp. Journal of Molecular Recognition, 1989. 2(3): p. i-i
49. Quinodoz S, Guttman M (2014) Long noncoding RNAs: an emerging link between gene regulation and nuclear organization. Trends Cell Biol 24(11):651–663
50. Tsai MC et al (2010) Long noncoding RNA as modular scaffold of histone modification complexes. Science 329(5992):689–693
51. Tirado-Magallanes R et al (2017) Whole genome DNA methylation: beyond genes silencing. Oncotarget 8(3):5629–5637
52. Merry CR et al (2015) DNMT1-associated long non-coding RNAs regulate global gene expression and DNA methylation in colon cancer. Hum Mol Genet 24(21):6240–6253
53. Kapusta A, Feschotte C (2014) Volatile evolution of long noncoding RNA repertoires: mechanisms and biological implications. Trends Genet 30(10):439–452
54. Johnson R, Guigo R (2014) The RIDL hypothesis: transposable elements as functional domains of long noncoding RNAs. RNA 20(7):959–976
55. Kapusta A et al (2013) Transposable elements are major contributors to the origin, diversification, and regulation of vertebrate long noncoding RNAs. PLoS Genet 9(4):e1003470
56. Tsirigos A, Rigoutsos I (2009) Alu and b1 repeats have been selectively retained in the upstream and intronic regions of genes of specific functional classes. PLoS Comput Biol 5 (12):e1000610
57. Gong C, Maquat LE (2011) lncRNAs transactivate STAU1-mediated mRNA decay by duplexing with 3′ UTRs via Alu elements. Nature 470(7333):284–288
58. Mariner PD et al (2008) Human Alu RNA is a modular transacting repressor of mRNA transcription during heat shock. Mol Cell 29(4):499–509
59. Hirsch ML (2015) Adeno-associated virus inverted terminal repeats stimulate gene editing. Gene Ther 22(2):190–195
60. Quentin Y (1992) Origin of the Alu family: a family of Alu-like monomers gave birth to the left and the right arms of the Alu elements. Nucleic Acids Res 20(13):3397–3401
61. Kriegs JO et al (2007) Evolutionary history of 7SL RNA-derived SINEs in Supraprimates. Trends Genet 23(4):158–161
62. Batzer MA, Deininger PL (2002) Alu repeats and human genomic diversity. Nat Rev Genet 3 (5):370–379
63. Quentin Y (1994) A master sequence related to a free left Alu monomer (FLAM) at the origin of the B1 family in rodent genomes. Nucleic Acids Res 22(12):2222–2227
64. Hacisuleyman E et al (2014) Topological organization of multichromosomal regions by the long intergenic noncoding RNA Firre. Nat Struct Mol Biol 21(2):198–206
65. Hacisuleyman E et al (2016) Function and evolution of local repeats in the Firre locus. Nat Commun 7:11021
66. Muller J et al (2002) Histone methyltransferase activity of a Drosophila Polycomb group repressor complex. Cell 111(2):197–208
67. Cabili MN et al (2011) Integrative annotation of human large intergenic noncoding RNAs reveals global properties and specific subclasses. Genes Dev 25(18):1915–1927
68. Zhou T, Kim Y, MacLeod AR (2016) Targeting long noncoding RNA with antisense oligonucleotide technology as cancer therapeutics. Methods Mol Biol 1402:199–213
69. Sharma VK, Watts JK (2015) Oligonucleotide therapeutics: chemistry, delivery and clinical progress. Future Med Chem 7(16):2221–2242
70. Weeks KM, Mauger DM (2011) Exploring RNA structural codes with SHAPE chemistry. Acc Chem Res 44(12):1280–1291
71. Smith KT, Workman JL (2009) Histone deacetylase inhibitors: anticancer compounds. Int J Biochem Cell Biol 41(1):21–25
72. Chang J et al (2012) Differential response of cancer cells to HDAC inhibitors trichostatin A and depsipeptide. Br J Cancer 106(1):116–125
73. de Ruijter AJ et al (2003) Histone deacetylases (HDACs): characterization of the classical HDAC family. Biochem J 370(Pt 3):737–749

74. Gregoretti IV, Lee YM, Goodson HV (2004) Molecular evolution of the histone deacetylase family: functional implications of phylogenetic analysis. J Mol Biol 338(1):17–31
75. Finnin MS et al (1999) Structures of a histone deacetylase homologue bound to the TSA and SAHA inhibitors. Nature 401(6749):188–193
76. Wagner JM et al (2010) Histone deacetylase (HDAC) inhibitors in recent clinical trials for cancer therapy. Clin Epigenetics 1(3–4):117–136
77. Mottamal M et al (2015) Histone deacetylase inhibitors in clinical studies as templates for new anticancer agents. Molecules 20(3):3898–3941
78. Takayama K et al (2013) Androgen-responsive long noncoding RNA CTBP1-AS promotes prostate cancer. EMBO J 32(12):1665–1680
79. Wang R et al (2012) Role of transcriptional corepressor CtBP1 in prostate cancer progression. Neoplasia 14(10):905–914
80. Sung YY, Cheung E (2013) Antisense now makes sense: dual modulation of androgen-dependent transcription by CTBP1-AS. EMBO J 32(12):1653–1654
81. Stirchak EP, Summerton JE, Weller DD (1989) Uncharged stereoregular nucleic acid analogs: 2. Morpholino nucleoside oligomers with carbamate internucleoside linkages. Nucleic Acids Res 17(15):6129–6141
82. Kole R, Krainer AR, Altman S (2012) RNA therapeutics: beyond RNA interference and antisense oligonucleotides. Nat Rev Drug Discov 11(2):125–140
83. Lundin KE, Gissberg O, Smith CI (2015) Oligonucleotide therapies: the past and the present. Hum Gene Ther 26(8):475–485
84. Subramanian RR et al (2015) Enhancing antisense efficacy with multimers and multi-targeting oligonucleotides (MTOs) using cleavable linkers. Nucleic Acids Res 43(19):9123–9132
85. Adams BD et al (2017) Targeting noncoding RNAs in disease. J Clin Invest 127(3):761–771
86. McClorey G, Wood MJ (2015) An overview of the clinical application of antisense oligonucleotides for RNA-targeting therapies. Curr Opin Pharmacol 24:52–58
87. Kauppinen S, Vester B, Wengel J (2005) Locked nucleic acid (LNA): high affinity targeting of RNA for diagnostics and therapeutics. Drug Discov Today Technol 2(3):287–290
88. Crooke ST (1999) Molecular mechanisms of action of antisense drugs. Biochim Biophys Acta 1489(1):31–44
89. Vickers TA, Crooke ST (2015) The rates of the major steps in the molecular mechanism of RNase H1-dependent antisense oligonucleotide induced degradation of RNA. Nucleic Acids Res 43(18):8955–8963
90. Wu H et al (2004) Determination of the role of the human RNase H1 in the pharmacology of DNA-like antisense drugs. J Biol Chem 279(17):17181–17189
91. Gagnon KT et al (2014) RNAi factors are present and active in human cell nuclei. Cell Rep 6 (1):211–221
92. Torres AG, Threlfall RN, Gait MJ (2011) Potent and sustained cellular inhibition of miR-122 by lysine-derivatized peptide nucleic acids (PNA) and phosphorothioate locked nucleic acid (LNA)/2′-O-methyl (OMe) mixmer anti-miRs in the absence of transfection agents. Artif DNA PNA XNA 2(3):71–78
93. Fabani MM, Gait MJ (2008) miR-122 targeting with LNA/2′-O-methyl oligonucleotide mixmers, peptide nucleic acids (PNA), and PNA-peptide conjugates. RNA 14(2):336–346
94. Davis S et al (2009) Potent inhibition of microRNA in vivo without degradation. Nucleic Acids Res 37(1):70–77
95. Gustincich S, Zucchelli S, Mallamaci A (2016) The Yin and Yang of nucleic acid-based therapy in the brain. Prog Neurobiol. doi:10.1016/j.pneurobio.2016.11.001
96. Bishop KM (2016) Progress and promise of antisense oligonucleotide therapeutics for central nervous system diseases. Neuropharmacology. doi:10.1016/j.neuropharm.2016.12.015
97. Moreno PM, Pego AP (2014) Therapeutic antisense oligonucleotides against cancer: hurdling to the clinic. Front Chem 2:87

98. Crooke ST, Geary RS (2013) Clinical pharmacological properties of mipomersen (Kynamro), a second generation antisense inhibitor of apolipoprotein B. Br J Clin Pharmacol 76 (2):269–276
99. Lim KR, Maruyama R, Yokota T (2017) Eteplirsen in the treatment of Duchenne muscular dystrophy. Drug Des Devel Ther 11:533–545
100. Bassett AR et al (2014) Considerations when investigating lncRNA function in vivo. Elife 3: e03058
101. Nelson CA et al (2011) Inhibiting TGF-beta activity improves respiratory function in mdx mice. Am J Pathol 178(6):2611–2621
102. Halley P et al (2014) Regulation of the apolipoprotein gene cluster by a long noncoding RNA. Cell Rep 6(1):222–230
103. Castle JW et al (2015) Therapeutic ultrasound: increased HDL-cholesterol following infusions of acoustic microspheres and apolipoprotein A-I plasmids. Atherosclerosis 241 (1):92–99
104. Bebee TW, Dominguez CE, Chandler DS (2012) Mouse models of SMA: tools for disease characterization and therapeutic development. Hum Genet 131(8):1277–1293
105. Li F et al (2016) Novel insights into the role of long noncoding RNA in ocular diseases. Int J Mol Sci 17(4):478
106. Cooper C et al (2009) Increasing the relative expression of endogenous non-coding steroid receptor RNA activator (SRA) in human breast cancer cells using modified oligonucleotides. Nucleic Acids Res 37(13):4518–4531
107. Leygue E (2007) Steroid receptor RNA activator (SRA1): unusual bifaceted gene products with suspected relevance to breast cancer. Nucl Recept Signal 5:e006
108. Hube F et al (2006) Alternative splicing of the first intron of the steroid receptor RNA activator (SRA) participates in the generation of coding and noncoding RNA isoforms in breast cancer cell lines. DNA Cell Biol 25(7):418–428
109. Powell WT et al (2013) A Prader-Willi locus lncRNA cloud modulates diurnal genes and energy expenditure. Hum Mol Genet 22(21):4318–4328
110. Cruvinel E et al (2014) Reactivation of maternal SNORD116 cluster via SETDB1 knockdown in Prader-Willi syndrome iPSCs. Hum Mol Genet 23(17):4674–4685
111. Carpenter S et al (2013) A long noncoding RNA mediates both activation and repression of immune response genes. Science 341(6147):789–792
112. Rapicavoli NA et al (2013) A mammalian pseudogene lncRNA at the interface of inflammation and anti-inflammatory therapeutics. Elife 2:e00762
113. Faghihi MA et al (2008) Expression of a noncoding RNA is elevated in Alzheimer's disease and drives rapid feed-forward regulation of beta-secretase. Nat Med 14(7):723–730
114. Gharami K et al (2008) Brain-derived neurotrophic factor over-expression in the forebrain ameliorates Huntington's disease phenotypes in mice. J Neurochem 105(2):369–379
115. Xie Y, Hayden MR, Xu B (2010) BDNF overexpression in the forebrain rescues Huntington's disease phenotypes in YAC128 mice. J Neurosci 30(44):14708–14718
116. Khalil AM et al (2008) A novel RNA transcript with antiapoptotic function is silenced in fragile X syndrome. PLoS One 3(1):e1486
117. Ladd PD et al (2007) An antisense transcript spanning the CGG repeat region of FMR1 is upregulated in premutation carriers but silenced in full mutation individuals. Hum Mol Genet 16(24):3174–3187
118. Jeck WR et al (2013) Circular RNAs are abundant, conserved, and associated with ALU repeats. RNA 19(2):141–157
119. Memczak S et al (2013) Circular RNAs are a large class of animal RNAs with regulatory potency. Nature 495(7441):333–338
120. Hancock JM (2014) Circles within circles: commentary on Ghosal et al. (2013) "Circ2Traits: a comprehensive database for circular RNA potentially associated with disease and traits". Front Genet 5:459
121. Chen L et al (2015) Circular RNAs in eukaryotic cells. Curr Genomics 16(5):312–318

122. Dong YF et al (2016) Potential role of microRNA-7 in the anti-neuroinflammation effects of nicorandil in astrocytes induced by oxygen-glucose deprivation. J Neuroinflammation 13 (1):60
123. Chaudhuri AD et al (2016) MicroRNA-7 regulates the function of mitochondrial permeability transition pore by targeting VDAC1 expression. J Biol Chem 291(12):6483–6493
124. Zheng XB, Zhang M, Xu MQ (2017) Detection and characterization of ciRS-7: a potential promoter of the development of cancer. Neoplasma 64(3):321–328
125. Hansen TB et al (2013) Natural RNA circles function as efficient microRNA sponges. Nature 495(7441):384–388
126. Zhao J et al (2010) Genome-wide identification of polycomb-associated RNAs by RIP-seq. Mol Cell 40(6):939–953
127. Woo CJ et al (2017) Gene activation of SMN by selective disruption of lncRNA-mediated recruitment of PRC2 for the treatment of spinal muscular atrophy. Proc Natl Acad Sci U S A 114(8):e1509–e1518
128. d'Ydewalle C et al (2017) The antisense transcript SMN-AS1 regulates SMN expression and is a novel therapeutic target for spinal muscular atrophy. Neuron 93(1):66–79
129. Cabianca DS et al (2012) A long ncRNA links copy number variation to a polycomb/trithorax epigenetic switch in FSHD muscular dystrophy. Cell 149(4):819–831
130. Chen JC et al (2016) Morpholino-mediated knockdown of DUX4 toward facioscapulohumeral muscular dystrophy therapeutics. Mol Ther 24(8):1405–1411

Top Med Chem (2018) 27: 237–254
DOI: 10.1007/7355_2016_30

Published online: 17 March 2017

Messenger RNA as a Novel Therapeutic Approach

Matthew G. Stanton and Kerry E. Murphy-Benenato

Abstract The concept of mRNA as a therapeutic platform has historically been ignored owing to challenges in oligonucleotide delivery and, maybe more importantly, the perceived shortcomings of mRNA with regard to stability and immunogenicity. Advances in several areas have recently prompted a reexamination of such dogma. Significant improvements in oligonucleotide delivery have been realized over the past decade and their application to mRNA has enabled a more rapid path toward clinical development of this new modality. Similarly, recent discoveries in mRNA chemistry further enhance the attractiveness of this platform by attenuating innate immune activation and maximizing protein expression. With these advances, mRNA is positioned to become an important new therapeutic modality.

Keywords Delivery, Immunogenicity, Lipid nanoparticles, mRNA, Nucleotides, Polymers, RNA, Transcript therapy

Contents

1 Delivery 239
2 Innate Immunity 245
3 Conclusion 250
References 250

The advent of a new and potentially revolutionary technology is rare. Modern medicine is defined by a few such examples including rational small molecule drug design and protein therapeutics. These breakthroughs happen not in isolation, but often as a confluence of technical advances that overcome previously limiting barriers. For rational drug design this includes but is not limited to advances in

M.G. Stanton (✉) and K.E. Murphy-Benenato
Moderna Therapeutics, 200 Technology Square, Cambridge, MA 02139, USA
e-mail: matt.stanton@modernatx.com

synthetic organic chemistry, structural biology and computational science. Biopharmaceuticals became possible with the explosion of advances in recombinant DNA technology. While still early in its development and application, mRNA therapeutics hold such revolutionary promise and like other breakthroughs is enabled by advances in multiple areas. This chapter will focus on advances in RNA chemistry, delivery and attenuation of immune stimulation as distinct scientific fields that when brought together, enable mRNA as an exciting and viable therapeutic platform.

Before outlining the technical details of mRNA therapeutics, it's worth discussing the reasons for such intense interest in this space. Indeed, this interest is highlighted by the flow of investment capital over the past several years (for a good online review of the mRNA therapeutics space: [1]). Figure 1 outlines the general concept of mRNA therapeutics by focusing on the central dogma of biology, which is DNA codes for RNA which codes for protein. All therapeutic intervention must necessarily work at some level of this dogma. Small molecule and biopharmaceuticals focus on the protein and skip over the preceding nucleic acid components. This makes great sense in that the protein is the most proximal mediator of disease, however it comes with great cost and complexity. Small molecule therapies are discovered in a bespoke manner with substantial investment in protein production, screening for small molecule binders and subsequent optimization of leads. There are always challenges identifying molecules with the desired potency, target selectivity, and pharmacokinetic properties which enable effective translation in the clinic. Biopharmaceuticals, on the other hand, focus on recombinant technologies to produce therapeutic proteins. These products enjoy the benefit of selectivity but are encumbered by the technical costs of manufacture and the limitation of only being able to access extracellular targets. The obvious attraction of mRNA therapeutics lies in the ability to simultaneously address limitations of both small molecule and protein therapeutics. A common set of

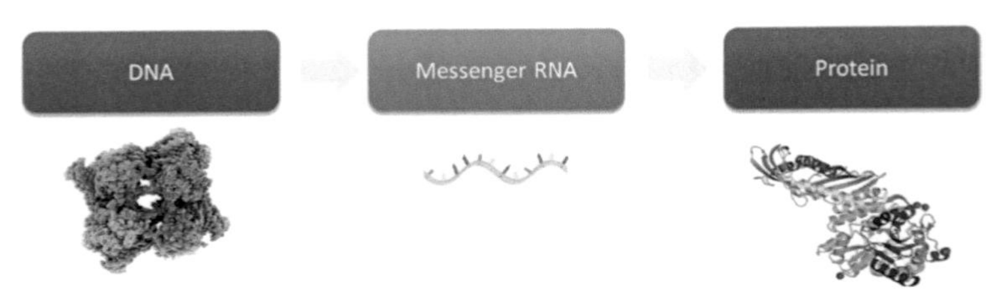

- mRNA as software for cellular machinery
  - Native post-transcriptional modifications
- Human body as protein factory
  - Large molecule targets at small molecule prices
- Transient dose-dependent expression of protein
  - Safety benefit. Pharmacological tuning
- Translated protein already in the cell. Signal peptides for trafficking inside or outside the cell

Fig. 1 The case for mRNA therapeutics

technical platform tools could enable the generation of any protein of interest in vivo, be it secreted, intracellular, or transmembrane. This holds the promise of large molecule therapeutics at a manufacturing cost that is more common to small molecules while also accessing previously undruggable targets. And unlike gene therapy, which theoretically has similar advantages, mRNA therapy is devoid of many safety concerns including genomic integration, immunologic response to viral vectors and risk associated with uncontrolled, sustained protein expression.

Given the potential of mRNA as a therapeutic modality, the question most likely to ponder is "why are we just now witnessing the emergence of this?" The remainder of this chapter will cover the two main hurdles to implementation of this technology and how these fields have evolved to address such concerns—RNA delivery and innate immunology. There was a third topic that was once considered a limitation to the realization of mRNA therapeutics and this was drug substance stability but this has been addressed as witnessed by the advent of mRNA vaccines and their clinical advancement and thus will not be covered here.

1 Delivery

A critical factor in mRNA therapeutics becoming a viable clinical option is the identification of a delivery vehicle that enables efficient and safe systemic delivery of mRNA to multiple tissues (Fig. 2). A delivery vehicle is required to protect the mRNA from nuclease degradation as well as minimizing the vehicle/cargo specific innate immune stimulation. Early in vivo examples of mRNA translation have utilized a variety of delivery methods such as direct injection of unformulated mRNA, commercially available transfection reagents, or cationic liposomes [2–6]. In fact, Türeci, Sahin, and coworkers have demonstrated positive human clinical data delivering mRNA based cancer vaccines in positively charged liposomes composed of cationic amino lipids DOTMA (**1**) and DOTAP (**2**) (Fig. 3) and the phospholipid DOPE (**3**) (Fig. 4) [7]. Through optimization of the surface charge of the liposome they were able to affect selective delivery of mRNA to antigen presenting cells. Looking ahead, to be able to translate the power of mRNA into a broadly applicable therapeutic, a delivery vehicle which targets different organs and tissue types will be required.

mRNA offers its own unique challenges for delivery due to its large size, instability in vivo, and immunogenic nature. Therefore mRNA requires a vehicle which will completely protect the material from degradation without contributing to innate immune activation. Additionally, an ideal mRNA delivery vector will target desired cell types specifically and will employ components that maximize endosomal escape and are efficiently cleared by the body. Early discovery efforts in the development of mRNA based therapies have taken advantage of the effort that has gone into the identification of clinically viable delivery vehicles for other nucleic acid based therapies, such as siRNA [8]. Of these delivery systems, one of the most advanced are lipid nanoparticles (LNPs) which are being utilized in

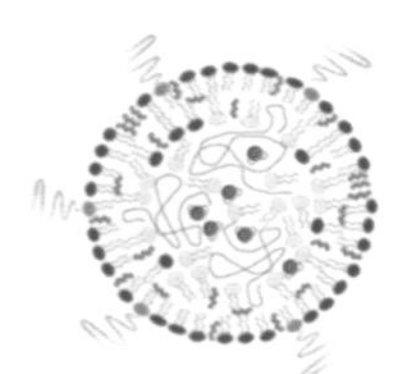

Attributes of Ideal mRNA Delivery Vehicle

- Protection from nuclease mediated degradation
- Desired opsonin profile to maximize delivery and minimize immune effects
- Maximal cell specific uptake
- Efficient endosomal escape
- Efficient clearance of delivery vehicle components

Fig. 2 Attributes of ideal mRNA delivery vehicle

DOTMA (1)

DOTAP (2)

Fig. 3 Cationic amino lipids

DOPE (3)

DSPC (4)

Fig. 4 Representative phospholipids

multiple clinical trials [9]. Lipid nanoparticles are multicomponent systems composed of an ionizable lipid (Fig. 5), a phospholipid (Fig. 4), cholesterol, and a PEG-lipid conjugate (Fig. 6) [10]. The structure of the ionizable lipid is a key factor in potency. Modification of the pKa of the ionizable amine can dramatically affect the surface charge of the particle and therefore its ability to associate with plasma proteins and the lipid bilayer of endosome walls. When delivered intravenously, LNPs selectively travel to the liver via an ApoE dependent mechanism, as shown through in vivo experiments in ApoE knock-out mice [11]. Depending on the ionizable lipid component, the LNPs are taken up by hepatocytes through an

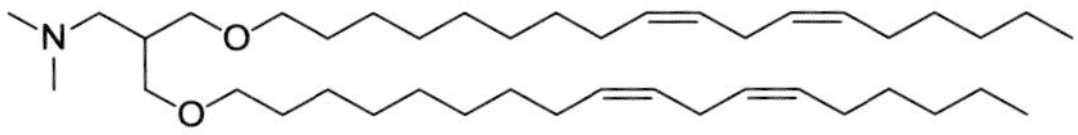

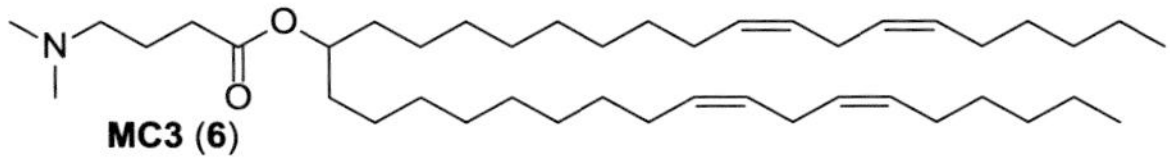

DLin-DMA (5)

MC3 (6)

Fig. 5 Ionizable amino lipids

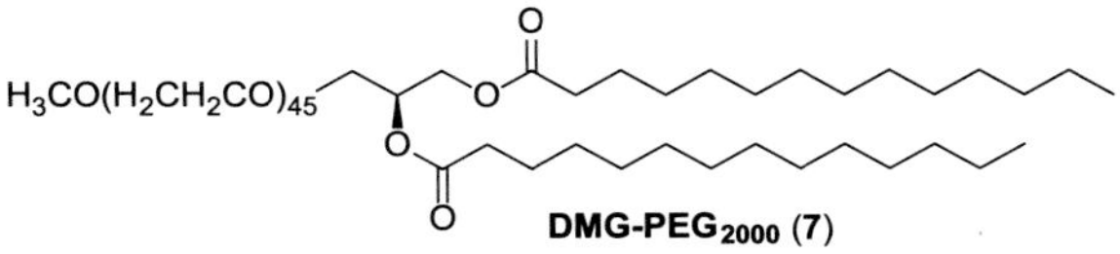

$H_3CO(H_2CH_2CO)_{45}$

DMG-PEG$_{2000}$ (7)

Fig. 6 Representative PEG-lipid conjugate

LDLr-dependent or micropinocytosis mechanism. The structure of the amino lipids tails play a role in endosomal escape, helping to drive phase transition from a bilayer to hexagonal H_{II} thereby enabling release of the mRNA. The phospholipid component increases the fusogenic character of the LNP lipid bilayer aiding endosomal escape. Cholesterol stabilizes the LNP membrane and the PEG-lipid shields the surface and stabilizes the particle.

The ability of LNPs, which have been optimized for siRNA delivery, to be repurposed for mRNA delivery has been verified by three different groups. In 2012 researchers at Novartis demonstrated intramuscular delivery of a self-amplifying RNA vaccine in a rodent model of RSV using D-Lin-DMA (**5**, Fig. 5) based LNPs [12, 13], a vehicle that has been employed by Alnylam in a Phase I study for the delivery of siRNA [9]. This was the first study that established amino lipid based LNPs (which had been optimized specifically for siRNA delivery to the liver) form stable particles with mRNA and effectively release the mRNA for protein translation in vivo.

In 2015 the Acuitas family of amino lipids were utilized for the systemic delivery of mRNA to mice via multiple routes of administration [14]. In this study intravenous delivered LNPs containing luciferase mRNA predominantly show protein expression in the liver, whereas intramuscular and subcutaneous injections result in more local expression. In vitro, the mRNA LNPs also utilize an ApoE dependent mechanism for cellular uptake. Following this report, scientists as CureVac GmbH employed the same lipid class and showed this class of LNPs were a viable delivery route for mRNA in both pig and cynomolgus monkeys. In both pigs and monkeys they observed efficacy with a single intravenous dose of human erythropoietin (hEPO) mRNA. In addition, in the monkey (0.037 mg/kg dose of hEPO mRNA) they observed no appreciable changes in TNF-α or INF-γ levels relative to pretreatment [15].

A group at Pfizer has also lent support to the concept of LNPs for mRNA delivery. The researchers demonstrated intravenous and intrathecal delivery of frataxin mRNA in MC3 effective based LNPs (**6**, Fig. 5, Alnylam's second generation clinical LNP formulation for their siRNA program). Alnylam's second generation clinical LNP formulation for their siRNA programs [16]. In this report, using mRNA encapsulated in an MC3 based LNP, they were able to show mRNA LNPs behave similarly in vivo to siRNA LNPs, delivering to the liver when dosed intravenously.

In addition to the classes of amino lipids exemplified by MC3 (single ionizable amine) Anderson and co-workers have pioneered the development of a series of polyamino lipid-like materials for siRNA delivery. The lipidoids, exemplified by C12-200 (**8**, Fig. 7), enter cells via a non-LDLr dependent mechanism [17]. Starting with the optimal C12-200 formulation for siRNA delivery, C12-200:DSPC:Chol:DMG-PEG_{2000} (50:10:38.5:1.5), Anderson and co-workers performed a systematic design of experiment (DOE) to optimize for mRNA delivery [18]. By modifying the LNP composition (e.g., phospholipid, lipid:mRNA ratio, molar composition of components) they were able to improve the overall particle characteristics (smaller size, improved encapsulation efficiency) and the in vivo efficacy. With the optimized mRNA LNP composition C12-200:DOPE:Chol:DMG-PEG_{2000}(35:16:46.5:2.5), a three- to sevenfold improvement in protein expression was observed depending on the reporter used. It is of note that the same improvement in in vivo efficacy was not observed when the optimized formulation was utilized for siRNA mediated protein knockdown, highlighting that an optimized siRNA vehicle is not necessarily going to

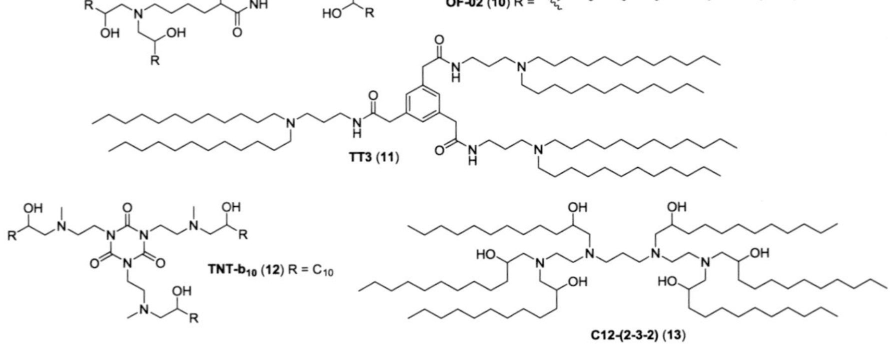

Fig. 7 Lipidoids

be optimal for mRNA delivery, and vice versa. The hepatocyte specificity and therapeutic efficacy of the C12-200 formulation has been further validated in collaboration with scientists at Shire [19].

In a separate report Anderson and co-workers demonstrated that an another LDLr-independent lipidoid which had been developed for siRNA delivery, the lysine based CKK-E12 (**9**, Fig. 7), could be applied to systemic mRNA delivery [20, 21]. Through variation of the lipid tails [lauric (**9,** CKK-E12) to linoleic (**10**, OF-02)] they were able to increase hEPO expression in mice relative to control. Additional experiments with luciferase mRNA demonstrated selective expression in the liver and minimal expression in the spleen with both CKK-E12 and OF-02 LNPs.

Using a systematic approach to LNP optimization, Dong and coworkers have built on the SAR generated with C12-200 (**8**) and developed a new lipidoid scaffold, exemplified by TT3 (**11**, Fig. 7) [22]. Beginning with a triacetamide benzene core, the linker between the amide and lauric lipid tails was modified. The optimized LNP composition of TT3:DOPE:Chol:DMG-PEG$_{2000}$(20:30:40:0.75) when delivered intravenously to a mouse selectively delivered luciferase mRNA to the liver and spleen and was shown to deliver human Factor IX in both wild type and Factor IX KO mice in a dose dependent manner.

Dong and co-workers have published an alternative scaffold for mRNA delivery based on a 1,3,5-triazinane-2,4,6-trione core [23]. By varying the lipid tail length off the core and utilizing systematic in vitro screening and optimization of formulation parameters they identified TNT-b$_{10}$ (**12**):DOPE:Chol:DMG-PEG$_{2000}$(34:40:35:0.75) as the optimal LNP formulation. Efficient translation of firefly luciferase mRNA was observed in the spleen of mice when delivered intravenously. When delivered intraperitoneal however, expression was diminished and no luciferase expression was observed when delivered subcutaneously.

In an investigation of the correlation between buffering capacity of an oligoalkylamine and its ability to complex mRNA and deliver intracellularly, Dohmen and co-workers synthesized a series of oligoalkylamine-based lipids to deliver mRNA intravenously [24]. In a screen of various alkyl amine cores, they found the 2-3-2 ethanolamine core was optimal (**13**, Fig. 7); it had the highest buffering capacity and highest in vitro transfection. In vivo LNPs formed with 2-3-2 lipid **13** and firefly luciferase mRNA had the highest level of bioluminescence in the liver of mice when given intravenously.

LNPs have great potential for the systemic delivery of mRNA, however alternative delivery modes may be necessary in the future in order to enable delivery to other tissues and by other routes of administration. mRNA polymer complexes are another non-viral delivery option which researchers have investigated for nucleic acid delivery (Fig. 8) [25]. Kataoka and co-workers applied their *N*-substituted polyaspartamide polyplexes to the cellular transfection of mRNA [26]. In this investigation they synthesized a series of *N*-substituted polyaspartamides, varying the number of aminoethyl groups in order to vary the protonation states of the different polymers. The studies demonstrated the mRNA polyaspartamide polyplex has low stability in the cytoplasm resulting in lower levels of protein expression.

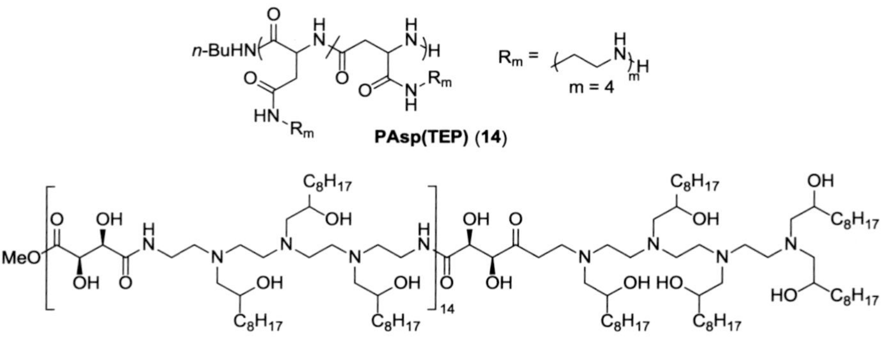

Fig. 8 Polymers for mRNA delivery

Therefore mRNA polymer complexes which had a slower rate of endosomal escape were optimal for sustained protein expression in cells. The polyaspartamide was incorporated into a PEG-block copolymer with cholesterol and nanomicelles were formed (PEG-PAsp(TEP)-Chol/mRNA) (**14**) [27]. The incorporation of cholesterol to the co-block polymer increased in vitro stability to nucleases. This was further validated in vivo and efficacy was demonstrated in a xenograph mouse model of pancreatic cancer by delivering mRNA encoding an anti-angiogenic growth factor (sFLt-1) intravenously.

Anderson and co-workers evaluated the structure activity relationship for nanoparticles consisting of poly(glycoamidoamine) (PGAA) brushes [28]. The PGAA polymers were synthesized from tartarate, galactarate, or glucarate. Differing lengths of polyamines were added, which were capped via epoxide ring opening to introduce hydroxy-containing alkyl chains, similar to those found in the lipidoids described above. The materials were evaluated in vivo for the delivery of both siRNA and mRNA and they found activity correlated, with TarN3C10 (**15**) being optimal for both. From the different analogs tested they found a number of key structural parameters for efficient delivery; three aminoethanol groups, tartarate as the PGAA polymer and shorter tail lengths. When TarN3C10-hEPO nanoparticles were delivered intravenously they produced high serum levels of hEPO, reportedly higher than C12-200 LNPs, however they do not report if C12-200 was formulated under optimized conditions described above. Using luciferase mRNA, they determined the nanoparticles deliver mRNA most efficiently to liver and spleen, in addition to a number of other tissues. A single dose tolerability study (1 mpk) with the TarN3C10 nanoparticles showed transient increases of some cytokines (G-CSF and MIG) and no effect on liver enzymes.

Being able to build on the work that has enabled siRNA and other nucleic acid based therapies to enter the clinic has allowed the power of mRNA based therapies to be realized. However the publications referenced here highlight there are still advances to be made and questions to be answered. It has been shown that optimal siRNA formulations may not be the best composition for mRNA. Also none of the referenced studies has addressed the question of whether these mRNA compositions are able to evade the innate immune system. However as evident by recent patent applications, a focus on delivery specifically for mRNA is an active pursuit by the scientific community [29–32].

2 Innate Immunity

A longstanding barrier to mRNA-based therapeutics has been concern about activating innate immune sensors with exogenously delivered RNA. The consequence of activation is somewhat complex but ultimately results in lack of expression of the desired transcript along with potential for unwanted toxicity. There are numerous RNA based immune sentinels as depicted in Fig. 9 [33]. Strategically, these sensors are designed to trigger a response based on either (1) modifications that distinguish foreign from self-RNA or (2) geography—the presence of RNA in locations that are restricted to self-RNA. In order to minimize the activation caused by the former, mRNA purity/quality is a key determinant. For the latter, modifications to the mRNA are key to eliminating activation.

Cytosolic RNA sensors appear to have evolved to discriminate self vs. non-self RNA on the basis of differences in molecular structure. Many viruses lack 5′ capping enzymes for instance, and rely on mammalian host capping enzymes to install a cap on a 5′ triphosphate substrate [34]. Accordingly, mammals have evolved two sensors capable of detecting foreign 5′ triphosphate mRNA—RIG-I and MDA5 [35, 36]. RIG-I is reported to sense 5′ triphosphate capped single stranded RNA of any length. 5′ triphosphate duplex RNA is reported to be detected

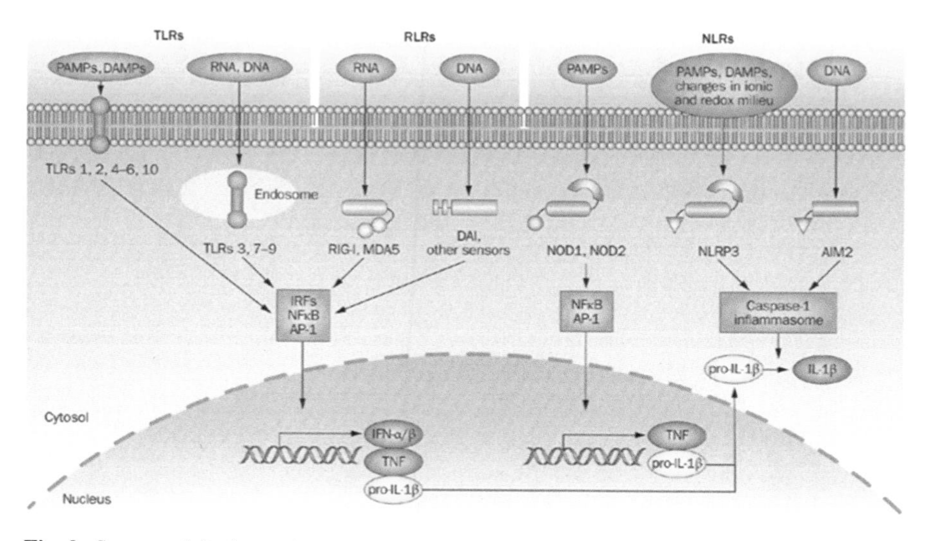

Fig. 9 Sensors of the innate immune system

based on length with RIG-I sensing shorter duplex RNA (≤1 kbp) and MDA5 activated by longer duplex RNA (>1 kbp) [35]. In addition to RIG-I and MDA5, more nuanced mechanisms of innate sensing exist such as the interferon-induced proteins with tetratricopeptide repeats (IFITs, Fig. 10) [36, 37]. These proteins are proposed to operate in an mRNA autonomous manner, effectively sequestering any mRNA that lacks a Cap1 like 2′ OMe group (Fig. 11) and limiting the expression of that particular mRNA [37]. As a result of this autonomous activity, there are no secreted biomarkers for activation of this pathway but in unpublished results from Moderna, we observe up to a 2-log difference in expression in mice for Cap1 vs. Cap0 mRNA. This highlights the potential for subtle structural differences leading to profound pharmacological effects. In this particular case, the potential for triggering an IFIT specific response is easily remedied by the use of enzymes that install the Cap1 2′ OMe functionality. The ability to avoid recognition by RIG-I and MDA5 is less straightforward. The generation of mRNA relies on the use of RNA polymerases such as those from T7 or S6 bacteriophage (Fig. 12). While these polymerases exhibit a wide range of desirable attributes, including overall efficiency, promoter specificity and low error rates, they are also prone to generation of non-desired products such as abortive transcripts, run-on transcripts and complementary, RNA templated transcripts [38–41]. Controlling for purity of the desired

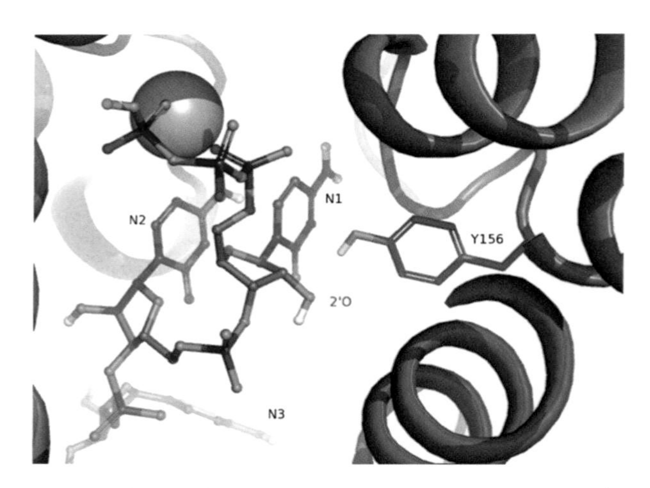

Fig. 10 X-ray structure of IFIT5 demonstrating Y156 interaction with Cap0 2′ OMe

Fig. 11 Structure of Cap0 and Cap1 mRNA

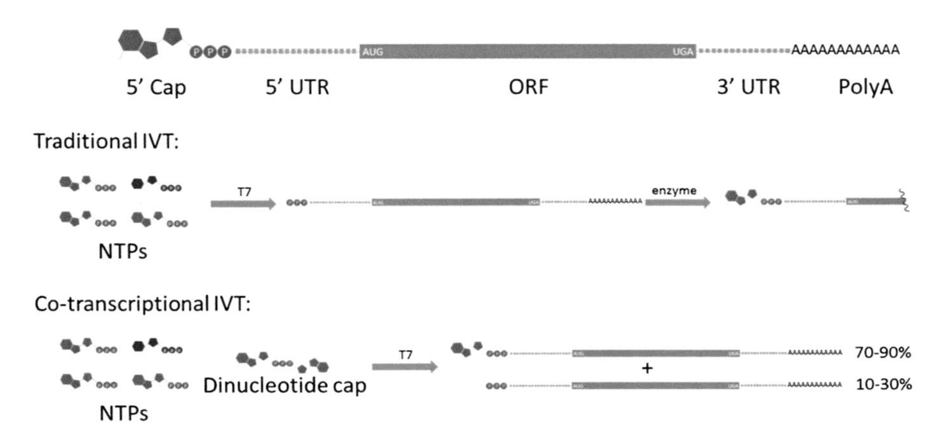

Fig. 12 mRNA structure and common IVT methods of preparation

polymerase product thus becomes a key element for minimizing RIG-I and MDA5 mediated innate immune stimulation. An example of the impact of enhanced purification methods was reported in 2011 by Kariko et al. where they demonstrated that HPLC purification of chemically modified mRNA reduced the secretion of TNF-α and IFN-α to background levels in dendritic cells transfected with mRNA [42]. Similarly, they observed background levels of activation of a host of genes associated with RNA sensing in dendritic cells with chemically modified, HPLC purified mRNA in contrast to unpurified control. Consistent with these results, expression of eGFP was significantly higher for purified vs. unpurified mRNA in the same cell lines. Cumulatively, these data highlight the importance of mRNA purity as it relates to reducing innate immune activation.

Conceptually, evading endosomally restricted pattern recognition receptors such as TLR3, TLR7, and TLR8 represents a much more significant challenge for exogenously introduced mRNA. As discussed in the previous section on mRNA delivery, entry to target cells almost invariably requires passage through TLR rich endosomes. The earliest examples aimed at addressing this challenge relied on the use of chemically modified nucleotides. In 2005, Kariko et al. reported the use of naturally occurring modified nucleotides to avoid activation of TLRs (Fig. 13) [43]. Their original hypothesis was that mammalian specific post-translational modifications might be a mechanism of discrimination relative to viral or bacterial RNA that lack such extensive modification. While this may be true for rRNA or tRNA, it remains highly speculative with regard to mRNA. Regardless, this study did provide the first empirical evidence that modification of RNA could have dramatic effects on TLR recognition as evidenced by reduction in cytokine and interferon secretion in dendritic cells transfected with modified RNA. The same authors followed up this publication with one a few years later where they applied the modification strategy to mRNA for transient protein expression [44]. When examining the effects of modification on expression in several cell-free expression

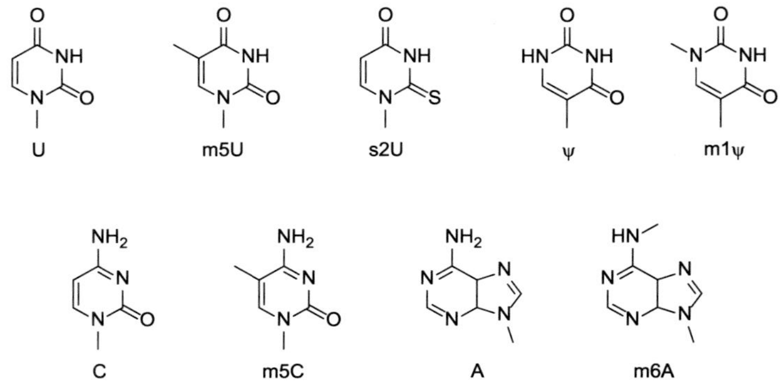

Fig. 13 Nucleobases and modified versions

systems they found that pseudouridine (ψ) gave the best expression followed by 5-methyl cytidine (m5C) and 5-methyl uridine (m5U). 2-thiouridine (s2U) and 6-methyl adenosine (m6A) were not tolerated with regard to expression. Pseudouridine-modified mRNA was chosen for further characterization in cells and in mice where it was demonstrated that expression was improved relative to unmodified mRNA. They also demonstrated improvement was independent of RIG-I, presumably in support of the TLR hypothesis. In 2012, the same group published the use of HPLC purified, ψ containing mRNA encoding for erythropoietin in both mice and rhesus macaques [3]. In the mouse studies, they demonstrated the expected pharmacological effects (increases in reticulocytes and hematocrit) while the rhesus study demonstrated expression of the target protein.

In 2008 and 2010 two papers were published that invoked the role of chemical modification of mRNA in dampening RNA dependent protein kinase (PKR) activation in vitro [45, 46]. PKR, when activated, phosphorylates eIF2α, the result of which is arrested protein translation. While the data provided in these manuscripts is compelling, both utilized mRNA that was not purified by advanced HPLC methods (these pre-dated the publication highlighting the importance of purity) and therefor one needs to be careful to interpret the results as impurities could be driving the PKR phenomenon. Nonetheless, it is worth noting the potential impact of chemical modifications on direct RNA-based activation of PKR.

In 2010, Warren et al. published the combined use of m5C and ψ modified mRNA to induce pluripotency and directed differentiation of human cells [47]. This represented the first utilization of multiple modifications simultaneously which led to increased protein translation and reduced innate immune activation (in cell culture). The authors also highlight that when employing co-transcriptional capping (vs. enzymatic based protocols) there appeared to be a benefit of treatment of the product mRNA with phosphatase to cleave the remnant 5′ triphosphate of the uncapped RNA (see Fig. 12 for diagram of co-transcriptional capping).

A relatively unique modification strategy was published in 2011 involving the use of non-optimal uridine and cytidine modification in sub-stoichiometric amounts

[5]. The authors describe substitution of 25% of uridines and cytidines with s2U and m5C, respectively. The empirical result was a measurable reduction in markers of innate immune activation including IL-12 and IFNα and an increase in protein expression. The use of sub-stoichiometric amounts of modified NTPs in the IVT reaction to prepare mRNA raises interesting questions around polymerase bias (with or without sequence specificity). The authors did not address the analytical characterization of %-modified nucleotide incorporated and/or any sequence specificity for its incorporation. This says nothing of the regulatory hurdle of characterizing a stochastic mixture of partially modified mRNA transcripts and demonstrating batch to batch equivalence. Nonetheless, this represents a novel conceptual approach to mRNA optimization.

The first published example of the use of N1-methyl pseudouridine (m1ψ) appeared in 2012 from the team at Moderna Therapeutics. This and a subsequent patent published in 2013 demonstrate improved translation when employing m1ψ relative to transcripts that utilize ψ or m5C/ψ while simultaneously demonstrating similarly low levels of interferon secretion (relative to unmodified) [48, 49]. Several years after Moderna's patents, academic groups began to publish on the value of m1ψ for mRNA expression. Pardi et al. published a manuscript focused on LNP mediated mRNA delivery and in it, employed m1ψ containing mRNA [50]. The focus was not on the modified mRNA and thus did not contain comparative analysis of other mRNA modifications. Andries et al. later reported in the same journal that m1ψ alone or in combination with m5C can improve translation of firefly luciferase in multiple cell lines, relative to ψ or ψ/m5C [51]. The authors also demonstrated this benefit in mice, but interestingly m1ψ alone gave by far the highest levels of luciferase expression both intramuscularly and intradermally. In an attempt to explain the benefit of m1ψ, it was shown that expression in HEK293 cells overexpressing TLR3 was minimally diminished for m1ψ containing mRNA relative to unmodified or ψ containing mRNA. Earlier this year, another group published improved expression with m1ψ (along with other modifications) in both cell-free and cell-based assays [52]. This group did not characterize the modifications in vivo nor did they explore the impact of modifications on stimulation of the innate immune system.

It is important to point out that not all investigators subscribe to the notion of modifications as necessary for mRNA therapeutics. Thess et al. recently reported the use of unmodified, sequence engineered mRNA encoding for erythropoietin and conclude that with HPLC purified, unmodified mRNA they can achieve expression that is improved over ψ modified mRNA and shows no signs of stimulation of innate immunity (as measured by secreted cytokines) [15]. They demonstrate this in both small (rodents?) and large animals (non-human primates). The challenge in interpreting these results rests in the notion of sequence engineering and the interplay between chemical modifications and sequence. In other words, we're left to take it at face value that a sequence optimized for expression of an unmodified mRNA is similarly optimal for a chemically modified mRNA. Given the vast sequence space for both untranslated and coding regions (employing synonymous codons) this remains a highly speculative assumption. It will be of interest to see how this concept evolves over time.

3 Conclusion

A critical component of mRNA optimization that has yet to be addressed in any detail publicly is the notion of sequence optimization/engineering. As mentioned above, Thess et al. did address this concept, but did not disclose in any detail the salient features of sequence optimization. With no public disclosures on this, we're left to speculate about the translatability of efforts to understand endogenous post-transcriptional gene regulation. What lessons can be learned regarding differences in endogenous UTRs and their impact on translation initiation and elongation? What impact would they have on mRNA decay? For the coding region, can important lessons learned be ascribed to synthetic mRNA? Presnyak et al. recently described in great detail the notion of codon optimality and the impact of this on expression and stability [53]. Will such lessons apply to chemically modified mRNA? And will there be cross talk between ORF and UTR sequence optimization? The answers to these important questions will likely form the basis of future disclosures on the continued optimization of mRNA as a therapeutic modality. We're currently in the early days of optimization of mRNA therapeutics but already there is a rapidly accelerating pace of discovery and disclosure, one that promises to continue with the efforts on delivery, maximization of expression while simultaneously minimizing innate immune stimulation and, prophetically, the efforts to improve features such as duration of protein expression. Importantly and not always obviously, medicinal chemistry has a significant role to play in this effort.

References

1. http://labiotech.eu/ultimate-review-how-could-mrna-overtake-all-other-biologicals-in-medicine/
2. Zangi L, Lui KO, von Gise A et al (2013) Modified mRNA directs the fate of heart progenitor cells and induces vascular regeneration after myocardial infarction. Nat Biotechnol 31:898–907
3. Karikó K, Muramatsu H, Keller JM et al (2012) Increased erythropoiesis in mice injected with submicrogram quantities of pseudouridine-containing mRNA encoding erythropoietin. Mol Ther 20:948–953
4. Phua KKL, Leong KW, Nair SK (2013) Transfection efficiency and transgene expression kinetics of mRNA delivered in naked and nanoparticle format. J Control Release 201:41–48
5. Kormann MSD, Hasenpusch G, Aneja MK et al (2011) Expression of therapeutic proteins after delivery of chemically modified mRNA in mice. Nat Biotechnol 29:154–157
6. Wang Y, H-h S, Yang Y et al (2013) Systemic delivery of modified mRNA encoding herpes simplex virus 1 thymidine kinase for targeted cancer gene therapy. Mol Ther 21:358–367

7. Kranz LM, Diken M, Haas H et al (2016) Systemic delivery to dendritic cells exploits antiviral defence for cancer immunotherapy. Nature 534:396–401
8. Yin H, Kanasty RL, Eltoukhy AA et al (2014) Non-viral vectors for gene based therapies. Nat Rev 15:541–555
9. Bobbin ML, Rossi JJ (2016) RNA interference (RNAi)-based therapeutics: delivering on the promise? Annu Rev Pharmacol Toxicol 56:103–122
10. Hope MJ (2014) Enhancing siRNA delivery by employing lipid nanoparticles. Ther Deliv 5:663–673
11. Akinc A, Querbes W, De S et al (2010) Targeted delivery of RNAi therapeutics with endogenous and exogenous ligand-based mechanisms. Mol Ther 18:1357–1364
12. Geall AJ, Verma A, Otten GR et al (2012) Nonviral delivery of self-amplifying RNA vaccines. Proc Natl Acad Sci 109:14604–14609
13. Hekele A, Bertholet S, Archer J et al (2013) Rapidly produced SAM® vaccine against H7N9 influenza is immunogenic in mice. Emerg Microb Infect 2:e52
14. Pardi N, Tuyishime S, Muranatsu H et al (2015) Expression kinetics of nucleoside-modified mRNA delivered in lipid nanoparticles to mice by various routes. J Control Release 217:345–351
15. Thess A, Grund S, Mui BL et al (2015) Sequence-engineered mRNA without chemical nucleoside modifications enables an effective protein therapy in large animals. Mol Ther 23:1456–1464
16. Nabhan JF, Wood KM, Rao VP (2016) Intrathecal delivery of frataxin mRNA encapsulated in lipid nanoparticles to dorsal root ganglia as a potential therapeutic for Friedreich's ataxia. Sci Rep 6:1–10
17. Love KT, Mahon KP, Levins CG et al (2010) Lipid-like materials for low-dose, in vivo gene silencing. Proc Natl Acad Sci 107:1864–1869
18. Kauffman KJ, Dorkin JR, Yang JH et al (2015) Optimization of lipid nanoparticle formulations for mRNA delivery in vivo with fractional factorial and definitive screening designs. Nano Lett 51:7300–7306
19. DeRosa F, Guild B, Karve S et al (2016) Therapeutic efficacy in a hemophilia B model using a biosynthetic mRNA depot system. Gene Ther 23(10):699–707
20. Dong Y, Love KT, Dorkin JR et al (2014) Lipopeptide nanoparticles for potent and selective siRNA delivery in rodents and nonhuman primates. Proc Natl Acad Sci 111:3955–3960
21. Fenton OS, Kauffman KJ, McClellean RL et al (2016) Bioinspired alkenyl amino alcohol ionizable lipid materials for highly potent in vivo mRNA delivery. Adv Mater 28:2939–2943
22. Li B, Luo B, Deng B (2015) An orthogonal array optimization of lipid-like nanoparticles for mRNA delivery in vivo. Nano Lett 15:8099–8107
23. Li B, Luo X, Deng B (2016) Effects of local structural transformation of lipid-like compounds on delivery of messenger RNA. Sci Rep 6:22137
24. Jarzębińska A, Pasewald T, Lambrecht J et al (2016) A single methylene group in oligoalkylamine-based cationic polymers and lipids promotes enhanced mRNA delivery. Angew Chem Int Ed 55:9591–9595
25. Lächelt U, Wagner E (2015) Nucleic acid therapeutics using polyplexes: a journey of 50 years (and beyond). Chem Rev 115:11043–11078
26. Uchida H, Itaka K, Nomoto T et al (2014) Modulated protonation of side chain aminoethylene repeats in N-substituted polyaspartamides promotes mRNA transfection. J Am Chem Soc 136:12396–12405
27. Uchida S, Kinoh H, Ishii T et al (2016) Systemic delivery of messenger RNA for the treatment of pancreatic cancer using polyplex nanomicelles with a cholesterol moiety. Biomaterials 82:221–228
28. Dong Y, Dorkin JR, Wang W et al (2016) Poly(glycoamidoaine) brushes formulated nanomaterials for systemic siRNA and mRNA delivery in vivo. Nano Lett 16:842–848
29. Heartlein M, Anderson D, Dong Y, DeRosa F (2015) Lipid formulations for delivery of messenger RNA. WO 2015061467 A1

30. Heyes J, Palmer LR, Reid SP et al (2015) Compositions and methods for delivering messenger RNA. WO 2015011633 A1
31. Byers C, Caplan SL, Gamber GG et al (2015) Leptin mRNA compositions and formulations. WO 2015095351 A1
32. Almarsson O, Lawlor C (2016) Lipid nanoparticle mRNA compositions. WO 2016118725 A1
33. Theofilopoulos AN, Gonzalez-Quintial R, Lawson BR et al (2010) Sensors of the innate immune system: their link to rheumatic diseases. Nat Rev Rheumatol 6:146–156
34. Picard-Jean F, Tremblay-Létourneau M, Serra E et al (2013) RNA 5′-end maturation: a crucial step in the replication of viral genomes. In: Romanowski V (ed) Current issues in molecular virology-viral genetics and biotechnological applications. http://www.intechopen.com/books/current-issues-in-molecular-virology-viral-genetics-and-biotechnological-applications/rna-5-end-maturation-a-crucial-step-in-the-replication-of-viral-genomes
35. Kato H, Takeuchi O, Mikamo-Satoh E et al (2008) Length-dependent recognition of double-stranded ribonucleic acids by retinoic acid-inducible gene-I and melanoma differentiation-associated gene 5. J Exp Med 205:1601–1610
36. Leung DW, Amarasinghe GK (2016) When your cap matters: structural insights into self vs non-self recognition of 5′ RNA by immunomodulatory host proteins. Curr Opin Struct Biol 36:133–141
37. Vladimer GI, Górna MW, Superti-Furga G (2014) IFITs: emerging roles as key anti-viral proteins. Front Immunol 5(9):1–9
38. Milligan JF, Groebe DR, Witherell GW, Uhlenbeck OC (1987) Oligoribonucleotide synthesis using T7 RNA polymerase and synthetic DNA templates. Nucleic Acids Res 15:8783–8798
39. Triana-Alonso FJ, Dabrowski M, Wadzack J, Nierhaus KH (1995) Self-coded 3′-extension of run-off transcripts produces aberrant products during in vitro transcription with T7 RNA polymerase. J Biol Chem 270:6298–6307
40. Nacheva GA, Berzal-Herranz A (2003) Preventing nondesired RNA-primed RNA extension catalyzed by T7 RNA polymerase. Eur J Biochem 270:1458–1465
41. Arnaud-Barbe N, Cheynet-Sauvion V, Oriol G et al (1998) Transcription of RNA templates by T7 RNA polymerase. Nucleic Acids Res 26:3550–3554
42. Karikó K, Muramatsu H, Ludwig J, Weissman D (2011) Generating the optimal mRNA for therapy: HPLC purification eliminates immune activation and improves translation of nucleoside-modified, protein-coding mRNA. Nucleic Acids Res 39:e142
43. Karikó K, Buckstein M, Ni H, Weissman D (2005) Suppression of RNA recognition by toll-like receptors: the impact of nucleoside modification and the evolutionary origin of RNA. Immunity 23:165–175
44. Karikó K, Muramatsu H, Welsh FA, Ludwig J, Kato H, Akira S, Weissman D (2008) Incorporation of pseudouridine into mRNA yields superior nonimmunogenic vector with increased translational capacity and biological stability. Mol Ther 16:1833–1840
45. Nallagatla SR, Bevilacqua PC (2008) Nucleoside modifications modulate activation of the protein kinase PKR in an RNA structure-specific manner. RNA 14:1201–1213
46. Anderson BR, Muramatsu H, Nallagatla SR, Bevilacqua PC, Sansing LH, Weissman D, Karikó K (2010) Incorporation of pseudouridine into mRNA enhances translation by diminishing PKR activation. Nucleic Acids Res 38:5884–5892
47. Warren L, Manos PD, Ahfeldt T, Loh Y-H, Li H, Lau F, Ebina W, Mandal PK, Smith ZD, Meissner A, Daley GQ, Brack AS, Collins JJ, Cowan C, Schlaegar TM, Rossi DJ (2010) Highly efficient reprogramming to pluripotency and directed differentiation of human cells with synthetic modified mRNA. Cell Stem Cell 7:618–630
48. Schrum, JP, Afeyan NB, Seiczkiewicz GJ, Bancel S, de Fougerolles A, Elbashir S (2012) US 20120251618 Delivery and formulation of engineered nucleic acids
49. de Fougerolles A, Roy A, Schrum JP, Siddiqi S, Hatala P, Bancel S (2013) US 20130115272 Modified nucleosides, nucleotides, and nucleic acids, and uses thereof

50. Pardi N, Tuyishme S, Muramatsu H et al (2015) Expression kinetics of nucleoside-modified mRNA delivered in lipid nanoparticles to mice by various routes. J. Controlled Release 217:345–351
51. Andries O, McCafferty S, De Smedt SC et al (2015) N1-methylpseudouridine-incorporated mRNA outperforms pseudouridine-incorporated mRNA by providing enhanced protein expression and reduced immunogenicity in mammalian cell lines and mice. J Controlled Release 217:337–344
52. Li B, Luo X, Dong Y (2016) Effects of chemically modified messenger RNA on protein expression. Bioconjug Chem 27:849–853
53. Presnyak V, Alhusaini N, Chen YH et al (2015) Codon optimality is a major determinant of mRNA stability. Cell 160:1111–1124

Index

A
Acetylation, 26, 153, 213
Acetyl choline esterase, 8
Acetylenes, 85
Acetyltransferases, 213
Actinomycin D, 10
ADAM10, 189
Alternative splicing, 135, 177, 180, 184, 189
Alzheimer's disease, 188, 208, 222
Amide *N*-methylation, 31
Amidines, 122
2-Aminobenzimidazoles, 117, 124
Aminoglycosides, 5, 82
Amino lipids, ionizable, 241
Amino-phenylthiazole, 123
2-Aminopyridine, 67
Amino-quinazoline, 125
Amyloid β (Aβ), 188, 222
Amyloid precursor protein (APP), 186
Amyotrophic lateral sclerosis (ALS), 187
Antagomir, 99
Antisense oligonucleotide (ASO), 2, 23, 82, 165
Antiviral drugs, 69, 111, 112, 120, 122, 193
AO19, 158
APOA1-AS, 225
ApoE, 240, 241
Apolipoprotein A1, 225
Apoptosis, 85, 90, 98, 100, 138, 184, 197, 217, 224
Argonaute (AGO), 81, 192
ASH1L, 228
ASO-10-27 (nusinersen), 146, 147
5-Aza-2′-deoxycytosine (5-AzadCyD), 228

B
BACE1, 189
Bcl-2, 197
Becker muscular dystrophy (BMD), 157
Benzimidazoles, 64, 67, 87, 98, 117, 124
Benzo[*g*]quinoline, 31
Biaryl guanidines, 117
Bioisosteres, 17, 29, 33
Biotin, 96
Bis-benzimidazole, 9, 64, 100

C
CamKIIA, 190
Cap analysis gene expression (CAGE), 209
Catalytic subunit of protein phosphatase 2A (PP2Ac), 186
cat-ELCCA, 96
Cell-penetrating peptides (CPPs), 162
Cerebellar degeneration related protein 1 (CDR1), 223
Charcot-Marie-Tooth type A, 216
Chemical biology, 1
Circular RNA (circRNA), 223
CiRS-7, 223
CKK-E12, 243
Coenzyme B12 (AdoCbl), 70
Combinatorial screening, 2D (2DCS), 5–7, 90, 98
Conformational search algorithms, 51
C9orf72, 187
C-terminal binding protein 1-antisense (CTBP1-AS), 217
Cyclosporine, 31
Cyproheptadine, 123

D
Dantrolene, 165
Delivery, 237, 239
Dengue virus (DENV), 111, 113, 119
2-Deoxystreptamine, 91
Diaminocyclopentanol, 68
Diaminopiperidines, 124
1,4-Diazepane, 128
Dicer, 81, 86–101
DiGeorge syndrome critical region 8 (DGCR8), 81
Dimer initiation sequence (DIS), 114
6,7-Dimethoxy-2-(1-piperazinyl)-4-quinazolinamine (DPQ), 69
Dinucleotide methyltransferases (DNMT), 214
DLin-DMA, 241
DMG-PEG$_{2000}$, 241
DMNT1-associated Colon Cancer Repressed lncRNA1 (DACOR1), 214
DMS footprinting, 179
DNA, 3, 18, 55, 70, 82, 209, 212, 238
  epigenetic modifications, 212–218
  G-quadruplexes, 126, 177–198
  methylation, 153, 220, 227
  tricycle (tcDNA), 160
Docking, 47, 50, 52
DOPE, 239
DOTAP, 239
DOTMA, 239
Drisapersen (PRO051), 163
Drosha, 81, 88–90, 98–101, 192
Drug design, 1
  structure-aided, 8
Drug targets, 111
DSPC, 240
Duchenne muscular dystrophy (DMD), 135, 141, 156, 164, 219
Dynamic combinatorial chemistry (DCC), 17, 18
Dynamic combinatorial library (DCL), 17, 19
Dynamic light scattering (DLS), 29
Dynamic linker, 39
Dystrophia myotonica protein kinase (DMPK), 9
Dystrophin-associated protein complex (DAP), 156
D4Z4, 227

E
Ebola virus (EBOV), 111, 120, 193
EBV nuclear antigen 1 (EBNA1), 193
ELISA, 34, 96, 151
Enhancer of nuclear retention element (ENE), 121
Ensemble docking, 60
Epigenetics, 207
Epstein–Barr virus (EBV), 193
Estrogen receptor modulator, 123
Eteplirsen (AVI-4658), 164
Ethidium bromide, 93
Exon inclusion, 141
Exon skipping, 135, 138, 140, 155, 163, 165

F
Facioscapulohumeral muscular dystrophy (FSHD), 221, 223, 227
Familial dysautonomia (FD, Riley–Day syndrome/hereditary sensory and autonomic neuropathy, HSAN) III), 135, 152
Filoviruses, 111, 113, 120
Firre (functional intergenic repeating element), 216
Flaviviruses, 111, 113, 119
Fluorescence indicator displacement (FID), 93
FMN-riboswitch, 49
FMR1, 222
Förster resonance energy transfer (FRET), 91
FR901464, 166
Fragile X mental retardation protein (FMRP), 186
Fragile X syndrome, 216, 222
Fragment-based drug discovery (FBDD), 61
Fragment screening, 47, 61
Frameshifting, 17
Frameshift site (FSS), 25, 114, 115
Framycetin, 93
FRAXE-associated mental retardation syndrome, 186
Free energy perturbation (FEP), 60
Fronto-temporal dementia (FTD), 187

G
Gapmers, 218
G-quadruplexes, 177, 179
Guanine riboswitch (GA), 60

H
Hepadnavirus, 114
Hepatitis B virus (HBV), 114, 121
Hepatitis C virus (HCV), 64, 85, 111–113, 117, 124

HepG2 hepatocellular carcinoma, 100
Herpesvirus, 114, 120, 193
High density lipoprotein (HDL), 225
High throughput screening (HTS), 1, 18, 59, 79, 87, 117, 123, 148, 198
Histone(s), 209–213, 217, 220, 225
Histone acetylation, 153, 213
Histone acetyltransferases, 213
Histone deacetylase (HDAC) inhibitors, 216, 217
Histone methyltransferase, 213
HIV, 17, 24, 111, 121
HIV FSS, 25, 27–36, 40, 116
  RNA, 27, 40
HIV TAR hairpin RNA, 90
H3K27me3, 213
Homesostasis, 177
Horseradish peroxidase (HRP), 96
HOXD10, 90
HOX transcript antisense RNA (HOTAIR), 210
HTLV-2, 33
Human erythropoietin (hEPO), 241
Human immunodeficiency virus (HIV), 17, 24, 111, 121
Huntington's disease, 3, 222
Hydrazone, 19, 39
Hypercholesterolemia, 225

I
Immunity, 220
  innate, 245, 249
Immunogenicity, 237
Influenza A virus, 69, 111, 118, 126
Interferon-induced proteins with tetratricopeptide repeats (IFITs), 246
Internal ribosome entry site (IRES), 67, 117, 185
Isoginkgetin, 168

K
Kanamycin, 6, 9, 90, 91
Kaposi's sarcoma associated herpesvirus (KSHV), 120
Kynuramine, 68

L
Large tumor suppressor 2 (LATS2), 92
Lethe, 222
Ligands, 177
LincRNA-COX2, 222
Lipid nanoparticles (LNPs), 237, 239
Lipidoids, 242
LMI070, 151
Locked nucleic acids (LNA), 160, 218
Luciferase, 33
Lysine acetylation, histones, 213
Lysine-specific demethylase (LSD1), 225

M
Macrolides, 48, 82
Madrasin, 169
Marburg virus, 111, 113, 120
Mass spectrometry (MS), 65
MC3, 241
Metal salicylaldimines, 20
Metastasis associated lung adenocarcinoma transcript 1 (MALAT1), 212
N-Methyl amides, 34
Methyltetrazine, 96
microRNAs/miRNAs, 79, 80, 191, 208
Microtubules associated protein 1b (MAP1B), 186
Morpholinos, 218
mRNA, 24, 38, 80, 85, 118, 137, 163, 237, 249
Muscular dystrophy, 1, 135, 221
  Becker (BMD), 157
  Duchenne (DMD), 135, 141, 156, 164, 219
  facioscapulohumeral (FSHD), 221, 223, 227
  myotonic, 1, 17, 23
Myostatin, 163
Myotonic muscular dystrophy, 17
  type 1 (DM1), 1, 3, 23
  type 2 (DM2), 1, 24

N
Natural products, 17
Neamine, 91
Neomycin, 37, 90–95, 100, 115
Neurodegeneration, 177, 186
  model, 223
Norstictic acid, 169
Nuclear magnetic resonance (NMR), 64
Nuclear paraspeckle assembly transcript 1 (NEAT1), 220
Nucleic acids, 1, 70, 82, 123, 179, 238, 244
  intercalators, 95
  locked (LNA), 160, 218
  peptide (PNA), 160
Nucleobases, 92, 123, 248

Nucleolin, 187
Nucleotides, 24, 69, 237, 247

O

Oligomers, 207
Oligonucleotides, 2, 82, 135, 158, 165, 207, 208, 218, 237
  antisense (ASO), 2, 23, 82, 165
  immobilization, 20
  therapeutics, 218
Orthomyxovirus, 113
Oxazoles, 124
Oxazolidinones, 48

P

p53, 196
Parallel Analysis of RNA Structure (PARS), 179
Parkinson's disease, 224
PEG-lipid conjugate, 241
Pentamidine, 10
Peptide nucleic acids (PNA), 160
Perixosome proliferator-activated receptor δ (PPARδ), 222
Phenazines, 117
Phosphorodiamidate morpholino oligonucleotides (PMO), 160, 218
Phosphorothioate (PS), 218
Pladienolides, 167
Poly(glycoamidoamine) (PGAA) brushes, 244
Polycomb Repressive Complex 2 (PRC2), 210
Polymers, 237
Porphyrins, 118, 188
Postsynaptic density protein 95 (PSD-95), 192
pre-mRNA, 2, 167
  splicing, 135
pri-miRNAs, 81
Protein-ligand docking, 53
Proto-oncogene serine/threonine-protein kinase 1 (PIM1), 185
PSD-95, 190
Pseudouridine, 248
Psoromic acid, 169
Pyridine-2,6-bis-quinolino-dicarboxamide (RR82), 195
Pyridostatin, 126
Pyrithiamine, 48

Q

Quinazolines, 66, 122, 125, 127

R

Rapamycin, 100
Resin-bound dynamic combinatorial chemistry (RBDCC), 21
Rev response element (RRE), 114
RG7800, 149
RGB1, 198
RHAU, 194
Ribavirin, 193
Riboswitches, 48, 60, 70
  FMN, 49
  guanine (GA), 60
  preQ1, 70
  thiamine pyrophosphate (TPP), 68
RIG-I, 245
Riley–Day syndrome, 152
RNA, 1, 47, 177, 237
  circular (circRNA), 223
  dependent protein kinase (PKR), 248
  enhancer (eRNAs), 209
  G-quadruplexes, 181, 191
  mRNA, 24, 38, 80, 85, 118, 137, 163, 237, 249
  miRNAs, 80, 191, 208
  natural antisense (NATs), 209
  noncoding, 111, 208
    long (lncRNAs), 180, 190, 207, 222
  polymerases, 246
  pre-mRNA, 2, 7, 135, 155, 167
  pri-miRNAs, 81
  short hairpin RNAs (shRNAs), 220
  short interfering RNA (siRNA), 92, 219, 225, 239–244
RNA-directed fragment libraries, 62
RNA-induced silencing complex (RISC), 81
RNA interference (RNAi), 214
RNA-ligand docking, 55
Roseoflavin, 48
RR110, 195
RR82, 195

S

Salicylamides, 20
Schizophrenia, 222
Scoring functions, 52
Selective 2′-hydroxyl acylation analyzed by primer extension (SHAPE), 4, 179, 216
SERPINE1, 215
Severe respiratory syndrome coronavirus (SARS CoV), 111, 119, 127
SF3b, 166

Short hairpin RNAs (shRNAs), 220
Short interfering RNA (siRNA), 92, 219, 225, 239–244
Small molecules, 79
  microarrays, 5, 87
Small nuclear ribonucleoproteins (snRNPs), 136
S nucleobase, 123
Spinal muscular atrophy (SMA), 135, 141, 142, 226
Spliceosome, 135–138, 153, 163, 166–169
Spliceostatin A, 167
Splice switching oligonucleotides (SSOs), 135, 141
Splicing, 2, 135, 196, 209
  alternative 135 , 177, 180, 184, 189
  druggable, 140
  enhancers/silencers, 136
  inhibition, 166
Steroid hormones, 222
Streptomycin, 178
Structure-based design, 47
Surface plasmon resonance (SPR), 31, 64, 70, 95, 128, 151
Survival motor neuron (SMN) protein, 142, 222
  SMN1, 226
  SMN2, 141, 223, 226
  SMN-C3, 149

T
Targaprimir-96, 100
Telomere(s), 177, 180, 194, 211
Telomere repeat-binding factor 2 (TRF2), 196
Telomeric RNA G-quadruplexes (TERRA), 69, 183, 198
Tetracyclines, 48, 82
Tetra(*N*-methyl-4-pyridyl)porphyrin (TMPyP4), 118, 120
TG003, 165
Therapeutics, 79
Thermoanaerobacter tengcongensis, 70
Thiamine pyrophosphate (TPP) riboswitch, 68
Thienopyridine, 122
6-Thioguanine (6TG), 165
Thioxanthone, 93
THRIL, 222
Thrombin binding aptamer (TBA), 183
Tobramycin, 70
TP53, 196
Transactivation response element (TAR), 114
Transcriptome, 3
Transcript therapy, 237
Transforming growth factor β2 (TGFβ2), 185
Transphosphoesterification, 137
Triaminoquinazoline, 66
Triaminotriazine, 10
Trichostatin acid (TSA), 228
Tricycle-DNA (tcDNA), 160
Triostin A, 22
Tumor growth, 100
Tumor suppressor genes, 179

U
Upregulation, 207

V
Vancomycin, 29
Vascular endothelial growth factor (VEGF), 185
Viruses, 111, 177
  antiviral drugs, 111
  dengue (DENV), 111, 113, 119
  ebola (EBOV), 111, 120, 193
  Epstein–Barr (EBV), 193
  hepatitis B (HBV), 114, 121
  hepatitis C (HCV), 64, 85, 111–113, 117, 124
  herpes, 114, 120, 193
  HIV, 17, 24, 111, 121
  infections, 177
  influenza A, 69, 111, 118, 126
  inhibitors, 111
  Marburg, 111, 113, 120
  West Nile, 111, 113, 119
  Zika, 111, 113, 119

W
West Nile virus, 111, 113, 119

X
Xanthone, 93

Z
Zika virus, 111, 113, 119

Printed in the United States
By Bookmasters